HANDBOOK GEOGRAPHY TEACHERS

Edited by
David Boardman

THE GEOGRAPHICAL ASSOCIATION
343 FULWOOD ROAD
SHEFFIELD S10 3BP

First published 1986
Reprinted 1988

© The Geographical Association 1986
ISBN 900 39598 2

The views expressed in this publication are those of the authors and do not necessarily represent those of the Geographical Association

Published by the Geographical Association, 343 Fulwood Road, Sheffield S10 3BP

Printed and bound by
Adlard and Son Limited,
Dorking, Surrey, and Letchworth, Hertfordshire

Contents

	List of Contributors	5
	Acknowledgements	6
	Preface	7
1.	Geography in the Secondary School Curriculum	9
2.	**Planning Teaching and Learning**	
	2.1. Planning with Objectives	27
	2.2. Steps in Planning	41
3.	**Teaching and Learning in the Classroom**	
	3.1. Approaches to Teaching and Learning	56
	3.2. Talking, Reading and Writing	68
	3.3. Games and Simulations	79
4.	**Resources for Learning**	
	4.1. Managing Resources	85
	4.2. Evaluating Textbooks	92
	4.3. Producing Resource Sheets	96
	4.4. Discussing Photographs	103
	4.5. Using Videotapes	108
5.	**Microcomputers in the Classroom**	112
6.	**Maps and Mapwork**	
	6.1. Map Reading Skills	123
	6.2. Relief Interpretation	130
	6.3. Atlases and Atlas Mapwork	139
	6.4. Satellite Imagery	145
7.	**Structure and Progression in Geography**	149

8.	**Diverse Abilities**		
	8.1.	Mixed Ability Groups	164
	8.2.	Bright Pupils	171
	8.3.	Slow Learners	176
9.	**Special Considerations**		
	9.1	Multi-cultural Society	181
	9.2.	Third World Studies	185
	9.3.	Sex Bias and Differentiation	193
	9.4.	Pre-vocational Education	198
10.	**Framework Fieldwork**		205
11.	**Combined Studies**		
	11.1.	Planning Combined Studies	219
	11.2.	Humanities in Practice	227
12.	**Evaluation and Assessment in Geography**		
	12.1.	Evaluation and Assessment 11–16	234
	12.2.	Coursework Assessment 14–16	250
	12.3.	Advanced Level Examinations	257
Appendix 1.			269
Appendix 2.			276
Appendix 3.			278
Appendix 4.			280
Index.			282

List of Contributors

And their posts at the time of writing

PATRICK BAILEY	School of Education, University of Leicester
TREVOR BENNETTS	Her Majesty's Inspectorate
DAVID BOARDMAN	Faculty of Education, University of Birmingham
MALCOLM BROWN	Gordano School, Bristol
DAVID CARTER	Department of Geography, Portsmouth Polytechnic
PETER FOX	Chilwell Comprehensive School, Nottingham
NEVILLE GRENYER	Winchester College, Hampshire
DAVID HALL	School of Education, University of Bristol
CLIVE HART	Geography 16-19 Project
MELVYN JONES	Department of Environmental and Recreational Studies, Sheffield City Polytechnic
SHEILA JONES	Colston's Girls' School, Bristol
RICHARD KEMP	Lord Williams's School, Thame
FRED MARTIN	The Grange School, Bristol
HOWARD MIDGLEY	Department of Geography, University of Loughborough, and City of Leicester School
MICHAEL MORRISH	Alleyn's School, Dulwich
ELEANOR RAWLING	Geography 16-19 Project
MARGARET ROBERTS	Division of Education, University of Sheffield
ROGER ROBINSON	Faculty of Education, University of Birmingham
HERBERT SANDFORD	College of St Mark and St John, Plymouth
JEFF SERF	Lordswood Girls' School, Birmingham
FRANCES SLATER	Institute of Education, University of London
TONY THOMAS	Field Studies Council, Shrewsbury
REX WALFORD	Department of Education, University of Cambridge
PATRICK WIEGAND	School of Education, University of Leeds
MICHAEL WILLIAMS	Department of Education, University of Manchester
DAVID WRIGHT	School of Education, University of East Anglia

Acknowledgements

The Geographical Association wishes to thank the following individuals and organisations for permission to reproduce material in this handbook.

Geoffrey Berry for Figure 1.2.

The Guardian for Figure 1.3.

Youth Hostels Association for Figure 2.1.

Michael Storm for Figure 2.13 from ILEA *Geography Bulletin,* No 6.

Eleanor Rawling for Figure 3.1.

John Earish for Figures 3.4 and 3.5.

Geography 16-19 Project and Longman Group Ltd. for Figure 3.7.

Margaret Roberts for Figures 3.9 and 3.10.

Peter Kelly for Figure 4.1.

Bristol Evening Post for the cartoon in Figure 4.3.

Development Education Centre, Birmingham, for Figure 4.4 from *The World in Birmingham.*

Maggie Murray for Figure 4.5 from *Living with the Land.*

United Feature Syndicate Inc. for Figure 6.1.

Space Department, Royal Aircraft Establishment, Farnborough, for Figure 6.12.

The Geographical Magazine for Figures 9.4 and 9.5.

The Times for Figure 9.10.

British Broadcasting Corporation for Figure 11.3 from the BBC Hulton Picture Library.

Longman Group Ltd. for Figure 11.4 from *Enquiries: Life in Developing Countries.*

Preface

This handbook has been written for specialist and non-specialist geography teachers in secondary schools and Postgraduate Certificate in Education students taking geography method courses. It is a companion volume to the Geographical Association's popular handbook *Geographical Education in Primary and Middle Schools,* and replaces the former handbook published by the Association under the title *Geographical Education in Secondary Schools.*

Teachers in secondary schools are currently facing challenges from a number of directions. Falling pupil rolls are leading to a reduction in the size of most schools, amalgamation of smaller ones and closure of some. Staff contraction and redeployment often mean that more subject teaching is being undertaken by non-specialists as schools attempt to maintain a broad and balanced curriculum. Financial stringency and decreased capitation are making it difficult for schools to purchase the books and other resources which teachers need to carry out their work effectively. At the same time poor employment prospects for school leavers are leading to demands from pupils, parents and employers for vocationally relevant courses. In a speech delivered to the Geographical Association on 19 June 1985 the Secretary of State for Education challenged geographers to be clear about what their subject is uniquely or best qualified to offer, since it is on this basis that their subject's claim for scarce curricular time should be made and judged.

Changes are also taking place in teacher training, which is being concentrated in a decreasing number of institutions. The Government White Paper *Teaching Quality* published in March 1983 was followed a year later by Department of Education and Science Circular 3/84 which set out the criteria for the accreditation of courses of initial teacher training. The lengthening of PGCE courses to 36 weeks, including at least 15 weeks of school experience and teaching practice, provides opportunities for the more thorough training of student teachers. The circular stresses the importance of giving adequate attention to the methodology of teaching the chosen subject specialism and of relating it not only to the school curriculum as a whole, but also to the everyday life and work of the community. It emphasises that all courses should prepare students to teach the full range of pupils whom they are likely to encounter in an ordinary school, with their diversity of ability, behaviour, social background and ethnic and cultural origins.

In the light of these different but related challenges for schools and training institutions, Publications and Communications Standing Committee of the Geographical Association recognised the need to produce a completely new handbook. Twenty-six authors were subsequently commissioned to write original contributions which together provide an overview of the various aspects of the geography teacher's work.

A handbook of this kind has to appeal to a wide readership and cater for a variety of needs. Generally speaking, the first six chapters may be regarded as foundation chapters which will be of most use to younger teachers and students preparing for their first teaching practice. After the introductory chapter, they focus directly on the day-to-day practical tasks of planning and implementing geography lessons with the aid of a wide range of resources.

The last six chapters are designed to assist teachers in their professional development. They discuss, often in a reflective rather than prescriptive manner, fundamental matters which are the professional concern of all geography teachers. Heads of departments and other experienced teachers may prefer to start with these last six chapters and subsequently use them, together with the first six, as a basis for discussion with their colleagues and with students undertaking their final teaching practice.

It is hoped that all geography teachers, however, will find something of interest in every chapter. Owing to restrictions on length, each chapter and section within it can only provide an introduction to the topic under discussion. References are listed and further reading is suggested where appropriate.

Every effort has been made to ensure that contributions were up-to-date at the time of writing. Readers are asked to bear in mind, however, that changes were taking place even during the process of publication, notably the introduction of the General Certificate of Secondary Education. Members of the Geographical Association will be regularly up-dated through the pages of the journals *Geography* and *Teaching Geography*.

I should like to express my sincere thanks to all of the contributors who have given freely of their time to write for the handbook. Without their willing co-operation the volume could not have been contemplated, let alone produced. Special words of thanks are due to Eleanor Rawling, Chairperson of Publications and Communications Standing Committee, for the support and encouragement which she and members of the committee gave me whilst the handbook was in preparation; to Rex Walford and Michael Williams for commenting on the draft of the first chapter; and to my colleague Roger Robinson for his suggestions whilst the volume was taking shape. I also wish to thank Rosemary Sutton for assisting me in my editorial task by word processing the manuscripts. Finally, my warmest thanks are due to Noreen Pleavin, Assistant Editor at GA Headquarters, for the care and thoroughness with which she saw the handbook through the various stages of publication.

David Boardman
Birmingham
July 1985

1. Geography in the Secondary School Curriculum

David Boardman

Why teach geography? What contribution does the subject make to the school curriculum? How do pupils in secondary schools benefit from studying geography? In recent years public discussion about education and falling rolls in secondary schools have made it increasingly necessary for geography teachers to pay close attention to these questions as they have sought to explain and justify the place of the subject in the school curriculum. This chapter provides an overview of developments in thinking about the secondary school curriculum and their implications for geography.

Geography in the educational debate

The 'Great Debate' on education was initiated by the then Prime Minister James Callaghan when he delivered a speech at Ruskin College, Oxford, in October 1976. He identified four main issues: the three 'Rs' in primary schools; the comprehensive school curriculum; school examinations; and provision for the education of 16-19-year-olds. Since then there has been an unprecedented flow of official publications on the school curriculum both from the Department of Education and Science and from Her Majesty's Inspectorate.

Early in 1977 the DES produced a consultative paper, *Educating our Children: Four Subjects for Debate* (DES, 1977a). This was widely distributed and used as the basis for a series of regional conferences organised in February and March. The four subjects for debate were: the school curriculum; the assessment of standards; the education and training of teachers; and school and working life. Then followed a Green Paper, *Education in Schools: A Consultative Document* (DES, 1977b), which suggested eight broad aims of education.

The Education Standing Committee of the Geographical Association prepared a contribution to the 'Great Debate' which was published in *Teaching Geography* (GA, 1977). This outlined the contributions of geography to the school curriculum and included sections on concepts, skills, content, attitudes, requirements for learning geography, progression and interdisciplinarity.

Her Majesty's Inspectorate also entered the debate in 1977 with the publication of *Curriculum 11-16. Working Papers by HM Inspectorate: A Contribution to Current Debate* (HMI, 1977) which became commonly known as the 'Red Book'. This argued the case for a common curriculum which introduces pupils during the period of compulsory schooling to eight 'areas of experience'. These were listed in alphabetical order as the aesthetic and creative, the ethical, the linguistic, the mathematical, the physical, the scientific, the social and political, and the spiritual. There followed a series of supplementary papers consisting of statements by HMI subject committees. Teachers of geography were disappointed to find that their subject was not included in the twelve subject areas that were discussed in the 'Red Book', although it was pointed out in the document that the selection was neither exhaustive nor exclusive, and emphasised that subjects which were omitted should not be regarded as of lesser importance in the curriculum. Nevertheless the widespread availability and free distribution of the

'Red Book' inevitably meant that its contents and its omissions were noted by curriculum decision makers in schools and local education authorities.

A statement by HMI geography committee, however, was written and issued as a separate leaflet, and was sent to LEAs in the hope that it would be copied for circulation to schools. It was included as one of the additional subject papers in a reprint of the 'Red Book' and was also published in *Teaching Geography* (HMI, 1978a). The statement began with a reference to the nature of geography, continued with an outline of content areas that need to be studied, and then considered conceptual knowledge, skills, attitudes, contribution to curricular areas and interdisciplinary work.

It is important to distinguish between statements produced by HMI and those which come from the DES. HMI statements are written mainly to stimulate and inform discussion in local education authorities and schools. DES statements are usually derived from the work of the Inspectorate but are also concerned with curriculum policy making and developments which the government wishes to promote through local education authorities. The distinction is essentially that between matters for discussion and matters for action. This distinction becomes particularly important when considering subsequent publications about the curriculum which appeared in 1980-81.

The Secretaries of State invited Her Majesty's Inspectorate to formulate a view of a possible curriculum. This was published under the title *A View of the Curriculum* in the HMI series 'Matters for Discussion' (HMI, 1980). The paper argued for a broad curriculum to be followed by all pupils up to the age of 16. It urged teachers responsible for subject teaching to identify explicitly the knowledge and skills that each subject is intended to promote and to examine the contribution of subject studies to the education of individual pupils. Although HMI emphasised that too much should not be read into what were intended to be purely illustrative selections and comments, it was inevitable that the subjects discussed tended to be regarded as a guide to those that might be included in the common curriculum.

Rex Walford and Richard Daugherty were critical of the series of propositions put forward in *A View of the Curriculum* (Daugherty and Walford, 1980). They considered that these propositions were a poor substitute for thorough discussion of the criteria on which to base decisions about the secondary school curriculum. In particular they were concerned about the implications of the support given in the document for the study of history but not for geography.

The DES proposals for consultation were contained in *A Framework for the School Curriculum* (DES, 1980a). This paper consisted of preliminary views on which comments were invited. Geography received only a brief mention on the last page. A year later came the publication of *The School Curriculum* (DES, 1981) which was sent direct to all schools. Unlike its predecessor, the revised paper was a policy document intended to provide guidance to local education authorities and schools in England and Wales on how the school curriculum could be further improved. Both the 'Framework' paper and its successor *The School Curriculum* listed six essential aims of education (Figure 1.1, p. 14). The fifth aim, 'to help pupils to understand the world in which they live, and the interdependence of individuals, groups and nations', would appear to assure geography of a place in the curriculum for all pupils. Geography was given rather higher visibility in *The School Curriculum* than in the 'Framework' paper. The part played by geography in the broadly similar programmes for the first three years of secondary schooling is recognised in *The School Curriculum*. The document recommends that all lower secondary geography courses should be designed so that pupils who study the subject no further 'are enabled to achieve something of value'. The importance of sustaining a broad curriculum in the fourth and fifth years is emphasised and it is recommended that all pupils should undertake 'some study of the humanities designed to yield lasting benefit'.

Criticism of *The School Curriculum* came from the House of Commons Select Committee when subjecting the document to analysis. Paragraph 22 had said: 'History, geography and economics serve to give the pupil an insight into the nature of society (including his own) and man's place in his environment'. The Select Committee commented: 'We regard this kind of statement as precisely of the kind which government (or for that matter any informed person) should avoid ... The lumping together of such different subject areas as history, geography and economics as if they were in some way interchangeable does no service to education, nor to the credibility of the authors' (House of Commons Education, Science and Arts Committee, 1982).

The School Curriculum was also criticised by Daugherty (1981) for its assertion that not only are there curriculum questions which all schools should be asking, but also that there are right answers to be found to such questions. The possibility that there could be several equally valid answers to the questions did not appear to have been considered. Daugherty argued that the strong case for a common curriculum for all the pupils in a particular school had been confused with the more dubious case for a uniform curriculum in all schools (Daugherty, 1981).

The formal response of the Geographical Association to *The School Curriculum* was contained in a letter which Rex Walford and Michael Williams sent to the DES and which was published in *Geography* (Walford and Williams, 1982). They expressed concern that 'when specific areas of the curriculum' are discussed, 'no reference is made to the humanities in general and to geography in particular'. They observed that in the eyes of parents and of teachers of subjects other than geography, 'geography would appear from your publication to be of minor importance, not worthy of a place in the curriculum'. They concluded that 'given the lack of strong support in curriculum statements from the Department of Education and Science, geography will have difficulty in helping our pupils leave school adequately prepared "to understand the local, national and international environments in which they live and which they will help to shape" '.

This published letter was accompanied by the reply from the DES which stated: '... the fact that geography is not one of the subjects discussed under the heading "specific areas of the curriculum" does not mean that its value has been underestimated. We would, however, expect geography to be in the forefront of the humanities available'. There is a recognition that the position of geography in the first three years of secondary schools will remain unchanged and that it will still be an important option in years four and five.

As a contribution to the Great Debate, a Geographical Association Working Party produced a brief statement, *Geography in the School Curriculum 5-16*, which was approved by Council and printed as a leaflet for distribution to every school in England and Wales (GA, 1981). This leaflet listed the special and general contributions of geography to the school curriculum. Later another working party produced a further statement, *Geography in the Curriculum 16-19*, which was published in *Geography* (GA, 1983). The role of the Geographical Association in the Great Debate has been recorded in detail by Williams (1985).

In his presidential address to the Geographical Association in 1984 Rex Walford offered a timely reminder: 'Unprotected by the umbrella of the core curriculum, geography will make or lose its way by the individual health of the subject in each school and by the efforts of its practitioners'. He emphasised the need for constant vigilance: 'The past decade gave us the luxury of debating the kind of geography we wished to teach; but in the eighties the focus has changed. The debate is now about whether geography should or should not be taught at all' (Walford, 1984).

Initiatives taken by the President and Joint Honorary Secretaries led to an historic and auspicious occasion on 19 June 1985 when the Secretary of State, Sir Keith Joseph, delivered an address to the Geographical Association. Sir Keith reminded the audience of government policy, set out in a White Paper *Better Schools* published in March, that the school curriculum should meet the four principles of breadth, balance, relevance and practicality; and differentiation according to ability and aptitude. In discussing the place of geography in the curriculum, Sir Keith recognised the contribution that geography can make to the balance of a broad curriculum and saw the subject as an important element in a broad education in the humanities for all pupils. He urged that the curriculum should bring geography and history into a relationship in which they reinforce each other, so that pupils acquire both a spatial map and a chronological map. He also recognised that in some ways geography is a bridge between the humanities and the sciences, since some aspects of the subject depend on fundamental and applied science. Sir Keith then made the major challenge of his speech: 'Geographers themselves have to be clear about what their subject is uniquely or best qualified to offer, since it is on this basis that their subject's claim for scarce curricular time should be made and judged' (Joseph, 1985).

In view of the clear need to explain and justify the place of geography in the school curriculum, an attempt is made in the next two sections to help readers to identify the special and general contributions which the study of the subject makes to the education of pupils in secondary schools. It draws upon the statements made by the Geographical

Association (1977, 1981, 1983) and by Her Majesty's Inspectorate (1978a), but it is not intended to be definitive or exhaustive. It is hoped, however, that it will form a useful framework which teachers will develop and expand, particularly when the position of geography in the curriculum is threatened by falling rolls and non-replacement of specialist staff, or when geographers are asked to contribute to various forms of combined courses.

The special contributions of geography
Knowledge

If pupils are to 'understand the world in which they live, and the interdependence of individuals, groups and nations', they will require a great deal of knowledge about the world. Through the study of geography pupils acquire a body of factual knowledge which enables them to know where countries are and to find places within them. Pupils learn how places in various contrasting parts of the world differ from one another, and how the ways of life in different cultures vary over the surface of the earth. This body of factual knowledge should help pupils to place in context the events in the world around them which are reported daily in the media.

Radio, television and newspapers provide pupils with knowledge which is essentially geographical in nature, but it is in geography lessons that they learn to organise separate pieces of information into a coherent whole. Events in the protracted miners' strike of 1984 were reported by the media almost daily. The fear of pit closures; the clashes on the picket lines in Derbyshire; the strength of support for the strike in Yorkshire and lack of it in Nottinghamshire; the convoys of lorries carrying coal to the major iron and steel plants at Ravenscraig, Scunthorpe and Llanwern; all could be located within a framework of the spatial distribution of the coalfields and major iron and steel plants of the United Kingdom.

The systematic branches of the subject are commonly used as the basis for organising school geography courses. Pupils normally study, for example, some of the geographical aspects of industry, agriculture, settlement, population and transport. Usually these are studied in relation to the physical environment but some specialised aspects of physical geography, such as landforms, hydrology and climate, may be treated separately. Most geography courses for 11-14-year-old pupils emphasise human geography, whilst the study of physical geography is mainly found in courses for older pupils.

Places selected for geographical study usually range from the local to the regional, national and global. In this way the interest, knowledge and understanding of pupils is extended from their immediate surroundings to more distant places. Pupils are likely to study their own neighbourhood before certain features of the town or city in which their school is situated. They then study selected aspects of the surrounding region and follow this with studies of contrasting parts of the United Kingdom: lowland and upland, urban and rural, industrial and agricultural. Geography courses usually include studies of selected areas of Europe, particularly countries within the European Economic Community, and other parts of the developed world, notably the USA and Canada, but the USSR and China are often neglected. Increasingly more attention is being given to the developing world, but India, Asia, Africa and South America receive uneven treatment in geography courses.

Studies of different places in the world provide pupils with a frame of reference within which they can attempt to understand contemporary events. They provide the points of contact which can help pupils to interpret events and situations in the world, whether they be civil unrest in Britain, agriculture in the EEC, war in the Middle East, drought in the Sahel or a hurricane in the Caribbean. Geographical studies on a global scale provide a specialist perspective within which pupils may place local, regional, national and international events. They also form the foundation upon which conceptual understanding can be built. Unless pupils can recall the appropriate basic information relating to events and situations, they will experience difficulty in understanding and discussing general ideas.

Recognition of the importance of background knowledge in geography is reflected in the Worldwise Quiz organised annually by the Geographical Association. Teams of pupils under 16 years of age throughout England and Wales compete in the quiz, which seeks to explore and test (in a light-hearted and enjoyable framework) that area of general knowledge in geography which might be expected to be background for most pupils by the time they leave school. The quiz book in which the questions are subsequently published emphasises that the quiz questions should not be taken to represent all that is geography, but they seek to represent that which should be 'useful knowledge' for any properly-educated future citizen (Geographical Association, 1984).

Ideas

The factual knowledge which pupils acquire in learning geography forms the foundation for their understanding of general ideas. Pupils should use the information they acquire about people and places in such a way that it helps them to recognise and explain the physical and human spatial patterns which exist over the surface of the earth, and to understand the processes which lead to the creation of these patterns. It is important that the acquisition of knowledge leads to the development of ideas in geography, because it is these ideas that help pupils to make some sense of the world around them.

The distinctive contribution of geography to the development of ideas lies in concepts associated with the location of phenomena in space and with the relationships between people and their environment. General ideas require an understanding of basic concepts. Terms such as river, town, farm and factory are gradually enlarged to include more abstract concepts such as drainage basin, settlement, agriculture and industry. These terms form the foundation for building the organising concepts of geography, such as spatial location and distribution, spatial organisation and interaction.

If learning is to be effective pupils will need to use their previous experiences and apply them to new situations. Thus younger pupils may learn to recognise, from the study of photographs or through first-hand experience in the field, different kinds of agricultural land use: arable, improved grassland and rough pasture. The pupils subsequently learn to recognise these types of land use by studying other photographs or areas in the field, identifying landscapes that are mainly arable, mainly pasture or mixed in land use. Similarly, when pupils have learnt to identify on a topographical map a landform such as an escarpment, they should then be able to show that they understand the concept by identifying similar landforms on other maps. In this way pupils develop a clearer and more complete notion of each concept and what it is that distinguishes one feature from another that is classified differently. Unless this conceptual development takes place, pupils will not be able to apply their learning to new situations.

The concepts which pupils are able to understand are clearly dependent on their age and on the stage of maturity that they have reached. Learning is influenced not only by chronological age but also by developmental age and by the breadth of the pupils' everyday experience. Most pupils entering secondary schools at the age of 11 years are still at the stage of what Piaget terms concrete operational thinking. Essentially this means that pupils need to see at first hand or physically handle objects if they are to understand the properties of these objects and the relationships between them. For example, if pupils observe different kinds of crops grown on a farm, or examine specimens in the classroom, they should be able to attach the correct labels to them. Similarly, if they observe the processes at work on the bend of a stream, or measure the depth of a stream at various points on the bend, they will not only learn the terms 'erosion' and 'deposition' but also acquire an understanding of the processes involved.

If an object or process is outside the range of experience of pupils they will have greater difficulty in fully understanding that object or process. The concepts that are presented to pupils at the concrete stage of reasoning, therefore, should as far as possible be closely related to their personal experiences. At this stage, for example, the basic concept of accessibility would preferably be introduced by means of a study of a local shopping centre in relation to the residential area which it serves. Similarly the concept of distribution patterns would be illustrated initially through a study of the houses, shops, offices, factories, roads and parks in the area around the school.

As pupils progress through the secondary school they reach what Piaget terms the stage of formal operational thinking. This means that they develop the ability to understand concepts which are not based on their personal observations or experiences. During adolescence pupils develop the ability to understand the more abstract concepts used in geography, such as accessibility, distribution pattern, spatial interaction, regional variation and global interdependence.

It is important to appreciate that some of the concepts used in geography with older pupils relate to complex processes, the full understanding of which demands a high level of abstraction. The extent to which pupils are able to understand abstract concepts depends upon their rate of intellectual development. Some pupils may not attain the stage of formal operational thinking until late in their secondary school careers. Many pupils will acquire only a partial understanding of the more abstract concepts used in geography and their understanding will be more closely related to specific situations. In such circumstances the abstract concepts that are used to explain the spatial structure of towns, for example, have to be based on a number of simpler concepts which the pupils are able to grasp through observation. The development of concepts of different types of houses, shops, offices and factories precedes their grouping into residential, shopping, commercial and industrial areas. This forms the basis for a study of the functional relationship between different areas, the patterns of movement of vehicles and pedestrians, and the changing spatial structure of the town.

A full discussion of concepts and generalisations frequently encountered in geography syllabuses is provided in a publication by Her Majesty's Inspectorate, *The Teaching of Ideas in Geography* (HMI, 1978b). This contains sets of generalisations relating to farming, landforms, manufacturing industries, natural resources, population, recreation, settlement, soils and vegetation, towns and cities, transport, and weather and climate. Each set forms a useful checklist although it is not related to specific age or ability groups.

Skills

Recent curriculum documents attach considerable importance to the acquisition of skills. The aims of education listed in *The School Curriculum* (Figure 1.1) specify physical skills in the first aim, knowledge and skills in the second, and the effective use of language and number in the third. According to the 'Red Book' 'no-one disputes the irrefutable case for basic skills and techniques' (HMI, 1977). In *A View of the Curriculum* teachers are urged to identify explicitly the knowledge and skills which each subject is expected to promote (HMI, 1980). The Schools Council Working Paper, *The Practical Curriculum,* gives more prominence to skills: 'We believe schools ought to emphasise the development of skills. Teachers could help themselves to be more effective if they saw the pursuit of certain types and levels of skills as one of the dimensions of curriculum planning. Nothing could do more than this to ensure that the efforts of different subject specialists contributed to the same ends' (Schools Council, 1981).

Discussion about a common curriculum in DES documents has been largely based on the contributions of the various subjects, although HMI suggested the 'areas of experience' to which subjects might contribute (HMI, 1977). An alternative view which has been suggested by Proctor

---

(i) To help pupils to develop lively, enquiring minds, the ability to question and argue rationally and to apply themselves to tasks and physical skills.

(ii) To help pupils to acquire knowledge and skills relevant to adult life and employment in a fast-changing world.

(iii) To help pupils to use language and number effectively.

(iv) To instil respect for religious and moral values, and tolerance of other races, religions and ways of life.

(v) To help pupils to understand the world in which they live, and the interdependence of individuals, groups and nations.

(vi) To help pupils to appreciate human achievements and aspirations.

Figure 1.1. Aims of education.
Source: *The School Curriculum,* DES, 1981.

(1984) is that the debate about the common curriculum should be based not so much on the contribution of each subject *in toto* but on the skills of communication. In the case of geography an important contribution would lie in the development of graphicacy in pupils. Graphicacy is one aspect of the subject which could be regarded as an indispensible component of the common curriculum.

The term graphicacy is used by geographers to describe the understanding and communication of spatial information that cannot be conveyed adequately by verbal or numerical means. Examples are the map of a shopping street, the plan of a farm, the location of a village and the route through a town. Graphicacy is the most distinctively geographical form of communication. It complements literacy and oracy, communication by means of the written and spoken word, and numeracy, communication by means of numbers.

Graphicacy is largely associated with the skills involved in reading and drawing maps, although it is not confined to them. The degree to which particular skills are mastered at any stage of geographical education depends to a large extent on the age and maturity of the pupils and the amount of practice and experience they have in using the skills. In geography lessons younger pupils learn a variety of skills involved in giving and using direction, reading and stating location, and using scale to measure distance. Pupils need considerable practice in using these skills before at a later stage they learn to recognise contour patterns on maps and attempt to visualise the landforms that they represent.

Intellectual and physical skills are closely interdependent. Intellectual skills require the possession of knowledge and understanding of ideas, and physical skills demand mastery and practice if pupils are to reach high levels of achievement. In order to read and interpret a topographical map, for example, pupils need some prior knowledge of landforms and drainage patterns, and they also have to be able to identify features shown by means of contour patterns and other conventional forms of representation.

A related graphical skill is the ability to recognise and interpret features shown on landscape photographs, taken both from ground level and from the air. Pupils have to learn to recognise physical features, such as common landforms, and human features, such as different types of land use, before they begin to appreciate the interrelationships between people and their environment. In doing so they may attempt to correlate an oblique air photograph with the corresponding area which is shown on a topographical map. This requires quite sophisticated intellectual and physical skills, because not only is the representation of features on a map different from that on a photograph, but also the scale of a photograph varies from foreground to background as a result of the effects of perspective. The drawing and annotation of a sketch from a landscape photograph is another example of the way in which physical and intellectual skills are combined in a single task.

Values

Knowledge, ideas and skills do not constitute all that is taught and learnt in geography lessons. Attitudes and values influence teaching and learning. Discussions of matters of social and environmental concern are affected by the views and opinions held by teachers and pupils. Geography lessons provide many opportunities for helping pupils to understand the nature of attitudes and values. The study of controversial issues necessarily involves a consideration of the ways in which the attitudes and values held by people influence the decisions they take about the environment and of the conflicts which often arise in the process. Equally important is the opportunity which the study of controversial issues provides for pupils to become aware of their own attitudes and values in relation to the environment. In the classroom it is usually most profitable to discuss attitudes and values in the context of a specific issue. This approach is adopted here: a particular issue is considered in some detail.

Consider the photograph of Wastwater in the English Lake District (Figure 1.2). This could be used by the teacher to impart knowledge about the shape and depth of the lake, the rocks and slopes of the fells rising sharply above the lake, and the massive screes coming down to the water's edge. The photograph could be used to develop ideas about the shaping of the valley by glaciation, the deepening of the valley floor now occupied by the lake, and the shattering of the rocks which contributed to the formation of the screes. The photograph might also be used to enable pupils to practise skills of landscape sketching, photograph annotation and correlation with a topographical

Figure 1.2. Wastwater.

map. All this teaching would be value laden. The photographer took the picture because he considered the scene to be a beautiful view on a summer's evening. A teacher who chose to show the photograph to pupils would be conveying an impression of beauty, stillness and isolation, and thus be encouraging them to develop a sensitive awareness of place. The teacher might share with the pupils the words of one of the many writers who have described Wastwater:

> 'To visit Wasdale is to experience the power of the English landscape at its greatest, its constantly changing light, and the varying clarity of its atmosphere, bringing infinite variety to the majesty and wilderness of the scene' (Berry, 1982).

Wastwater is a completely natural lake from which four million gallons of water are abstracted daily without any artificial control of the level of the water in the lake. In 1979 an application was submitted for the construction of a weir across the outlet of Wastwater in order to raise the water level and increase the volume to be abstracted from four to eleven million gallons daily. During a lengthy public enquiry held between January and May 1980 the principal objection made by numerous bodies was that the increased abstraction would result in variations in the level of the lake. At times of drawdown in periods of dry weather areas of mud and stones normally submerged would be exposed, creating a 'rim' effect round the lake. The inspector who conducted the public enquiry concluded in his report that the proposed weir would change the appearance of Wastwater in a way that was totally out of keeping with the environmental scene (Berry, 1982). The Secretary of State for the Environment who accepted this conclusion and rejected the proposal for the weir made a political decision which was strongly influenced by the values and attitudes of people concerned with protecting the landscape and conserving the environment.

The proposal for the weir and increased abstraction of water was submitted by British Nuclear Fuels Ltd., owners of the Sellafield (formerly Windscale) nuclear reprocessing plant and adjacent Calder Hall nuclear power station located on the

Cumbrian coast less than ten miles from Wastwater (Figure 1.3). Again this photograph could be used to provide information about the appearance of the plant, and to illustrate ideas about the location of nuclear plants and nuclear power as a source of energy. It could also be the starting point for a discussion about values and attitudes.

In November 1983 the environmental group Greenpeace investigated the pipeline through which the solvent used to flush out radioactive sludge from waste tanks at Sellafield was discharged into the sea. The Greenpeace boat and equipment were contaminated with adherent radioactive material. A few days later radioactive contamination was found on the beaches on both sides of Sellafield and a 15-mile section of the coast was closed to the public for several months. Two government reports, one from the Department of the Environment and the other from the Health and Safety Executive, both published in February 1984, strongly criticised the management at the plant for permitting the discharge of radioactive waste that led to contamination of the sea and beaches. British Nuclear Fuels Ltd. was subsequently prosecuted over the incident.

A Yorkshire television documentary, 'Windscale — The Nuclear Laundry', screened in the same month that the contamination occurred (November 1983), drew attention to the abnormally high incidence of childhood cancer in the village of Seascale, a mile from the Sellafield complex. The government commissioned a separate enquiry under the chairmanship of Sir Douglas Black, a medical scientist, whose report in July 1984 confirmed that the incidence of leukaemia in children under ten years of age in Seascale was about ten times the national average. But the report stressed that 'an observed association between two factors does not prove a causal relationship' and recommended further studies on the health records of leukaemia sufferers under 25 in West Cumbria and a separate study on the health records of all children born to mothers in Seascale since 1950.

The events of 1983-84 were a severe blow to the nuclear power industry which spends a great deal of money on an expensive public relations exercise designed to convince people of the safety of nuclear power stations. An impressive amount of attractively presented literature is freely available from the United Kingdom Atomic Energy Authority and

Figure 1.3. Sellafield.

the Central Electricity Generating Board. On 17 July 1984 a driverless diesel locomotive travelling at 100 mph struck a nuclear fuel flask which the CEGB had placed across the track on a derailed wagon. This spectacular crash was staged, at a cost of £1.6 million, to reassure the public about the safety of the transport of irradiated fuel from nuclear power stations.

Clearly the geographical location and spatial distribution of nuclear power stations and reprocessing plants cannot be discussed only in terms of expanses of flat land and proximity to good supplies of cooling water. Neither is it sufficient to explain that, because of the possibility of a leak of radioactive material or of an explosion, nuclear plants should be sited at considerable distances from towns and cities. People live in villages, too. The location and distribution of nuclear plants are the result of political decisions which are influenced by the values and attitudes of the people who make them.

Social as well as environmental issues are heavily value laden. Pupils should be encouraged to reflect on their own attitudes to social issues such as those associated with the inner city and urban deprivation, and should learn to recognise the values which underly these attitudes. They should consider the attitudes and values of other people to the same issues and attempt to balance the advantages and disadvantages of different responses to these issues. As pupils grow older and their intellectual maturity develops, they should critically examine social, economic, technological and political perspectives on controversial issues.

Consideration of attitudes and values is particularly important in geography lessons about the Third World, the North-South divide and the relationship between the richer, better fed developed world and the poorer, less well fed developing world. The complex nature of the relationship between wealthier and poorer countries requires careful examination of such issues as the implications of programmes of international aid. Reflection upon attitudes and values is essential if pupils are to understand and respect the cultures and ways of life of other peoples in both the developed and developing world.

The general contributions of geography

Apart from the special contributions which geography makes to the school curriculum in the areas of knowledge, ideas, skills and values, it shares with other subjects a number of general contributions to the curriculum. Geography provides plenty of opportunities for pupils to practise their literary skills. Pupils obtain a great deal of information from reading a wide variety of geography textbooks and other resource material. They thereby acquire knowledge and ideas about places at the local, regional, national and global scales. Modern geography textbooks supplement the text with photographs, diagrams and other illustrations which help pupils to understand what they read. Most textbooks also contain questions to test the pupils' comprehension of passages they read. Other geographical knowledge is obtained by reading newspaper reports, commercial brochures, travel books, and sometimes novels and poetry.

There are also opportunities for pupils to develop their use of spoken language in geography lessons. The sensitive use of open rather than closed questions can encourage pupils to discuss and express views on a variety of social and environmental issues. Work in small groups enables pupils to explore together a wide range of questions, problems and issues. Simulations encourage pupils to present oral evidence on matters on which there is more than one viewpoint. Geographical games involve pupils in joint decision-making activities.

A great deal of written work is undertaken in geography lessons. The demands made upon pupils should go further than taking notes from the blackboard or making notes from textbooks. There are plenty of opportunities for comprehension, judgement, creative writing and descriptive narrative if pupils study photographs or view slides, films and videotapes of people and places. Fieldwork provides pupils with first-hand experience of places which form the basis of their written accounts. Enquiry approaches to fieldwork encourage pupils to formulate hypotheses, test generalisations, write up their results and draw their own conclusions, thus using their powers of analysis, synthesis and evaluation. Geography has much to contribute to the development of a policy of language across the curriculum advocated by the Bullock report (DES, 1975).

The recording and presentation of data enables pupils to practise their mathematical skills in geography lessons. Pupils may convert raw data into temperature and rainfall charts, draw linear graphs to show the relationship between two phenomena, or construct pie charts to show percentage groupings. Older pupils may sometimes use simple statistical tests such as rank correlation or chi square on data collected in the field or obtained from secondary sources. Geography is an ideal subject for helping pupils to apply skills learned in mathematics to another area of the curriculum, as recommended by the Cockcroft report (DES, 1982a).

Whilst literacy and numeracy are taught by English and mathematics teachers, responsibility for the development of graphicacy in pupils rests largely with geography teachers. Graphicacy is the distinctively geographical form of communication, particularly in so far as it relates to the reading and drawing of maps and the study and interpretation of photographs of townscapes and landscapes. Maps are used in history and economics lessons in particular, and in other subjects to a lesser extent. Photographs of landscapes and townscapes may be used in a wide variety of lessons. Skills of map reading and photograph interpretation have many practical uses in everyday life.

Various social skills may be encouraged in geography lessons. Opportunities are provided through small group discussion, the joint planning of an investigation in the field or classroom, and participation in role-playing simulations and geographical games. All of these activities can help pupils to listen to viewpoints which are different from their own and develop an understanding of the attitudes and values held by individuals and communities. Discussion of environmental problems and issues can lead to a deeper understanding of the processes of decision making and contribute to the pupils' social and political education.

It will be apparent from the above that the knowledge, ideas, skills and values which pupils learn in geography lessons can deepen their understanding and appreciation of the work they undertake in other subjects of the school curriculum. The links are likely to be strong with other subjects in the Humanities, particularly English, history, religious education, economics, and social and political studies. Certain aspects of modern geography also have close affinities with the physical, life and earth sciences, as well as with mathematics, statistics and computer studies.

There are clear advantages to be gained from establishing and maintaining links between geography and other subjects of the curriculum. Studies of some of the major social and environmental issues and problems faced by the world today can only be tackled effectively by approaches which are interdisciplinary in nature. Life in urban areas, preparation for employment and unemployment, education for a multicultural society, and the interdependence of the developed and developing world, provide a few examples where joint approaches from different subjects are advantageous.

In some schools links between subjects have been formalised through the establishment of collaborative schemes of work in courses of combined studies, integrated studies, social studies or environmental studies. Whilst these developments are to be welcomed in so far as they further the pupils' understanding of issues and problems which are interdisciplinary in nature, it is important to ensure that the distinctive contributions of the various subjects are retained. Otherwise there is a danger that the curriculum which is experienced by pupils will be weakened through the loss of some of the special contributions which are provided by different subjects. Both the Geographical Association (1982) and Her Majesty's Inspectorate (1978a), whilst recognising the value of links between subjects, emphasise the importance of careful planning if combined courses are to be successful. The theoretical and practical considerations involved in designing and teaching integrated courses are fully discussed by Williams (1984).

Geography in examination courses

16 +

The decision to introduce the General Certificate of Secondary Education (GCSE), taken by the Secretary of State in 1984, followed fourteen years of debate. It was in 1970 that the governing body of the Schools Council recommended that there should be a single examination system at 16+. The feasibility of a common system was investigated by means of experimental 16+ examinations in several subjects, including geography, administered by consortia of GCE and CSE boards between 1971 and 1976. The evidence from these experimental examinations again enabled the Schools Council in 1976 to recommend a common system to the then Secretary of State, Shirley Williams. She acknowledged that a common system could have considerable advantages but decided that further enquiry was needed to remove uncertainties. A steering committee was established under the chairmanship of Sir James Waddell to make an intensive and systematic study of the outstanding problems. The Waddell report, published in 1978, concluded that a common system was desirable and educationally feasible.

After the change of government in the following year, the next Secretary of State, Mark Carlisle, accepted that a common system at 16+ was desirable and in 1980 authorised preparatory work to be undertaken on a new examining system. The GCE and CSE Boards' Joint Council was asked to draw up national criteria for all subjects of the curriculum. In 1982 the new Secretary of State, Sir Keith Joseph, said that the single 16+ system would only proceed if he was satisfied with the national criteria. Meanwhile the Schools Council was abolished and replaced by two bodies: the School Curriculum Development Committee and the Secondary Examinations Council. When the SEC was asked for its advice on the feasibility of a single system in 1983, it replied that the system was educationally feasible and desirable. The Secretary of State eventually announced the decision to merge O-level and CSE into a single system on 20 June 1984.

Common syllabuses for pupils of a wide range of ability are not new because the feasibility studies of the 1970s had been developed into joint systems of examining by the consortia of boards responsible for their administration. By the early 1980s two common systems in geography, one administered by the Joint Matriculation Board and the four northern CSE boards, and the other by the University of Cambridge Local Examinations Syndicate and East Anglian Examinations Board, had become large examinations taken by many thousands of candidates. At the end of their courses candidates received either an O-level or a CSE according to their performance in the examination.

A considerable impetus to joint systems of examining also came from two Schools Council projects. The Geography for the Young School Leaver (GYSL) Project was established at Avery Hill College, London, in 1970 with the original aim of considering the contribution that geography could make to the education of pupils of average to below average ability between the ages of 14 and 16, and to produce schemes of work and supporting resources that could be used either in a subject or in an interdisciplinary framework. The teacher's guides to the project emphasise the changes in geography as a discipline which had led to a move from descriptions of individual and unique phenomena to a search for recurring processes and patterns. This requires the investigation of basic spatial concepts such as location, distance, accessibility and areal association, as well as a consideration of spatial problems of an environmental, economic, social and political nature.

The GYSL Project's three published themes, *Man, Land and Leisure, Cities and People* and *People, Place and Work* (Schools Council 1974-75) formed the basis of many Mode 3 CSE schemes taught throughout the country. Many of the GYSL ideas and resources were found to be appropriate for use with more able pupils, and in 1978 an O-level syllabus and scheme of assessment was introduced and administered jointly by the Welsh Joint Education Committee and Southern Universities Joint Board on behalf of all GCE boards. The Avery Hill O-level scheme was based on the three published themes and a fourth devised by teachers in a school or group of schools. The content of this fourth theme, called a further curriculum unit, in physical, regional or applied geography, complemented those in urban, economic and recreational geography in the published themes. An evaluation of the influence of the GYSL Project has been provided by Boardman (1985).

The Geography 14-18 Project, based at the University of Bristol from 1970 to 1975 and at Leeds Polytechnic from 1978 to 1980, initiated a programme of curriculum development for more able pupils designed to provide an intellectually exacting study of geography. In 1974 the Project introduced a new O-level examination system which fostered school-based curriculum development and was administered by the University of Cambridge Local Examinations Syndicate on behalf of all boards. A core syllabus defined the kind of geography which schools should teach through their own more detailed syllabuses and enabled pupils to use important skills, ideas and models. Groups of schools working as local consortia planned their own curricula using illustrative examples from both physical and human geography. A comprehensive guide to school-based curriculum development, with particular reference to the Project's O-level scheme, has been provided by Tolley and Reynolds (1977), and five sets of exemplar materials have been published (Schools Council, 1978-80). Most teachers who adopted the Project's O-level scheme also designed parallel Mode 3 CSE schemes based on the core syllabus.

A full account of the work of the Geography 14-18 Project will be found in Orrell (1985).

In both the GYSL Project and the Geography 14-18 Project, therefore, the examination syllabuses were extended to cover a wide ability range. This enabled teachers who adopted the projects to teach a common syllabus to all of their fourth and fifth form pupils. They no longer had to divide pupils into different groups at the beginning of their courses. The two projects showed that it is possible for pupils to follow a common course which leads to examinations at different levels at 16+. The final stage in this development was reached in the early 1980s when both projects produced their own joint GCE/CSE examination papers which were in effect pilot 16+ papers.

The GCE and CSE Boards' Joint Council for 16+ National Criteria published the recommended criteria for geography in January 1983 after widespread consultation. The introduction to the revised document containing the national criteria for GCSE geography courses published by the DES in March 1985 begins with a definition of the subject:

Geography is concerned to promote an understanding of the nature of the earth's surface and, more particularly, the character of places, the complex nature of people's relationships and interactions with their environment and the importance in human affairs of location and the spatial organisation of human activities.

The statement then affirms that geography is a valuable medium for education in a social context which includes such characteristics with a geographical dimension as:

(a) a rapidly increasing world population with expanding consumer demand and consequent resource implications;

(b) the existence of multicultural communities and societies;

(c) the existence of marked contrasts in the level of economic and technological development between and within nations;

Knowledge and Understanding

1 to develop a sense of place and an understanding of relative location;

2 to develop an awareness of the characteristics and distribution of a selection of contrasting physical and human environments;

3 to develop an understanding of some of the processes which affect the development of environments;

4 to promote an understanding of the spatial effects of the ways in which people interact with each other and with their environments;

5 to encourage an understanding of different communities and cultures within our own society and elsewhere in the world, together with an awareness of people's active role in interacting with environments and the opportunities and constraints those different people face in their different environments.

Skills

1 to develop a range of skills through practical work, including investigations in the field, associated with observation, collection, representation, analysis, interpretation and use of data, including maps and photographs.

Values

1 to develop a sensitive awareness of the environment;

2 to encourage an appreciation of the significance of the attitudes and values of those who make decisions about the management of the environment and the use of terrestrial space;

3 to develop awareness of the contrasting opportunities and constraints facing people living in different places under different physical and human conditions.

Figure 1.4. Aims of GCSE geography courses.
Source: *GCSE. The National Criteria: Geography,* DES, 1985.

(d) the occurrence of rapid social, economic and technological change in developed and developing countries;

(e) an increase in environmental hazards and growing concern about the deteriorating quality of some environments.

The aims of 16+ geography courses are shown in Figure 1.4. Four groups of examining boards in England and one in Wales have the task of designing GCSE syllabuses which meet the national criteria for geography. Differentiated examination papers or differentiated questions within papers cater for the range of ability in the target population, and there is a component of school-based coursework assessment. A feature of the GCSE is the move towards criteria-related grading, which means testing absolute standards rather than judging the performance of individual pupils relative to that of other pupils. The top three grades in the GCSE represent standards at least as high as the corresponding grades in the former GCE O-level, and the next four grades replace the former CSE grades.

17+

The expansion of comprehensive schools and the growth of the open access sixth form have been accompanied by a rapid increase in the numbers of students who are staying on for one further year of full-time education after reaching the age of 16. The increase has accelerated as youth employment prospects have deteriorated and the continuing growth in the number of students has led to an urgent debate on the nature of the curriculum which best meets their needs.

The Certificate of Extended Education (CEE), introduced as a pilot scheme in 1972, was designed as a one-year course for students who sought a further year of full-time education before they attempted to secure employment. Most of the students in this category had obtained grades 2 to 4 in the CSE. The CEE consisted mainly of courses in single subjects, of which geography was one, and was actively promoted by some CSE boards. Although courses became popular in some regions during the 1970s and early 1980s, the CEE was not, at the national level, widely adopted by schools nor widely understood by employers.

A Green Paper, *Examinations 16-18* (DES, 1980b) gave a clear indication of the intention of the government to establish a nationally validated 17+ course structure. It was proposed that 'courses and examinations for this group should concentrate on assisting the transition from school to work', and that 'any new examination for the identified target group must be vocationally orientated'. The Green Paper referred favourably to the Mansell report, *A Basis for Choice* (FEU, 1979) which considered the education of young people of average ability or attainment who wished to remain in full-time education for a further year after age 16. Developments relating to 17+ courses since 1980 have been strongly influenced by this report.

The Mansell Report proposed that pre-vocational courses should contain three elements: core, vocational and job-specific studies. Core studies should occupy the greater proportion of the available time and are defined as 'those studies to which all students of this age should have right of access, and that learning which is common to all vocational preparation'. Vocational and job-specific studies are designed respectively as those which are particular to a specific vocational sector and a specific job.

A paper *17+ A New Qualification* (DES, 1982b) sets out DES policy on the introduction of the new pre-vocational courses. These courses are intended for both schools and colleges of further education and they lead to the award of a national qualification at 17+ known as the Certificate of Pre-Vocational Education (CPVE). The DES paper describes the new qualification as 'a key element in a set of government policies designed to improve the vocational relevance of education and training for young people'. It is important to recognise that the purpose of CPVE courses is to give a vocational bias to a balanced programme of general education. Accordingly any student in the 16-17 age group might benefit from following such a course. In practice it is likely, however, that many students will be those who might formerly have taken CEE or similar courses. There are no minimum entry requirements to CPVE courses and a wide range of ability can be anticipated among students.

Although the Mansell Report did not make any reference to geography, there is a specific mention of the subject in *17+ A New Qualification*.

Geography is also included in the final proposals for the CPVE published in 1984 in preparation for the introduction of courses in 1985. The Geography 16-19 Project has given considerable thought to the contribution of geography to 17+ courses and produced a discussion document on *Geography and Pre-Employment Courses in the Sixth Form* (Hart, 1981). This was followed by more specific suggestions and a detailed example in *The Geographical Component of 17+ Pre-Employment Courses* (Hart, 1982).

The Examinations Working Party of the Geographical Association also produced a document, *The Contribution of Geography to 17+ Courses* (Geographical Association, 1982). In this the argument for a significant contribution from geography in 17+ pre-employment courses is summarised as follows:

(a) geographers are concerned with the way people use space: in particular they study the spatial effects of socio-economic and physical processes;

(b) the world of work and employment is a significant user of space;

(c) it follows that pre-employment students should study the spatial effects and consequences of the actions of such a significant space user;

(d) in doing so they will develop both general and subject-specific skills;

(e) they will also be involved in discussion of values, attitudes, political influences and beliefs in considering the impact of work activities on the environment and the quality of life;

(f) such experiences are essential if education is to be seen as a genuine opening up of awareness and understanding, whatever the ability or employment aspirations of the students concerned.

The kinds of practical work which might be undertaken by 17+ students are illustrated by means of examples relating to the changing accessibility by public transport of public facilities in inner urban areas. Further examples of ways in which geography teachers can contribute to a common core in the Certificate of Pre-Vocational Education have been provided by Hart (1984) and Hodkinson (1985). The overall structure of the CPVE, with its three elements of core, vocational and additional studies, is explained in the document published by the Joint Board (1985) in preparation for the introduction of the new courses.

18+

The General Certificate of Education in two or three subjects at Advanced Level remains the primary goal of students who undertake two years of full-time study after the age of 16, whether in secondary school sixth forms or in separate sixth form colleges or in colleges of further education. The A-level geography syllabuses offered by all of the boards have been revised to take into account changes in the approach to the subject as an academic discipline. The variety of approaches adopted reflect several well established geographical traditions: spatial, regional and ecosystems.

All A-level syllabuses include a study of the various processes which help to form different physical environments and emphasise in varying degrees the inter-relationships between the physical and human elements of those environments. Syllabuses also require a study of the variety and complexity of factors which contribute to spatial variations in physical and human phenomena. Some syllabuses use a systems framework as the basis for studying the links between physical and human geography, whilst others focus more directly on the processes which lead to regional differentiation.

Scientific rigour is encouraged at A-level through the use of theoretical models and quantitative techniques in making generalisations about the patterns and processes which can be detected over the surface of the earth. The logical positivist approaches to spatial analysis which were prevalent in the mid-1970s, however, were being tempered by softer humanistic and behavioural enquiries by the mid-1980s. There was an acknowledgement of the need for A-level geography courses to show a concern with broader social issues arising, for example, from contrasts between areas of growth and prosperity on the one hand, and areas of decline and deprivation on the other.

The problem of the diversity of syllabus content is not confined to geography. When the Secretary of State announced his decision to retain A-level in 1980, thus assuring its future throughout the decade, he also suggested that consideration be given to 'the clarification and rationalisation of syllabuses'. Inter-Board Working Parties were established for each of the subjects most widely studied at A-level with a brief which included 'identifying where possible an appropriate common core for each subject'.

The Geography Working Party argued that it was neither desirable nor feasible to identify a core of syllabus content for all A-level syllabuses. To a large extent this was because at least three strong traditions or paradigms were reflected to different degrees in geography syllabuses: man-environment relationships, regional synthesis and spatial science. Thus the Working Party did not attempt to specify detailed content but instead produced a broad framework against which the content of any particular A-level syllabus could be measured. The answer to the question 'What ought a student to have gained as a result of taking an A-level course in geography?' was expressed as a set of seven general principles to guide syllabus design (GCE Boards, 1983). These are as follows:

(a) an awareness of certain important ideas in three areas; in physical geography; in human geography; in the interface between physical and human geography;

(b) an appreciation of the processes of regional differentiation;

(c) knowledge derived from a study of a balanced selection of regions and environments, linked with a broad understanding of the complexity and variety of the world in which the student will become a citizen;

(d) an understanding of the use of a variety of techniques and the ability to apply these appropriately;

(e) a range of skills and experiences through involvement in a variety of learning activities both within and outside the classroom;

(f) an awareness of the contribution that geography can make to an understanding of contemporary issues and problems concerning people and the environment;

(g) a heightened ability to respond to and make judgements about certain aesthetic and moral matters relating to space and place.

An explanation of the principles and a discussion of their interpretation constitute most of the report of the Working Party (GCE Boards, 1983), a summary of which was published in *Teaching Geography* (Daugherty, 1982).

A distinctive approach to A-level geography has been initiated by the Geography 16-19 Project which was established by the Schools Council in 1976 and based at the University of London Institute of Education. The project has been involved in a reconsideration of the objectives, content and teaching methods of geography courses for students in the 16-19 age range in schools, sixth form colleges, and colleges of further education. The curriculum framework devised by the Project is based on questions, problems and issues arising from the interrelationships of people with the various environments in which they live. The framework recognises that the interaction of people in and with their varied environments provides certain distinctive spatial locations, distributions and associations, and that various flows or movements of goods, people, ideas and objects take place.

The Geography 16-19 Project uses an enquiry-based approach to learning. The process of enquiry involves discussion of the questions, problems and issues, consideration of the relevant aspects of geographical theory, and arrival at a solution or a range of possible solutions. The use of enquiry methods facilitates an appreciation of the organising concepts and ideas of geography. Consideration of important questions about environmental problems and the quality of life in the modern world is designed to develop an understanding both of the nature of geography as a subject and of the methods of working used by geographers. The enquiry-based approach and the study of issues and problems also provide opportunities for students to consider realistically questions of attitudes and values as they arise. The structure of the Project's A-level course, which is administered by the University of London Board, is different from all others in that it consists of a series of modules, each representing about six weeks of teaching time in length. Each module is a unit of study which is self-contained in its logical development of concepts and in its methods of enquiry but it has recognisable links with other modules. An overview of the Geography 16-19 Project has been provided by Naish (1985) and examples of the enquiry approach by Rawling (1981).

Repeated attempts have been made to broaden the relatively narrow sixth form curriculum in which students follow courses in two or three A-level subjects. In the mid-1970s the proposals for Normal and Further examinations led to the construction of syllabuses which created interest and produced comment before they were quietly buried. The consultative paper *Examinations 16-18* (DES, 1980b) reiterated the concern about the lack of breadth in A-level studies and suggested that the introduction of Intermediate level examinations deserved serious consideration. Comments made on this suggestion indicated that it had a certain measure of support. Early in 1984 the DES produced a further consultative paper on Advanced Supplementary (AS) level examinations (DES, 1984) which proposed that an AS-level syllabus should cover not less than half the amount of ground covered by the corresponding A-level syllabus. Students would continue to take A-levels in the subjects in which they wished to specialise but would be encouraged to study a wider range of subjects by taking one or more at AS-level. Geography was included in the initial list of ten subjects for which the Secretary of State considered it would be right to develop AS syllabuses. It remains to be seen whether this attempt to broaden the sixth form curriculum meets with more success than its predecessors.

Conclusion

At a time of falling rolls, staff reduction, curricular contraction and financial stringency, geography has shown remarkable resilience as a subject in the school curriculum. Almost all secondary school pupils between the ages of 11 and 14 study geography, either as a separate subject or as part of a combined studies course, and about half of all pupils continue to study geography between the ages of 14 and 16. Nevertheless there are no grounds for complacency. For a number of years geography had to struggle to maintain its place in curricular thinking at the Department of Education and Science. The Secretary of State has himself challenged geographers to clarify what their subject is best qualified to offer to pupils during the compulsory years of education if it is to retain its share of scarce curricular time. There is little doubt that in the coming years geography teachers will continue to be asked to justify the place of their subject in the curriculum. This chapter has attempted to provide the context for that debate.

References

Berry, G. (1982) *A Tale of Two Lakes,* Friends of the Lake District.
Boardman, D. (1985) 'Geography for the Young School Leaver', in Boardman, D. (ed) *New Directions in Geographical Education,* Falmer Press.
Daugherty, R. (1981) 'Geography and the school curriculum debate', in Walford, R. (ed) *Signposts for Geography Teaching,* Longman.
Daugherty, R. (1982) 'A common core for A-level?', *Teaching Geography,* 8, 2, 77-9.
Daugherty, R. and Walford, R. (1980) 'Government proposals for the school curriculum in England and Wales', *Geography,* 65, 3, 232-5.
Department of Education and Science (1975) *A Language for Life* (The Bullock Report), HMSO.
Department of Education and Science (1977a) *Educating our Children: Four Subjects for Debate,* DES.
Department of Education and Science (1977b) *Education in Schools: A Consultative Document* (A Green Paper), HMSO.
Department of Education and Science (1980a) *A Framework for the School Curriculum,* HMSO.
Department of Education and Science (1980b) *Examinations 16-18: A Consultative Paper* (A Green Paper), HMSO.
Department of Education and Science (1981) *The School Curriculum,* HMSO.
Department of Education and Science (1982a) *Mathematics Counts* (The Cockcroft Report), HMSO.
Department of Education and Science (1982b) *17+ A New Qualification,* HMSO.
Department of Education and Science (1984) *AS Levels,* DES.
Department of Education and Science (1985) *GCSE. The National Criteria: Geography,* HMSO.
Further Education Unit (1979) *A Basis for Choice* (The Mansell Report), HMSO.
GCE Boards (1983) *Common Cores at Advanced Level,* GCE Boards.
GCE and CSE Boards Joint Council for 16+ National Criteria (1983) *Recommended 16+ National Criteria for Geography,* GCE and CSE Boards.
Geographical Association (1977) 'Geography in the school curriculum: a contribution to the Great

Debate', *Teaching Geography*, 3, 2, 50-1.
Geographical Association (1981) *Geography in the School Curriculum 5-16*, Geographical Association.
Geographical Association (1982) *The Contribution of Geography to 17+ Courses*, Geographical Association.
Geographical Association (1983) 'Geography in the curriculum 16-19', *Geography*, 68, 2, 149-53.
Geographical Association (1984) *Worldwise Quiz Book No 1*, Geographical Association.
Hart, C. (ed) (1981) *Geography and Pre-Employment Courses in the Sixth Form*, Geography 16-19 Project.
Hart, C. (ed) (1982) *The Geographical Component of 17+ Pre-Employment Courses*, Geography 16-19 Project.
Hart, C. (1984) 'Common core — no cause for concern', *Teaching Geography*, 9, 4, 148-50.
Her Majesty's Inspectorate (1977) *Curriculum 11-16. Working Papers by HM Inspectorate: A Contribution to Current Debate* (The Red Book), DES.
Her Majesty's Inspectorate (1978a) 'Geography in the school curriculum: a discussion paper from Her Majesty's Inspectors'. *Teaching Geography*, 4, 2, 76-8.
Her Majesty's Inspectorate (1978b) *The Teaching of Ideas in Geography: Some Suggestions for the Middle and Secondary Years*, HMI Series: Matters for Discussion, 5, HMSO.
Her Majesty's Inspectorate (1980) *A View of the Curriculum*, HMI Series: Matters for Discussion, 11, HMSO.
Hodkinson, P. (1985) 'Piloting CPVE — a geographer's experience', *Teaching Geography*, 10, 4, 152–7.
House of Commons Education, Science and Arts Committee (1982) *The Secondary School Curriculum and Examinations*, HMSO.
Joint Board (1985) *The Certificate of Pre-Vocational Education*. Joint Board for Pre-Vocational Education.
Joseph, K. (1985) 'Geography in the school curriculum' *Geography*, 70, 4, 290–9.
Naish, M. (1985) 'Geography 16-19', in Boardman, D. (ed) *New Directions in Geographical Education*, Falmer Press.
Orrell, K. (1985) 'Geography 14-18', in Boardman, D. (ed) *New Directions in Geographical Education*, Falmer Press.
Proctor, N. (1984) 'Geography and the common curriculum', *Geography*, 69, 1, 38-45.
Rawling, E. (1981) 'Decision making in geography' and 'New opportunities in environmental education', in Walford, R. (ed) *Signposts for Geography Teaching*, Longman.
Schools Council (1974-75) *Geography for the Young School Leaver: Man, Land and Leisure; Cities and People; People, Place and Work*, Nelson (for the Schools Council).
Schools Council (1978-80) *Geography 14-18 Units: Population; Urban Geography; Transport Networks; Industry; Water and Rivers*, (General editor: Tolley, H.), Macmillan Education (for the Schools Council).
Schools Council (1981) *The Practical Curriculum*, Working Paper 70, Methuen.
Tolley, H. and Reynolds, J. B. (1977) *Geography 14-18: A Handbook for School-Based Curriculum Development*, Macmillan Education.
Walford, R. (1984) 'Geography and the future', *Geography*, 69, 3, 193-208, reprinted in King, R. (ed) (1985). *Geographical Futures*, Geographical Assocation.
Walford, R. and Williams, M. (1982) 'Recent involvement of the Geographical Association in the curriculum debate', *Geography*, 67, 1, 71-5.
Williams, M. (1984) *Designing and Teaching Integrated Courses*, Geographical Association.
Williams, M. (1985) 'The Geographical Association and the Great Debate'. *Geography*, 70, 2, 129-37.

2. Planning Teaching and Learning

2.1. Planning with Objectives

David Boardman

What should I teach this class? How should I tackle this topic? Where do I start to plan this lesson? Which resources are available to me? How will the pupils respond? What measure of success can I expect? These are questions that teachers ask themselves every day. All are important questions that need to be asked when planning to teach.

Planning a lesson

Let us imagine that we are planning a lesson for a class of 11-year-old children in their first term at secondary school. During the term all of the children are expected to master a series of basic map reading skills. So far they have had lessons on direction, location and scale. Their next lesson is to be a 70-minute double period on the symbols used on Ordnance Survey maps.

If we have limited time to spare for planning we might decide to give the pupils copies of the key sheets to map symbols. The pupils might spend the lesson copying the symbols into their exercise books, colouring them and labelling them. For homework the pupils might be set the task of learning the symbols and their meanings by heart for a test. The pupils might not see a map at any time during the lesson.

This lesson would not necessarily lack a plan. The content of the lesson, Ordnance Survey map symbols, is clearly defined. The method of teaching is to involve the pupils actively in drawing, colouring and writing during the lesson. An assessment of what the pupils learn will be provided by a test. Even the objective of the lesson is implicit: the pupils are to be able to recognise map symbols and know their meanings.

A question that does not appear to have been asked during the planning of the lesson, however, is *why* the pupils are to spend their time drawing and colouring map symbols. What is it hoped that the pupils will gain as a result of the lesson? In other words what is the purpose of the lesson?

It is always important to think carefully about the objectives of a lesson before teaching it. We cannot hope to teach anything effectively unless we know *why* we are teaching it. We need to decide what kinds of educational benefits the pupils will derive from studying a particular topic. Careful thought about objectives will often affect everything else that is attempted in a lesson.

When objectives are being formulated for a lesson important considerations are the age of the children and the stage of intellectual development that they have reached. Most children at the age of 11 years need concrete experiences to assist their thinking. They should have an opportunity to see and handle objects if their lessons are to be truly meaningful experiences for them. The thinking of children of different abilities at the same age varies considerably and will be very noticeable in a mixed ability class.

This has direct relevance to the method by which the children are taught. The teacher has to consider whether the symbols on an Ordnance Survey map mean very much to 11-year-old children. If they have not experienced a train journey, for example, the difference between an ordinary station and a principal station may not be apparent to them. Children who have never walked along a canal towpath may not always appreciate the difference between a canal and a river. For children who have never visited the sea, such symbols as those depicting cliffs, flat rocks, sand and shingle probably have little meaning for them.

If we are aware that pupils may not have seen some of the features represented by symbols on maps we will take steps to ensure that there are visual illustrations of them in the classroom. These may be photographs of the features mounted on to card and displayed on the wall, or they may be 35 mm slides projected on to a screen if the classroom has blackout and projection facilities.

Abbreviations used on the map for individual buildings may also have to be explained. It should not be assumed, for example, that the pupils will automatically realise that 'Cath' is a cathedral as well as a girl's name, or that 'Cemy' stands for a cemetery and not a particular kind of house. The letter P indicates a post office and not a public house, which is shown by PH, and the letters PC stand for public convenience, not a police constable, in a rural area.

For the purposes of illustration, let us imagine that we decide to use a class set of extracts from the Ordnance Survey 1:50,000 map of East London (sheet 177). The area shown on the map extract includes part of the north-south trending valley of the River Lea with its reservoirs and the nearby valley of the River Roding followed by the M11 motorway. Between the two valleys lies Epping Forest and the urban areas of Loughton, Chingford and Woodford. A map extract such as this contains vast quantities of information and it is important to focus the pupils' attention on one aspect at a time.

After revising with the pupils the method of giving four figure grid references, we might follow the convention of asking the pupils to identify, for example, the kinds of roads that cross selected grid squares. Similarly, after revision of six figure grid references, the pupils might be asked to state the particular features that are found at specific points. Having noticed that the map extract shows a youth hostel in Epping Forest, however, we could decide to use this as the basis for the next part of the lesson.

We could explain to the class that the Youth Hostels Association aims to help all, especially young people of limited means, to a greater knowledge, love and care of the countryside and that the YHA does this mainly by providing simple and cheap accommodation for people during their travels. Some pupils may have stayed at a youth hostel and so could tell the rest of the class where they had been and what they did. The handbook published annually by the YHA would provide a useful additional resource because it contains a location map and details of each of the 260 hostels in England and Wales. We could read to the class the description of Epping Forest Youth Hostel and the attractions of the area (Figure 2.1).

We could then ask the pupils to imagine that they are staying at the hostel for a day and to plan a route for a walk which takes in some of the interesting features shown on the Ordnance survey map extract. They could draw a sketch map to show their route and write a brief description of what they would hope to see during the day. Whilst undertaking this task the pupils would be making constant references to the key sheet to map symbols. At the same time the task would add interest to the lesson and help the pupils to appreciate the usefulness of map symbols. For homework the pupils could be asked to use their creative talent to draw their own maps of an imaginary area, showing on the maps by means of symbols and a key some roads, individual buildings and places of interest.

The way in which this lesson can be set out as a lesson plan is shown in Figure 2.2. Here the objectives have been grouped into ideas, skills and values. It will be noted that the ideas are stated in terms of the ways in which the pupils will benefit from the lesson. The section on content reminds the teacher about the essential points which need to be covered in the lesson, whilst the section on method indicates what the pupils will do during the lesson and for homework.

There remains the question of how the teacher can evaluate the success or otherwise of the lesson. During the lesson it will be possible to make some subjective judgement about the extent to which the pupils are actively engaged in their work and interested in what they are doing. Other judgements will be based on the pupils' success in finding

PLANNING TEACHING AND LEARNING

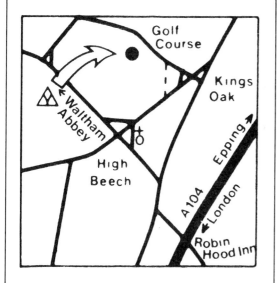

EPPING FOREST

Single storey timber building on edge of 6,000 acre forest which was once a Royal Hunting Forest, only 15 miles from London. 4m S of Epping town at High Beach, W from King's Oak Hotel. Hostel is first building on left down Wellington Hill. From Loughton station (Central Line) proceed N along Station Road, Forest Road, Earls Path, cross A104 at Robin Hood Inn, fork right for King's Oak Hotel

OS 167, 177 GR 408983 Bart 15, 16

Attractions: Forest walks, Queen Elizabeth's hunting lodge (natural history museum) 2m. Connaught Water 2m. Horse riding $\frac{1}{2}$m. Waltham Abbey (Norman) 3m. High Beach Conservation Centre $\frac{1}{2}$m

Walthamstow-Epping-Harlow (alight Loughton High Road 2m), Waltham Cross-Harlow, Waltham Cross-Loughton (alight Volunteer Inn, Upshire $\frac{3}{4}$m), Green Line 702 hourly

Loughton 2m (underground), Chingford 3$\frac{1}{2}$m (BR)

Next hostels: Harlow 11m, St Albans 17m, London 20m, Saffron Walden 29m

Figure 2.1. Youth hostel location map and details.

symbols on the map and in planning their route. The homework will be evaluated by judging the quality of the maps drawn by the children and the accuracy of the symbols used on their maps.

The lesson outlined here is not intended to be a definitive lesson on how to teach map symbols. There are different ways of teaching this topic and other equally effective lessons could be devised. The lesson is also taken out of context. In practice it would form part of a sequence of lessons on map skills. Each lesson would be related to the preceding one, where possible revising skills already learnt, and would pave the way for the next lesson, which would introduce new skills.

A planning model

The preceding example has indicated four considerations in lesson planning: objectives, content, methods and evaluation. In other words it is helpful to distinguish (1) what we are hoping to achieve; (2) the ground we are planning to cover in order to achieve it; (3) the kinds of activities that are likely to be most effective in helping us to achieve our goal; and (4) how we will determine whether we have achieved the goal. These four elements in planning are sometimes shown in linear form as in Figure 2.3.

This simple model requires us to specify our objectives, plan the content and the methods which will lead us towards them, and then to measure the extent of our success. The model assumes that this is the logical order for planning and will be used here as a starting point in order to consider briefly each of the four components in turn. Later it will be shown that it is better to think of the four components as being interrelated.

Aims should be distinguished from objectives. Aims are usually regarded as very general statements of goals and purposes. Thus a geography syllabus might include among its aims 'to develop an awareness of the interrelationship between people and their environment'. Such aims usually apply to the whole of the work for a term or year and are too general to provide clear guidelines for a teacher in the day-to-day task of planning work for pupils. When planning individual lessons and units of work, therefore, it is necessary to develop more precise statements of goals from these general aims.

Class	1st year	**Age**	11-12 years
Topic	Map symbols	**Time**	70 minutes

Objectives

Ideas
— Objects in the real world can be shown as symbols on maps
— Symbols may represent point, line or area features
— The amount of generalisation on maps increases as scale decreases
— Some symbols are disproportionate to the size of the objects they represent

Skills
— To identify symbols using a key sheet
— To plan a route on a map
— To draw and label symbols
— To insert symbols on a sketch map

Values
— Appreciation of the usefulness of symbols
— Enjoyment of reading and using symbols

Resources and Equipment

— Set of O.S. 1:50,000 map extracts, part of East London, Sheet 177
— Set of key sheets to map symbols
— Slides of selected features represented by symbols
— YHA Handbook

Content

1. Line symbols, eg. roads, railways, rivers, canals. Colours are conventional, eg. red, brown, yellow for roads; blue for rivers, canals and other water features — and also motorways. Lines representing roads exaggerate their true width.

2. Point symbols, eg. individual buildings, triangulation pillars. Symbols may show features in plan form, eg. squares for isolated farms; or in side elevation, eg. lighthouses and beacons; some use a combination of both, eg. churches; some are purely symbolic, eg. railway stations.

3. Area symbols, eg. built-up areas (orange), woodland (green), water (blue). Note that white areas are the most common of all!

Method

1. Revise scales of maps studied in earlier lessons. Show slide of local church. Sketch on blackboard how church would be shown on 1:2,500, 1:10,000, 1:25,000, 1:50,000 maps. Explain the need for symbols to highlight churches on smaller scale maps.

2. Revise four figure grid references. Pupils to refer to key sheet to map symbols and find out what kinds of roads cross the following squares:

    4597   Motorway (M11)
    4197   Main road (A104)
    4399   Secondary road (B172)
    4096   Minor road

(continued)

(continuation)

3. Revise six figure grid references. Pupils to find out what the symbol at each of the following represents:

    408 978   Church with spire
    437 984   Camp site
    406 985   Golf course
    408 983   Youth hostel

Show slides of examples of each of above.

4. Pupils to find out what initial letters at each of the following stand for:

    413 973 PH   Public house
    406 985 CH   Club house
    413 983 PC   Public convenience
    427 995 MS   Mile stone

Show slides of examples of each of above.

5. Read from YHA handbook description of Epping Forest Youth Hostel and attractions of the area. Exercise: Imagine that you are staying at the hostel (408 983). Plan a route for a day's walk taking in several interesting places. Possible places to visit include:

    Church at High Beach (408 978)
    Queen Elizabeth's Hunting Lodge (394 950)
    Loughton Camp (418 975)
    Epping Forest Conservation Centre (413 981)

Draw a sketch map to show your route and label it. Add a key, scale line and north point. Describe briefly what you would hope to see during the day.

6. Homework: Draw a map of an imaginary area and on it show by symbols the following: a main road, two secondary roads, a camp site, a caravan site, a youth hostel and a village with a post office, a public house, a church with spire and a school. Add a key, scale line and north point.

Evaluation

Grid references needed more revision and took longer than expected. All pupils then found features at given references. Most pupils showed great interest in planning routes and produced some good sketch maps and descriptions. Homework maps were variable but included some very creative layouts.

Figure 2.2. First year map reading lesson plan.

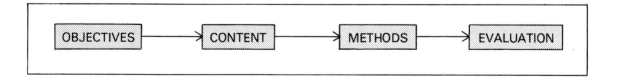

Figure 2.3. A linear planning model.

These more specific statements are termed objectives.

The use of objectives is based on the premise that if an activity is to be planned it should be directed towards some clear goal or purpose. The intentions of a particular activity need to be identified in advance because it is hard to see how we can attempt to teach something without knowing what it is that we are hoping to achieve. It is important to have a clear idea of where we are going in order to have a basis to guide and direct our work in the classroom. Objectives require clarity of thought and are best expressed in a way that tells us what the *pupils* will do or achieve during a lesson.

The second component of the model, content or subject matter, is often regarded as the most important element in planning geography lessons. This is because content is to a large extent determined by the resources that are available to the teacher, particularly textbooks and audio-visual aids. It is natural, therefore, that teachers should find out what is available when they start planning. There is nothing wrong in starting to plan by deciding on content provided that objectives are subsequently clarified. Otherwise teaching may become little more than the coverage of a body of factual information.

The third component of the model, methods, is closely linked to content. The teacher has to decide *how* to teach a particular topic to the pupils and then plan the learning activities in which they will engage. These should usually consist of tasks in which the pupils are actively involved, for example, questions to answer, data to analyse or problems to solve. The teacher has the responsibility for organising and structuring lessons so that learning takes place. The value of different methods of teaching lies in the extent to which they facilitate the achievement of the objectives of a lesson.

Evaluation, the final component, is often linked to assessment, from which it should be distinguished. Assessment involves some form of measurement as a means of ascertaining the quality of the pupils' work. It is usually expressed as marks or grades awarded in a test or examination. Evaluation is a broader term, however, and involves placing some overall value on the work which has been undertaken, usually with a view to modifying subsequent work if evaluation indicates that this is desirable. Judgements which are made as part of evaluation are necessarily often subjective in nature, being related, for example, to the way in which the pupils respond to a task and display an interest in it.

This brief discussion will have indicated that the four components of the planning model cannot be treated in isolation from one another. As was suggested earlier, the linear model, which progresses from the specification of objectives to decisions about appropriate content and suitable teaching methods and ends with the evaluation of the outcome, is too simple and inadequate for most lesson planning. Although the linear sequence suggests a logical order, in practice it is rare to keep to the sequence indicated because each component of the model is continually being modified by the others.

The constant interaction between the components means that the model has to be adapted to show these links and an interactive planning model is shown in Figure 2.4. Here the double-headed arrows indicate, for example, that objectives not only influence content, methods and evaluation, but may also be influenced by them. Similarly, when deciding on methods it is necessary to refer to objectives as well as content, whilst evaluation should be considered in relation to objectives as well as content and methods, and so forth.

Another advantage of thinking about a planning model as interactive is that it permits any of the components to be modified in the light of experience. Indeed, the model encourages constant modification if this is found to be desirable. Sensitive teachers make adjustments to their objectives, to the content of their lessons and to their teaching methods in the light of the feedback that they receive from their classes as the pupils attempt each piece of work. The feedback may indicate, for example, that the pupils do not understand the work, so that the teacher has to explain a task again, perhaps by using a different method, or perhaps by simplifying the content of the lesson so that it is comprehensible to the pupils. This in turn may mean that the original objectives of the lesson have to be modified.

It has only been possible here to outline briefly the 'objectives' model of curriculum planning as it is generally known. A full discussion of the model, each of its components and the ways in which they interact with one another will be found in general books on curriculum planning such as Barnes (1982) and Kelly (1982) and, in relation to geography, Graves (1979).

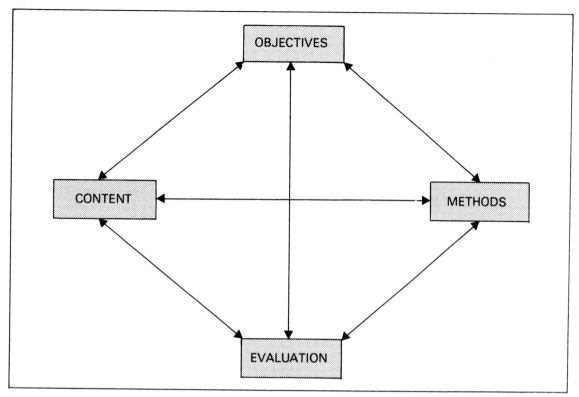

Figure 2.4. An interactive planning model.

The use of objectives

Objectives relating to knowledge and skills in geography can generally be stated more precisely than those relating to ideas, values and attitudes. Often it is also easier to state precise objectives for teacher-directed work than for pupil-centred learning. This can be illustrated by reference to the contrasting ways in which photographs may be used in the classroom.

Let us imagine that we are planning a lesson on residential environments in the inner city as part of an urban geography course for a class of fourth form (14-15-year-old) pupils. We find that textbook photographs of the inner city tend to emphasise old terraced houses, often in a poor state of repair. We might decide to take some photographs of types of dwellings which are in good condition such as those shown in Figure 2.5 in order to focus on the positive rather than the negative aspects of the inner city.

If we adopt a formal, didactic method of teaching we would probably show the slides to the class with clear objectives in mind: the pupils are to recognise the various types of residence, describe their characteristics and learn their origins. We might explain, for example, why the rows of small terraced houses were built in Sparkbrook in the late nineteenth century (photo A). We might then describe the operation of the 'envelope scheme' by which the local authority refurbishes the exteriors of old houses such as those in Balsall Heath (photo B). We should probably follow this with a reference to the wholesale demolition of rows of houses which preceded the construction of the tower blocks of flats such as those in Lee Bank (photo C). Finally we might explain that the large houses in Edgbaston are found in the inner city because of their location on the privately owned Calthorpe Estate (photo D). Throughout this exposition the pupils would be the passive recipients of information.

A Sparkbrook

B Balsall Heath

C Lee Bank

D Edgbaston

Figure 2.5. Inner city residential environments.

We could use the same photographs, however, in a quite different way so that the pupils could make a personal response to them. The content would be the same but the objective and method would both be different. Our objective might be to encourage the pupils to express their feelings about the photographs of these residential environments. The method might be to present the pupils with a list of pairs of adjectives such as those in Figure 2.6. We would show the first slide (photo A), give the pupils time to look at it carefully, and then ask them to record their feelings about the area by placing a dot at the appropriate point on the broken line separating each pair of adjectives. By joining up the dots each pupil would produce a profile of his or her feelings about the area. The procedure would be repeated for each of the other slides (photos B, C, D), the dots for each one being joined up in different colours or by distinctive lines. In this way the pupils would be making their own value judgements about each area.

The next stage of the method might be for the pupils to form small groups and compare the profiles that they have drawn for each photograph. They look for similarities and differences in their profiles and ask one another why they have similar or different feelings about the same area. They discuss the advantages and disadvantages of living in each area and try to imagine what it is like to live in each residence. In this way the pupils would be learning from each other rather than from the teacher. At the same time they would be trying to identify with people who live in different parts of the inner city.

As an alternative to taking photographs and showing them in the classroom, we might organise a piece of experiential fieldwork. In this case the objective might be to encourage the pupils to respond to the environment through their senses. We could take the pupils to the four places and at each one ask them to sit quietly and spend a few minutes concentrating on their senses and feelings. They concentrate on each sense in turn: sight, hearing, smell and touch. They record their reactions by making lists of key words, such as clean or dirty (sight), quiet or noisy (hearing), pure or foul (smell), smooth or rough (touch). They also draw a sketch of each place, annotating it to indicate what they regard as its distinctive features.

This humanistic approach to the study of residential environments in the inner city is designed to help the pupils to develop a heightened consciousness of their own 'private geographies' (Fien, 1983). The pupils are in effect learning from the everyday experiences of the environment in which people live. In school geography courses we tend to spend a great deal of time presenting formal geographical knowledge and ideas to our pupils. The everyday experiences that comprise the private geographies of the pupils tend to be neglected. Humanistic approaches place more emphasis on developing these private geographies and on the ways in which our pupils perceive, experience and respond to the environment.

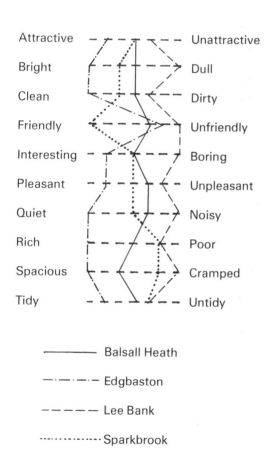

Figure 2.6. Profiles of inner city environments.

	Ward	Percentage of households with more than one person per room	Percentage of households with no car
1	Acock's Green	4.9	52.5
2	All Saints	14.0	69.5
3	Aston	16.0	71.1
4	Billesley	5.3	51.1
5	Brandwood	3.8	43.2
6	Deritend	14.3	72.6
7	Duddeston	8.9	80.7
8	Edgbaston	3.7	49.3
9	Erdington	4.3	52.0
10	Fox Hollies	5.5	57.4
11	Gravelly Hill	6.3	53.3
12	Hall Green	2.8	35.7
13	Handsworth	12.6	62.7
14	Harborne	2.8	44.7
15	King's Norton	3.5	46.8
16	Kingstanding	8.2	62.3
17	Ladywood	5.1	81.2
18	Longbridge	5.4	51.5
19	Moseley	4.3	46.2
20	Newtown	10.2	78.3
21	Northfield	3.0	44.6
22	Oscott	4.0	42.4
23	Perry Barr	3.5	37.7
24	Quinton	3.2	42.2
25	Rotton Park	12.1	66.5
26	Saltley	13.3	64.0
27	Sandwell	8.3	40.0
28	Selly Oak	3.3	45.0
29	Shard End	4.1	49.4
30	Sheldon	3.4	43.8
31	Small Heath	15.8	62.3
32	Soho	20.6	65.7
33	Sparkbrook	18.2	70.0
34	Sparkhill	15.2	57.6
35	Stechford	6.8	55.9
36	Stockland Green	4.2	56.7
37	Washwood Heath	9.1	62.5
38	Weoley	4.0	48.7
39	Yardley	4.6	44.3
40	Sutton Coldfield No 1	0.9	22.0
41	Sutton Coldfield No 2	0.8	18.1
42	Sutton Coldfield No 3	2.4	25.6

Figure 2.7. Data for the wards of Birmingham. Source: 1981 Census.

It is perfectly feasible to state objectives in general terms for humanistic approaches to geography, as was done here: 'to respond to the environment through the senses'. It would clearly be inappropriate, however, to specify precise objectives for work of this kind. The content and method are determined by the teacher, but each response is unique to the individual pupil. It is thus impossible and indeed undesirable to specify the exact nature of that response in advance. To attempt to do so would be to stifle creativity and produce stereotyped responses which would not be genuinely personal. Some of the important outcomes of humanistic approaches, such as appreciation, empathy and sensitivity, cannot be measured and so their evaluation depends on the subjective judgement of the teacher.

The formulation of objectives in this way meets the criticism that is sometimes made about the prespecification of objectives in the humanities. Thus Stenhouse (1975) has argued that attempts to predetermine the outcomes of study in such areas as the literary and fine arts are alien to the nature of the activities in which the pupils are engaged. Restrictions on space prevent a detailed consideration of the arguments about the nature of objectives in such circumstances; these will be found in general books on the curriculum such as those by Barnes (1982) and Kelly (1982). It is sufficient to state here that the debate seems not to be about the use of objectives as such but the narrowness with which they are sometimes specified.

Planning a unit of study

It is often necessary to plan work which extends over two or more lessons rather than one lesson at a time. The older the pupils, and the greater the weekly allocation of time to geography, the more likely it is that planning will be carried out in this way. These longer pieces of work may be described as units of study.

Let us imagine that we are planning a unit on spatial variations in a city as part of a course in urban geography for sixth form students. We decide to supplement the textbook with evidence provided by data from the 1981 Census. We visit the public library to look up data from the Small Area Statistics for the city in which the school is situated (in this case Birmingham) and obtain a map of the ward boundaries. The data are found to be too detailed and complex for our purposes but are manageable when converted into percentages. The content of the unit is determined by selecting the two sets of data shown in Figure 2.7. We decide to allow two lessons and an intervening homework for the unit.

In the first lesson our main objective is for the students to draw a map from the data giving the percentage of households with more than one person per room. The 42 wards are placed in rank order and the list is divided into quartiles, that is, four groups with equal numbers of wards (10 or 11) in each. The students then plot these four groups of data on to a base map of the wards of the city (Figure 2.8). The method is to use different colours or types of shading to complete a choropleth map similar to that shown in Figure 2.9.

Our next objective would be to analyse the map. We should need to explain that data on the percentage of households with more than one person per room provide a measure of overcrowding. The students would examine the map to see whether it shows any particular pattern. They would note the concentration of overcrowding in the wards near the city centre and the decrease which takes place steadily towards the north and east, more rapidly to the south and as far as the city boundary on the west.

A related objective would be to attempt to explain the distribution shown on the map. Why do variations of this kind occur in a city? Why is there more overcrowding in the inner city than in the suburbs? Why is the decrease in overcrowding more rapid in one direction than in another? The teacher might help the students to think of the social conditions which lead to a high percentage of households with more than one person per room, such as the presence of families with a large number of children, or the multiple occupancy of residences by single young people. Conversely, the students might consider possible reasons for a low percentage of households with more than one person per room, such as the presence of wealthy families who can afford large houses with many rooms, or of elderly people who continue to live in the same house after their children have grown up and left home.

For a homework exercise the students might be asked to draw another map using the second set of data, showing the percentage of households with no car. This would provide a test of the students' ability to plot data and draw a map on their own.

Figure 2.8. Base map for plotting data.

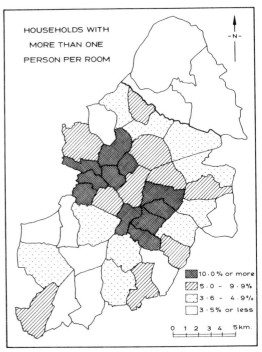

Figure 2.9. Choropleth map (1) compiled from data.

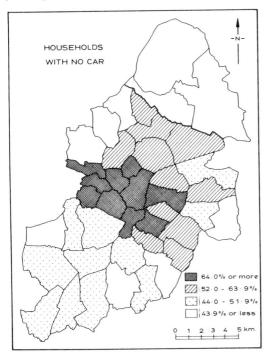

Figure 2.10. Choropleth map (2) compiled from data.

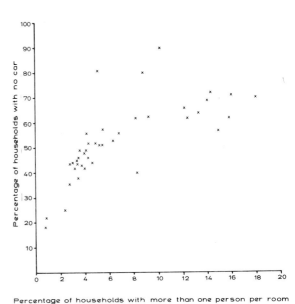

Figure 2.11. Scattergraph plotted from data.

The result should be a choropleth map similar to that in Figure 2.10.

The second lesson might open with an analysis of the pattern shown on this map and an attempt to explain it. The higher percentages of households with no car clearly occur in the inner wards of the city and the lower percentages in the suburbs. Why do fewer people own cars in the inner city than in the suburbs? If it is assumed that the data provide an indication of level of income, why do more people with lower incomes live in the inner city than in the suburbs? Might the reasons for the variation in car ownership sometimes be related to personal circumstances, such as age or health, or to personal choice, such as deciding to do without a car? What are the implications for the provision, cost and frequency of public transport facilities?

The next objective might be for the students to compare the two sets of data. A certain similarity between the two maps is apparent from visual comparison. A more precise analysis, however, is obtained by drawing a scattergraph. The students plot the data for overcrowding on the horizontal axis and the data for car ownership on the vertical axis (Figure 2.11). They then attempt to draw a 'best fit' line (which need not be a straight line) on the scattergraph to show the general relationship between the two variables. This enables the students to decide whether there is a positive or negative correlation, or no correlation, between the two variables. Some students might also use a statistical test such as the Spearman rank correlation coefficient to compute a precise measure of the degree of correlation.

The objectives of the unit are shown in Figure 2.12. Analysis of spatial distributions and associations such as these makes it possible to study the city in terms of human welfare and social well-being. Examination of the localisation of disadvantage, as shown in each of the two maps, is extended to the interrelationships between different kinds of deprivation, as suggested by the scattergraph. The objectives of this kind of analysis accord with the welfare approach to geography, which focuses on the study of 'who gets what where' and is concerned with issues of social

---

Spatial variations in a city

Ideas
— The percentage of households with more than one person per room provides a measure of overcrowding.
— The percentage of households with no car provides an indication of level of income.
— Spatial variations in overcrowding and in car ownership may be discerned in the wards of a city.
— The variations may be related to social conditions.
— There may be a correlation between the two variables.

Skills
— To transform census data into choropleth maps.
— To plot a scattergraph and draw a 'best fit' line.
— To analyse and explain the patterns on maps and graph.
— To calculate the Spearman rank correlation coefficient.

Values
— Appreciation of census data as a source of evidence.
— Empathy with people in different social circumstances.
— Concern for inequality and social injustice.

Figure 2.12. Objectives of a sixth form unit of study.

justice, human well-being and the quality of life. As Bale (1983) has observed, an infusion of the welfare approach into school geography courses adds to quantitative analysis a degree of relevance, humanism and political education.

Conclusion

When planning teaching and learning it is important to know why a particular lesson or unit of work is to be taught, what content will be selected, which methods will be used, and how the outcomes will be evaluated. The model of curriculum planning which is built around objectives, content, method and evaluation provides a useful framework for helping to ensure that teaching and learning are purposeful and effective. The components of the model interact with one another, however, and changes in one have implications for the others. The model should be used flexibly, therefore, and the various components should be subject to constant reappraisal in the light of experience.

References

Bale, J. (1983) 'Welfare approaches to geography', in Huckle, J. (ed) *Geographical Education: Reflection and Action*, Oxford University Press.

Barnes, D. (1982) *Practical Curriculum Study*, Routledge and Kegan Paul.

Fien, J. (1983) 'Humanistic geography', in Huckle, J. (ed) *Geographical Education: Reflection and Action*, Oxford University Press.

Graves, N. J. (1979) *Curriculum Planning in Geography*, Heinemann.

Kelly, A. V. (1982) *The Curriculum: Theory and Practice*, Harper and Row.

Stenhouse, L. (1975). *An Introduction to Curriculum Research and Development*, Heinemann.

2.2. Steps in Planning

Frances Slater

Teaching is very much an activity which involves personal encounters between people — adults and children. The nature of our responses in personal encounters in classrooms depends on many variables, including our personalities, our understandings and capacities as human beings, and the relationships that we are able to develop with our pupils. We might say at one level that our capacity to respond to the personal and intellectual encounters of the classroom partly if not wholly rests on the stage of development we have reached in that lifelong process of coming to be. Since in school teaching our responses must often be made spontaneously in the heat of the moment, under pressure of immediate demand, we may sometimes be found wanting as human beings. At other times we respond and match pupil needs. There is under the pressure of classroom demand little possibility, it must be acknowledged, of taking stock.

Learning episodes and lesson plans

The necessity for this kind of response pattern in the classroom can make teaching exhilarating, satisfying, exhausting, stressful and apparently unrelated to planning. We are usually very pleased with ourselves when suddenly in relation to a pupil's question or answer we see the difficulty being experienced. We are able to deviate from our overall plan, slip into another gear, engage momentarily in further listening and explanation, and experience a pupil deepening his or her understanding and ours. We have a sense of ourselves as teachers, and a feeling that we coped rather well at that particular point. Some of us do say to ourselves, "Well I couldn't have planned that and yet that's the best bit of teaching I did today. I actually know I contributed to someone understanding better". Who would disagree?

Yet, I ask "Hasn't that opportunity to clarify and deepen understanding come about within a general structure?" I see it as an episode embedded within a wider matrix. Perhaps the episode within a structure is analogous to having a general plan of taking a walk in the park in the spring and deciding upon a route. That walk along a route comes to be made up of shorter, very memorable, pleasant episodes of coming upon a clump of crocuses, finding a bird's nest, stopping to look at the greening tips of buds. These episodes occur within the general purposeful activity of taking a walk. I think therefore that it is important to distinguish between the *general lesson plan* or teaching plan, the overall strategy guided by a purpose, and the short individual *learning episodes* which may occur as part of classroom interaction in the course of the lesson. We set out on a walk or a lesson in a general fashion and find ourselves engaged in particulars. And I have suggested that our humanity as well as our minds come powerfully into play in the handling of both since a helpful, sensitive response is needed quickly.

In the classroom the quality of our personal responses in these episodes will count alongside our intellectual response. We feel pleased at our ability to explain, the pupil is pleased at our sensitivity in perceiving her or his difficulty and having sufficient grasp of the idea to explain or in some way take it further. We do not gain by dubbing the planning side sterile, though, when we get satisfaction out of the episodes. A lesson needs planning *and* sensitive responding. A lesson is the result of a carefully thought out general teaching plan which may be interspersed with spontaneous learning episodes.

Quite understandably and realistically, student teachers often tend to regard their standing out front and up front as an exercise in capturing and holding the attention of an audience. They know they have to project themselves, to appeal, to entertain as well as to teach and maintain control. The acting analogy is not new in descriptions of teaching. Indeed, it is probably quite an imperfect one. Part of teaching *is* acting — or that is what it feels like; the pressure is on, the footlights are full beam.

The educational and life experiences of people training to be teachers seems to encourage them to see the on-the-spot performance as teaching. Perhaps it is not surprising. We seem to pass examinations on the basis of an on-the-spot performance. Once we have passed we forget about all the preparation for which in one sense we are never rewarded. And let us not forget either the peer group pressure brought to bear to deny that we work hard and prepare for the final performance in any way. It seems to me that these attitudes spill over into teaching and certainly cause confusion in relation to how lesson planning is perceived, what aspects of teaching are defined, what activities are accepted as part of the job of teaching.

Teaching as performance and planning

Beginning teachers will surely grasp the similarity between their lesson planning and actors learning their lines and remembering directions. The lesson plan is the text and it also specifies the backdrop, the props, the context for the performance and presentation. Teaching as much as dramatising is a planned performance and presentation. Teaching is both planning *and* a performed presentation with learning often taking place for individuals in distinct episodes within that planned performance.

It is interesting to note that in teaching the allowable interaction between what is planned and what is spontaneous between actor and audience is much greater than in a dramatic performance. We cannot question an actor personally. We *can* question a teacher. At first, when inexperienced, this seems yet another demand which teaching makes on personality and intellect. The plan will not by definition be detailed enough to deal with spontaneous episodes and the Catch 22 of teacher training is that experience during a first teaching practice is yet limited. The teacher is more vulnerable than the actor. This fact only seems to make the necessity for planning all the stronger. At the same time, it makes the experience of the performance and presentation vis a vis the audience all the more powerful, too.

I have little doubt that just as we must distinguish between the general lesson plan and spontaneous learning episodes, it is equally necessary to make a distinction between planning and the performance or presentation in order to appreciate more clearly the contribution one makes to the other, to see the connections between the two. We can plan how we are to perform, how we are to present ourselves and ideas. A careful plan does not detract from a good presentation; it complements it. Over time these distinctions become clear to anyone frequently watching beginning teachers and trying to help them. And my justification of the distinctions is grounded in this experience. I certainly would not have been so clear about them after some years of teaching because I had not the same opportunity to see other people performing or to reflect on my own performance.

If some are hostile to lesson planning because they do not see its connection with performance and presentation, and others because learning episodes seem to be the essence of teaching, yet other people constrain their lesson planning because of management difficulties with classes. How then am I to write about lesson planning in view of the perceived and actual difficulties teachers experience in dealing with their often very demanding audience? I do not believe in one very important sense that written words can dissolve management difficulties for another person. When we write, when we talk to students about control and management, we are using the wrong central concepts. It is more helpful to ask students what they know, what they are aware of, as necessary and helpful to the process of building their own personal and group relationships.

Building personal relationships is yet another variable in the very complex set which must be addressed in the process of developing as a teacher. Lesson plans will not work consistently well if this side of the equation is missing or goes wrong. And it may often be that not enough time has passed to develop the relationships. Commitment, sincerity, a sense of humour and fairness, a consistency of behaviour and attitude, a sympathy and sensitivity to others, are some of the qualities we like in our own personal relationships. How we develop, convey, project these qualities in our dealings with people is very complex. How we develop and communicate these and other qualities in building up a relationship with a class is equally complex.

I would assert that thoroughly planned lessons assist in building relationships because they are evidence of commitment and show a respect for other people by using their time responsibly and usefully. One sees plenty of evidence to confirm that pupils respond to careful teaching. It is easy to set out suggestions about praising pupils, giving them a sense of their achievement, being firm and fair, getting to know them out of the classroom setting and so on. And all such tips work to some degree. Yet one knows that it is an amalgam of such things, together with the personal characteristics and undefinable qualities unique to each of us, which goes into a melting pot and over time, sometimes a long and difficult time, fuses to form a satisfactory relationship with a class or not. It is easy to pinpoint the unique element when it is an appeal based on charisma, but not all teachers are charismatic. Yet the noncharismatic succeed with their particular humanity and come across to the pupils. The success, I suspect, is made up of appropriate little acts and exchanges which reveal to the children a person with a grip on what it means to be human, with an ability to respond to the demands made on his or her personality, character and intellect. In 'Daring to be a teacher' Robin Richardson (1983) contemplates the awesomeness of the task.

Control and lesson style

These thoughts and recommendations recognise the perceived and actual difficulties and constraints which student teachers and teachers often place on variety in their lesson planning. In 'Some tentative thoughts on a taboo topic' Michael Storm (1979) has very usefully pointed out that frequently the first order question with some classes and for some teachers is "How am I to control this group, or this individual"? Selecting content or devising learning strategies become second order concerns. In his table of 'Some possible management/content/style interactions' (Figure 2.13), he shows, nevertheless, how management problems affect lesson planning. For example, reading across the third row, 'Resources', suggests that the more difficult a class, the less varied and the less demanding our teaching may be. We all appreciate the essential truth of this. We know classes which settle down when we make them copy something from the board. Thus occupied there is less chance of misbehaviour. In some schools it seems that such an activity is the applied solution lesson after lesson, year after year — copying from the board, the overhead projector, the worksheet. If this is the situation one is faced with, then one can at least try to ensure that the pupils copy down the most worthwhile, up-to-date material available whilst one attempts to work towards a relationship which may permit moving to a more diverse planning strategy. Alternatively, as I often suggest to students, planning short 5 and 10 minute interludes between the copying down routine may be possible. But we must also face up to the fact that it is too easy to let our fears about how a class *may* behave control and constrain our planning. Student teachers often justify unimaginative lessons by saying, "I had to keep them in order". I often feel that this is the situation they fear rather than the situation which would develop. No general pronouncements can be made on this tangled web. One can only urge people to examine their judgments, and to *revise* their judgments as time goes on, as confidence develops and experience expands.

Management problems have led teachers in some schools to using a worksheet/audio-visual strategy. Worksheets are nicely designed, content is up-to-date and the best of films and videos are brought in to introduce or reinforce the main theme and ideas. Teacher-led class discussion and questioning and small group discussion are not used, or used only sparingly. Instead, most often, the teacher goes round the class continually giving individual attention. It is at this time that learning episodes take place and individual personal relationships are built up. When the quality of the geography is high one can feel that management problems are not entirely dictating the curriculum experience. What one needs to think about more in order to expand the possibilities is the distinction between planning the *content* and planning the educational *process*.

Such a distinction also recalls the need to appreciate the difference between prior planning and the performance or presentation. The concept of performance or presentation needs to expand beyond the fairly individualistic conception of teaching as 'my performance' or 'my presentation' and to be integrated with an appreciation of planning an educational process. Not only do we need to see lesson planning as at least a two stage process of ordering and presenting information but also as a plan of acting which incorporates opportunities for pupils to explore the information conceptually, work with it, use it, rework it and make it their own. This in essence is the educational process which we must handle thoughtfully if we are to assist in the development of a product.

SITUATIONS	→ ONE →	→ TWO →	→ THREE →
CLASS some key attributes	hostility; disruption; pervasive frivolity; internal conflicts; confrontation. Rejection of what school has to offer.	tolerance; industry; intermittent frivolity; co-operation; argument. Using most of what school has to offer.	enthusiasm; involvement; occasional frivolity; co-operation; debate. Reaching out to resources beyond school.
CLASS qualities aimed for	acceptance of authority; recognition of need for order; regular attendance; respect for buildings and books in school.	diligence; neatness; improved work-rate and motivation; punctuality; respect for materials loaned.	articulacy; selectivity; inventiveness; independent learning; sensible planning of time; group co-operation.
RESOURCES characteristic uses	blackboard (copying from); occasionally textbook (copying from); homework largely impracticable.	worksheets and textbook; some use of AVA, esp.ohp; atlases; homework set.	diversity of worksheets; printed and AVA sources, esp. slides; library significant; 'non-school' sources (eg. press).
FIELDWORK	extremely hazardous (if not inconceivable).	guided tour; 'scrapbook' hazards.	independent observation and data-collecting.
EXAMINATIONS	largely irrelevant or unrealistic.	expectations of adequate grades.	diversity of exams considered (various modes) and possibly employed.
TEACHER characteristic preoccupations	physical survival; retention of sanity.	preparation of worksheets; curriculum debate within school; departmental cohesion.	assessing and extending resources; stimulating more able; extra-school networks; subject journals.
TEACHER achievements	containment; security of materials and institution.	quiet working atmosphere; large quantities of work produced by pupils.	high rates of purposive pupil participation in discussion; a lasting intellectual and imaginative concern for *places* developed and sustained.

Figure 2.13. Some possible management/content/style interactions. Source: Storm (1979).

Step one	Select and organise content
Step two	Decide how to present the information, ideas, skills etc.

Figure 2.14. Lesson planning Model 1.

Initial steps in planning

Let us begin with the very simplest two step model of lesson planning which I can offer (Figure 2.14). Later I shall make suggestions about building opportunities for learning in a process sense into lesson plans.

It is often difficult for student teachers to realise that once they have (1) sorted out the content they are going to teach, they then have to think very long and hard about (2) how the content is to be presented to a class. Students are often not given a curriculum document to guide them on teaching practice. They are usually told to teach, for example, Water or Bolivia, Rivers or Cities in India, Shopping Centres or New Towns, Population or Networks, for two or three weeks. Student teachers are unlikely to have studied all of these topics in their degree work and the task of getting to grips with the information, finding books on the topic, organising the content in a logically explanatory or logically descriptive unfolding sequence, can be demanding work. Students' earliest and intuitive lesson plans consist of a listing and an elaboration of content. It follows from this that organising the content into a good set of notes to be given to the pupils is one possible teaching strategy for getting information over. Or it may be that a good film or video is seen as the answer to the problem. Yet again a way in may be through a particularly clear and interesting description or a set of statistics illustrating the point to be understood. These latter suggestions show the student teacher grappling with the problem of presentation. 'What data are to be used in order to illustrate the facts and the ideas embodied in the general understanding I want to develop in the pupils?'

There are a number of other decisions which have been taken in the course of (1) selecting and organising content and (2) deciding how to present the information. These need to be made more explicit. This can be done by elaborating on Model 1 to develop a second model (Figure 2.15).

In particular, we need to think more about how content is selected at the level of lesson planning. Let us consider the planning of a lesson on the

Step one	Identify key questions.
Step two	Decide what answers, what generalisations to work towards.
Step three	Gather and select appropriate data and resources.
Step four	Sort out how to present the content and use the data: What learning tasks? What teaching strategies? Is there a balance and range of tasks? Is the level appropriate?
Step five	Examine the activity/activities for likely educational objectives. Accept, reject, modify the activity/activities.
Step six	Devise assessment and evaluation procedures.

Figure 2.15. Lesson planning Model 2.

Amazon Forest for second year (12-13-year-old) pupils as part of a series of lessons on land and resources in South America. The content may be indicated by themes (People and natural resources), topics (Forests), regions (the Amazon Forest), key ideas ('The tropical forest can be utilised by people in a number of ways, all of which have costs and benefits, advantages and disadvantages') or issues ('How should the Amazon be used or developed?'). Sometimes teachers will assist student teachers by indicating an aim or set of aims in relation to a topic or theme. This helps and earlier in this chapter the objectives model of lesson planning was outlined. Key ideas also help since if we know the generalisation or general understanding towards which we are building, working out the 'bits' which have to be taught in order to reach the goal is easier.

My preferred strategy for selecting content at lesson planning level is to use *key questions*. For example, if the set issue is the development or management of the Amazon Forest, how do you decide what precisely to teach over a number of lessons? Most geographers will comfortably decide to use and plan around a key question 'Where is the Amazon Forest?' and to look at 'What it is?' in vegetation and ecological terms. Other possible key questions are 'How is it changing?' 'What difficulties does it present for people?' 'How is it used by different people?' 'How is it to be developed and managed?' 'What difficulties do recent settlers of the Amazon experience?' 'Whose rights are to be considered?' 'How much of the forest should be cut down?' 'What should government policy be in relation to the Amazon?' 'Ought it to be preserved?' These, my key questions, help me to select the content I need in order to come to grips with the Amazon Forest today. I always advise students to turn themes or topics which they are given into a series of key questions as a way of beginning to sort out content and presentation. Key questions are a powerful selection device in my opinion. The steps in lesson planning are shown in Model 2 (Figure 2.15) and illustrated by means of a particular lesson described in detail in Figure 2.16. This should now be read through very carefully in conjunction with the associated resources (Figures 2.17-2.20).

Class	2nd year mixed ability	**Age**	12-13 years
Topic	Change and planning in the Amazon Forest	**Time**	70 minutes

This is one of a series of lessons over a term on land and resources in South America to a second year mixed ability class.

The previous lesson introduced pupils to the Amazon Forest through

(1) A description and diagram of its vegetation make up.

(2) The climate of the region taught through climate graphs of Manaus and London. The pupils answered questions drawing out comparisons between Britain's woodland and the Amazon and Britain's climate and the Amazon.

(3) A short account of the story of Juliane Koepeke who crash-landed in the tropical rain forest of South America some years ago. Pupils were asked to imagine that they had crash-landed in the Amazon and that they had been asked to write a newspaper account of their experience.

Step One Identify key questions	*The key questions* for this lesson are: How is the Amazon Forest being changed? How is it being opened up? How is the Amazon Forest being developed? How should the Amazon Forest be developed?

(continued)

PLANNING TEACHING AND LEARNING 47

(continuation)

Step Two Decide what answers, what generalisations to work towards.	It is decided to bring out the *general ideas* that the Amazon Forest is being opened up by the destruction of the forest and the building of roads, the settlement of farmers from other parts of Brazil, the development of a timber industry and mining of mineral resources. Equally, it is decided that pupils, after working through a planning exercise on developing the Amazon Forest should be challenged to think about such development and to consider another possibility — leave the forest alone. The general idea to be explored is that the development of the Amazon Forest does not have a simple answer. How it should be developed is a debatable question.
Step Three Gather and select appropriate data and resources.	1. A suitable BBC video *The Amazon Frontier* is chosen for showing to help answer the question, 'How is it being opened up?' 2. A mock newspaper account of Amazon Development is devised as an additional resource to introduce the video (Figure 2.17). 3. A worksheet/planning brief and map are devised for a planning exercise to answer the question 'How is the Amazon being developed?' (Figures 2.18 and 2.19). 4. A newsflash is made up for a homework written exercise (Figure 2.20).
Step Four Sort out how to present the content and use the data. What learning tasks? What teaching strategies? Is there a balance and range of tasks? Is the level appropriate?	It is decided to present the content via the reading of a mock newspaper account, viewing a video, a planning exercise and a homework exercise comprising new information and questions on it. The class read the Amazon Development article (Figure 2.17) as a briefing for the video. The debriefing is based on the children's immediate impressions and feelings about the size, sight, colour of the forest etc. related to anything in their experience; and with reference to Brazil's need of development and how the Amazon is being developed according to the video. The planning activity is carried out as organised on the resource sheets and map (Figures 2.18 and 2.19). The homework activity is carried out as indicated on the resource sheet (Figure 2.20). The teaching strategy developed can in actual fact be labelled as a mismatch strategy or discrepant data strategy (Tolley and Reynolds 1977, Slater 1982). The pupils are given data and asked to solve what seems to be a straightforward problem and then their viewpoint is challenged. In this way an educational process has been organised. There is a good range of tasks in the lesson from reading an account, observing and comprehending a video, to making a choice and setting out a plan in map form. The pupils are asked to give reasons for their choice and then to rethink the choice available. There are reading, observing, recalling, comprehending, writing, choosing, organising/synthesising tests as well as mapping, justifying and rethinking tasks.

(continued)

Step Five Examine the activity/activities for likely educational objectives. Accept, reject, modify the activity/activities.	General educational aims of promoting literacy and graphicacy are present in this lesson. Numeracy could be built in by quantifying and assigning costs to the development options. Reading and writing skills have to be practised. When the class is divided into small groups discussion is encouraged as the planning task takes place. Another opportunity for talk would be to have some pupils announce and explain their choices to the class as a whole in the last 10 minutes or so of the lesson before the homework is set. The content of the lesson contains worthwhile geographical and environmental education ideas and pupils are likely to have to think about the management of resources, some of the constraints, and some of the opportunities. Map reading and map making are among the educational objectives as well as choosing and decision making. The newsflash for homework brings in the conservation viewpoint.
Step Six Devise assessment and evaluation procedures.	The pupils' work is read and marked but no special test is devised for the lesson. At the end of the series of lessons the pupils are asked, on a show of hands, to indicate how enjoyable and how useful they found the lessons and parts of them (eg. descriptions, videos, planning exercise, etc). Their opinions are recorded on the board.

Acknowledgement. The lesson as outlined here is an amalgam of an initial planning idea devised by Anna Hayes of Holland Park School, London, and developed with PGCE students. The worksheets reproduced here are based on those by David Humphreys now of Bedford School.

Figure 2.16. Second year lesson plan: The Amazon Forest.

AMAZON DEVELOPMENT
A Great Step Forward!

Brazil is a developing country with vast resources in farming, forestry and mining that remain largely unexploited. In order to compete with or even join the world's richer nations Brazil must make the most of its natural resources, develop industries of its own based on these resources, and leave its image of poverty and backwardness behind. *Development* of the huge Amazon Basin could provide the vital impetus in Brazil's push for growth and prosperity.

Until recently development in the Amazon Basin was rarely planned. The main inhabitants were small tribes of Indians who cleared the forest, grew crops for a few years and then moved on to do the same elsewhere. This *shifting cultivation* provided them with most of their food.

Since 1970 the Brazilian Government has followed a programme to *'open up the Amazon Forest'*. Roads are being built over vast distances making it possible for more people to inhabit the forest. Along these roads families from over-crowded cities in the South and from the drought-torn, poverty-stricken North-East have moved in to farm the land and start a new life.

As the Trans-Amazon roads have been built, minerals in large quantities have been found, such as bauxite, iron ore, tin, copper and lead, even small quantities of 'black gold' (oil). The mining of these minerals, although involving the loss of large areas of the forest, are providing very valuable *income* (money) for the government.

In some places the government plans to build large new towns, housing people who will be employed in timber mills and paper factories using wood from the vast number of trees in the Amazon Forest. *But any development must be carefully planned.*

Figure 2.17. Newspaper report: Amazon Development.

AMAZON FOREST DEVELOPMENT PLANS

The Amazon Forest is a natural resource that must be managed. You have to decide how best to develop a small part of the Amazon Forest and *plan* how *you* would manage the land. You have three choices:

Plan A. Setting up a timber industry.

Plan B. Clearing land for farming by settlers from other parts of the country and building homes for the settlers in new villages.

Plan C. Exploiting the mineral resources of bauxite which is made into aluminium.

Choose *one* of these and on the map try to show how you would plan the use of the land for the particular choice you have made. The sheet for each plan will help you to decide the kinds of things that you should include on your map. Read the sheet carefully and try to use the information it gives you. As well as completing the map you must say *why* you have opted for the plan you have chosen and also give your reasons for rejecting the other two plans.

Points for all

1. When drawing up your plan try to think about the best locations for the timber mill, new village, railway line, etc. For example, if you choose a timber mill you will need to transport the logs to the mill and then transport the wood to the towns.
2. Complete your map and label it clearly.
3. Complete the key below the map.
4. Explain why you have chosen your plan.
5. Give your reasons for rejecting the other two plans.
6. Give more details of your plan when you have thought it out. Using your map describe how the plan will look when it is finished.

Plan A Timber industry

Timber mill (water is needed in the production process)

Areas cleared of trees

Areas scheduled for clearance

Forest left alone

Areas being replanted with trees (you need to plan ahead)

Houses for workers and their families

Services for the new village (eg. shops, petrol station, etc.)

Railway (for transport overland by train to other towns)

Roads (eg. to the mill from the village; from the village to the main road)

Power lines (and lines of communication, eg. TV and telephone)

Are there any other things that you would like to include in your plan? For example, perhaps you think it would be beneficial to have a jetty or wharf built so that goods, raw materials and people can travel by river; perhaps you would like to build a factory related to the timber mill, such as a factory producing paper, which will use a lot of water in the production process.

Plan B Farming and opening up the land for new settlers

Land cleared and scheduled for clearance

Land being farmed

Forest left alone

Farmhouses

New village

Services for the new village (eg. shops; petrol station)

Railway (for overland transport by train to other towns)

Access roads (eg. between the village and farmsteads; between the village and farmsteads and the main road)

Power lines

Are there any other things that you think should be included in your plan? If so, please include them, for example, other activities that the new settlers might do; facilities for river transport, etc.

Plan C Exploiting the bauxite resources

Loading depot (to transfer the bauxite to some form of transport)

Site offices and buildings

Houses for workers at the mine, ie, a new village

Services for people in the village (eg. shops, petrol station)

Forest left unaffected by these new developments

Railway (for overland transport of bauxite and people)

Access roads (to the mine and new village from the main road; between the mine and village)

Power lines (and lines of communication, eg. TV and telephone)

Are there any other things that you would like to include in your plan? Perhaps you think it would be beneficial to build a jetty or wharf so that you can use the river for transport. Perhaps you think it is a good idea to process the bauxite near the mine in order to reduce transport costs. If so, show on the map where you would build the processing factory.

Figure 2.18. Resource sheets for Amazon Forest planning exercise.

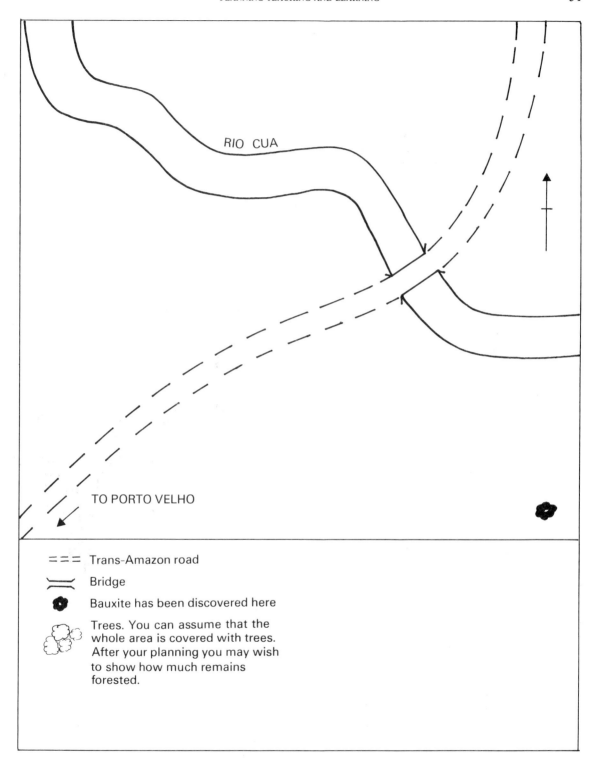

Figure 2.19. Map for Amazon Forest planning exercise.

THE AMAZON FOREST

Leave it alone!

As planners go ahead with their schemes for mining, forestry and agriculture in the Amazon Forest, news has reached us of strong reasons against cutting down any more forest. We ask you to read the report below and to think again about the plan you made in school today.

The Amazon Forest has about one million different kinds of plants and animals in it. Scientists call it a valuable storehouse of plants and animals. Many present-day plants and animals have developed or evolved from very old plants and animals in the rain forest. Now one or more kinds of plant become extinct every day as forested land is cut down and used for mining, forestry or agriculture. We are killing off plants on our planet faster than ever before in history.

Scientists say that it is unwise to go on destroying the plant and animal kingdoms in the Amazon Forest. A large number of different plants gives us a large storehouse. A large storehouse of plants for breeding will give us better plants for our future than a small storehouse.

In the planning exercise on the Amazon Forest which you did today you chose to plan a mining, forestry or agricultural development. You now have this new information.

Write a paragraph to a friend in your class saying *either* (1) how you would change or partly change your plan now, *or* (2) why you would not change your plan at all.

Figure 2.20. News flash for homework exercise.

Further steps in planning

Study of the lesson plan in Figure 2.16 shows that we have now moved beyond the first steps set out in Model 1 (Figure 2.14) consisting of (1) the use of key questions to select and order content, and (2) an appreciation of the generalisations towards which we are working, to making decisions about (3) what resources, what data to use and (4) how to organise the learning task, what strategy to adopt, as indicated in Model 2 (Figure 2.15). In the example group work is the chosen strategy. We have become aware hopefully of the complexity of the task and of how taking it step by step we manage the tasks and make it comprehensible to ourselves.

A further consideration as set out in Model 2 is (5) to analyse the activity for its likely aims and objectives. What is being achieved by teaching this activity? What are the likely educational outcomes? Are these valid, worthwhile outcomes to hold as aims and objectives? Do I need to revamp the activity in order to narrow the aims, broaden the aims, to include others? I have found that students are often more able to get a grip on aims and objectives if they are asked to sort them out at this stage. In a sense I have no quarrel with the objectives model of lesson planning described in the first part of this chapter. It is a matter of procedure, a matter of practice. I suggest using the concepts of aims and objectives to clarify what has been planned. In that way an analysis of aims and objectives becomes a way of evaluating the worthwhileness of the lesson activity. Students have often told me that they think up the aims after they have planned the lesson. Since this was my own habit I could understand the strategy and approve it, whereas I could not approve scrapping planning altogether!

The final consideration in Model 2 is (6) to give some thought to how to assess what you have hoped the pupils have learned. The homework exercises I have suggested in my example I see as a way of helping pupils to rework what has been presented and learned in class. It gives them a chance to sort through the ideas again and in marking the homework we will gain an indication of their level of understanding and the success of that particular teaching strategy.

An awareness of values

Figure 2.16 goes through a plan at some length. I think it will have been clear that once the key questions have been articulated we have logically obtained an idea of the generalisations, the understandings, the points of view, the ideas we want the pupils to explore, and the values we want them to think about. Let me emphasise the planning for values education and looking at different points of view in geography lessons a little more. For example, after a consideration of the recently planned uses of the Amazon, the question, *'Whose rights are to be considered?'* could be tackled. Such a lesson is dealing with points of view, not a generalisation as such, so the generalising or understanding we are moving towards concerns the validity of the reasons generated for the rights of one group. Sometimes the generalisation in a topic on settlements, for example, may be to do with the size of settlements and their spacing or the size of settlements and their functions. The nature of the generalisation depends very much on the paradigm of geography being used for teaching. A possible suitable activity which in actual fact is a values education strategy uses a mock newspaper extract. Entitled 'Amazon Tribe in Danger', it reads as follows:

> The largest remaining isolated tribe of Amazon Indians in Brazil is facing destruction. Anthropologists and missionaries working with the Indians are protesting against plans by a government agency to create 221 reserves for the Yanomami tribe which inhabits large areas of the forest.
>
> Protestors say that the reserves would be too restricted to sustain the Indians. They are farmers and hunters who move their villages every few years and thus need large areas of land to survive. Their system is much more suitable to the area than modern farming methods of new settlers which usually rapidly exhaust the fragile soil.
>
> Some of the worst accusations concern reports of how developers have killed Indians in the Amazon to make sure they did not interfere with plans for the area.

The following is the procedure:

1. Hand out newspaper extract 'Amazon Tribe in Danger', to be read through by teacher or around the class. Some recall of previous understanding of shifting agriculture, forestry and mining, the new colonists, genetic diversity, etc.

2a. Divide the class into two groups. One half are asked to take the view that these people want to be protected and *should* be protected against the proposal by the government agency. The other half are asked to hold the opposite view.

2b. Pupils individually write down three possible reasons for the view they are holding.

2c. Then grouped in like pairs each pair is to decide on the best four reasons they have come up with.

3. Pairs of opposite views join together in fours and discuss and select the best two reasons supporting each view.

4. The teacher lists some or all of the first ranked reasons on each side. A class vote, a free vote is then held on whether the Indians should be protected against the proposal by the government agency.

5. For homework the task is to write a report of the class work on the topic as if for the local newspaper or write an editorial on the issue.

This is one of four strategies for values education in geography which are discussed by Fien and Slater (1981). On a point of fact, the Brazilians have now created an Indian Park for the Yanomami which does include most of their tribal area. There is concern, however, that mining prospectors may enter this territory.

Planning for mixed ability

The worksheet/audio-visual strategy I described earlier is used with mixed ability classes in the first and second situations categorised by Storm (Figure

2.13) and generally speaking it works tolerably well. The class is handled so that each pupil works through the worksheet on her or his own with individual help if needed from the teacher at least two or three times in the course of the 70 minute period. The chance for learning episodes to occur between teacher and pupil is there.

An important feature of the worksheet/audio-visual strategy I have observed is its use of progressive questioning and as such it follows one of the three strategies for teaching mixed ability classes suggested by Richard Kemp in Chapter 8 of this volume. The questions become increasingly more difficult and more open ended in the sense that there is more opportunity for input from the pupil in the form of general knowledge, value judgements and decisions. It is expected that not all pupils will finish the work and that they will vary widely in the quality of their responses.

If the question of the rights of Indian tribes is to be tackled as part of a sequence of lessons on the Amazon Forest for a mixed ability class in Storm's situations 1 or 2 then some revamping may be necessary. As an activity, co-operative class discussion may be difficult and the teacher, while using the newspaper extract as an introductory resource, might need to provide two lists — one of reasons supporting Indian protection, the other of reasons against Indian protection. The pupils could be asked to rank both sets of reasons in order of strength of reason in their opinion; to add any other reason they think of; to write a sentence explaining whether they were for or against; and having decided, to justify their reasons in the form of a letter to the editor of the school newspaper; they are then to look at the reasons supporting the view they did *not* take and in two lists to say what is strong and weak about the reasons.

The homework might take the form of the moral dilemma described earlier. Moral dilemmas are very useful for dealing with attitude and value laden issues. For example, each pupil is a member of a committee of three advising the Brazilian government on its Amazon policy. The problem is to decide whether to construct a new road off the Trans-Amazon highway. The new road would be part of a wider plan to clear forest and set up a new farming settlement for poor people. It would mean taking land away from several Indian tribes. One member of the committee has decided to vote 'yes', the other 'no'. How will you cast your vote? Give your reasons.

The role of language in learning

I wish now to take up further the point made earlier about planning for the educational process, planning for learning and to expand the concept of process. We decide on content. We decide on how to present the content. How are we to give pupils some opportunity for *working over* the content so that *they make it their own* and expand their understanding of it in the same way as the boys in the well known porous rock example made new meaning for themselves out of old meaning (Schools Council, 1973)?

Assuming a co-operative class, I chose a group work strategy for developing ideas about the rights of Indians. Why? How is this related to learning as a process? We all have an intuitive grasp of the meaning of the phrase 'talking to learn'. We all know how talking over and through ideas helps to clarify and sharpen meaning. The role of language in learning is discussed by Margaret Roberts in the next chapter and other references should be consulted (eg, Barnes et al, 1969; Sutton, 1981).

I simply want to stress here that opportunities for exploring concepts and discussing ideas need to be built into lesson plans as distinct and necessary episodes. These may be episodes in which the teacher does not always share but they *are* episodes nevertheless, which allow pupils to sort out concepts and ideas, to rub ideas against one another, to polish understanding. Brainstorming, buzz groups, structured discussion as in the example above are teaching strategies, not gimmicks. Such teaching strategies are essential to an engagement in an educational process which is but a step along the road to achieving an educational product, but one which is being continually transformed. The spontaneous pupil-initiated learning episodes will not cease to occur and in fact will occur more abundantly when episodes that take into account the role of language in learning are incorporated deliberately into lesson plans.

The 'talking to learn' school has not only identified the role of language in learning but has also suggested for very good reasons that teachers should (1) provide a variety of audiences for spoken and written work so that there is a decrease in the number of times the teacher is audience and evaluator; and (2) give pupils a range of writing or talking purposes — sometimes suggest they write a letter to a friend, younger or older than themselves, to a newspaper or as if to prepare a speech.

Another way of expanding the range is to provide the pupils with a role. I deliberately chose to suggest a homework exercise as a newspaper report or editorial after the group discussion as a way of helping pupils to practise organising ideas and arguments. Likewise, I chose a moral dilemma as homework for a mixed ability class as a way of giving them a role and a different perspective from which to view the problem. I need hardly add that the structured discussion engages pupils in examining attitudes and values, as does the moral dilemma. Again, I deliberately selected these two very useful strategies as examples which can be built into lessons when value laden issues lie at the core of the content.

Selecting teaching strategies is as important as selecting content. I am back to my simple two stage model of lesson planning and stage two is now divided into (a) and (b). (1) Select content, (2a) think about presentation and (2b) select teaching strategies which frequently allow talking or writing opportunities and so take into account the role of language in learning.

Conclusion

I have sought to define a case for lesson planning in relation to people's preferences and their notions of what constitutes teaching; to describe models containing essential steps in lesson planning, given a variety of management situations and a wide ability range in any one class; to give examples of teaching strategies useful to values analysis and discussion and congruent with the concept of education as process; and finally to raise the issue of the significance of language in learning and its centrality in the learning process. I know that the suggestions and principles I have set out are feasible; I hope that my readers are convinced that they are desirable and useful.

References

Barnes, D. et al, (1969) *Language, the Learner and the School*, Penguin.

Fien, J. and Slater, F. (1981) 'Four strategies for values education in geography', *Geographical Education*, vol 4 pp 39-52, reprinted in Boardman, D. (ed) (1985) *New Directions in Geographical Education*, Falmer Press.

Richardson, R. (1983) 'Daring to be a teacher' in Huckle, J. (ed) *Geographical Education: Reflection and Action*. Oxford University Press.

Schools Council (1973) *From Information to Understanding: What Children do with New Ideas*. Schools Council and University of London Institute of Education Joint Project: Writing Across the Curriculum 11-13 years, Ward Lock.

Slater, F. (1982) *Learning through Geography*, Heinemann.

Storm, M. (1979) 'Some tentative thoughts on a taboo topic' *Geography Bulletin*, ILEA, No 6, pp. 5-7.

Sutton, C. (ed) (1981) *Communicating in the Classroom*, Hodder and Stoughton.

Tolley, H. and Reynolds, J. B. (1977) *Geography 14-18: A Handbook for School-Based Curriculum Development*, Schools Council/Macmillan.

3. Teaching and Learning in the Classroom

3.1. Approaches to Teaching and Learning

Eleanor Rawling

'... the truth is that the best resource is a sensitive teacher'   (Eric Brough, 1983)

How we teach

How do you approach your teaching in geography? In what ways do you help your pupils to learn? Do you always use the same approach? Or do you vary your classroom strategies according to the characteristics of the topic or of the pupils? Have you ever asked yourself why you do things this way? Are you satisfied with the responses of the learners?

Until we have seriously asked ourselves questions like these, a discussion of teaching approaches will be meaningless. Many discussions of teaching approach take place as if the only consideration were the choice of classroom methods and techniques from those currently available. One may choose, for instance, to involve pupils in a role play, to propound theories by means of an illustrated lecture, or to send them out to 'experience' the local town for themselves. It is clear at a superficial level that each method involves a different organisation and results in a different kind of pupil response — but perhaps there is more to it than that ...

The four photographs (Figure 3.1) present four different teaching-learning situations and the comments beneath each one draw attention to the piece of conversation or thought which best represents that situation. One way of analysing the four situations is to consider each under the following headings:

   pupil activity and role;

   role and involvement of the teacher;

   general character and atmosphere of the learning situation;

   intended outcome or end product of the event.

Try writing some comments on each photograph, using these headings as a guide.

Figure 3.2 presents one version of a possible list of such comments. This analysis presents four very different teaching approaches, and refers to a whole range of characteristics relating to all aspects of the teaching-learning situation. Teaching approach analysed in this way is much more than just the particular classroom strategy chosen. Indeed, from the photograph it is not always clear which actual strategy is being used. Teaching approach is also a matter of pupil-teacher relationship, of social grouping, of classroom management and inevitably of the aims and objectives implicit in the exercise. For instance, the relative formality of the

TEACHING AND LEARNING IN THE CLASSROOM

(a) Teacher: "Yes, that's an interesting feature. You might find it useful now, to examine the photograph carefully and to refer back to the list of characteristics which we discussed in class. This might help you to identify the feature."

(b) Pupil: "I wonder what it's like living near to all these derelict wharves. The river smells oily and it makes a funny splashing sound as it laps against those posts."

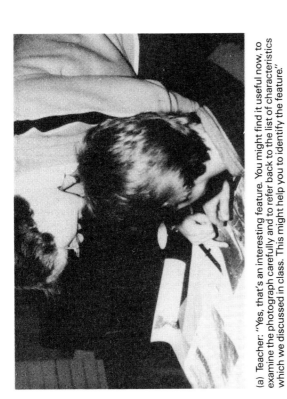

(c) Teacher: "I want you to draw your own version of this diagram and then to make some notes about the main characteristics of the feature shown."

(d) Pupil: "This book which the teacher suggested contains some useful information. I think it will help me to answer the question I've been studying."

Figure 3.1. Four different teaching and learning approaches.

classroom situation in photograph (c) and the apparent emphasis on the class as a group contrast markedly with the more personal and informal level of activity shown in (b). Similarly, the intended end product of the activity in (c) might be assumed to be the acceptance and understanding of a received body of knowledge, although the intended outcome of the experience being gained in (b) might well be of a more abstract and exploratory nature.

One of the difficulties in analysing photographs is that it is necessary to 'guess' at the context of the work and at the atmosphere generated. A more meaningful analysis of teaching approach will be obtained if you analyse your own or a colleague's classroom approaches in this way. The resulting analysis might surprise you; it will certainly provide a good basis for consideration of the ideas in the rest of this chapter.

A continuum of teaching-learning approaches

One way in which to consider different teaching-learning approaches is to envisage a continuum such as that shown in Figure 3.3. The diagram suggests that there is a range of methods available to teachers and that their choice and use relate to specific types of pupil learning activity. Exposition, narration and reception learning, for instance, occupy one extreme of the teaching-learning continuum and are characterised by a relatively low level of pupil autonomy, since the teacher is totally in control of the situation. Moving along the continuum, other possible teaching strategies and learning activities are identified. These include the close direction of question analysis and problem solving activities, the provision of advice and guidance in open-ended discovery situations, and finally, at the other end of the continuum, encouragement and support given to creative activity. In this latter situation, the pupil retains the dominant responsibility for learning taking place. The situations analysed in the four photographs could be located on the continuum; photograph (c) possibly representing the left hand end of the diagram — exposition and reception, photograph (a) providing an example of directed problem solving, photograph (d) illustrating individual and possibly open-ended discovery, and (b) showing a pupil involved in individual creative reflection. It must be emphasised at this point that the continuum diagram should not be interpreted as a rigid classification according to which every teacher can be categorised. Nor is it intended that any methods of approach are seen as better or worse than others. A teacher sensitive to the needs of the pupils is likely to make use of strategies characteristic of any point on the continuum at some time. All of the photographs (a) to (d) might then be illustrating different aspects of the work of one class.

Another important point relates to the use of the term 'enquiry'. Whether the teaching-learning situation is one of exposition and reception or of creative activity and support, it may be described as 'enquiry' if the activity is orientated towards answering questions and opening up problems and issues. More usually, however, 'enquiry-learning' tends to be equated with pupils 'learning by doing' and so with approaches from the right hand end of the continuum.

The continuum diagram is valuable because it provides assistance in making the link between teaching approach and learning activities. Where, as in photograph (c), the teacher expounds a theory or a set of information directly to pupils, the intended learning activity is inevitably that of listening, receiving information, memorising it and, perhaps, applying it later. When the teacher encourages a pupil to enquire into a topic on her own, as in photograph (d), offering support and advice if necessary, but no rigid structure, the pupil will automatically be required to use a different range of skills to collect, analyse and process the information needed, and to organise her own programme of work. It is important that teachers perceive this link because it has direct implications for the outcome or end product of teaching-learning activity. Pupils who spend all their time receiving wisdom in a relatively passive way from the teacher will never find the opportunity to practise other skills such as decision-making or planning their own programme of work. Similarly, pupils who are always following up their own individual enquiries may never be introduced to some particular piece of geographical knowledge that does not impinge on their work, and might rarely find the opportunity to distil important ideas from a well presented lecture.

The style of classroom approach is then closely related to the outcomes required in terms of pupil achievement and pupil attitude. Examples to illustrate this important point can be given from work with different age groups.

TEACHING AND LEARNING IN THE CLASSROOM

	Photograph 1(a)	Photograph 1(b)	Photograph 1(c)	Photograph 1(d)
PUPIL ACTIVITY AND ROLE	Working individually Using a range of resources Referring to teacher for help and advice Organising own work but within set limits	Observing and reflecting No obvious formal learning activity No evidence of teacher guidance Outside classroom	Working as a class Listening, watching, taking notes Teacher is main source of information Seated formally	Working individually Using reference book and noting contents Book suggested by teacher provides information Relative freedom to select appropriate evidence
TEACHER ROLE AND INVOLVEMENT	Teacher offering support and guidance within a relatively structured task Maintaining informal but well-defined pupil-teacher relationship	Pupil apparently organising own experience in an informal outdoor situation No obvious task direction	Teacher acting as main source of information and director of activities Maintaining formal pupil-teacher relationship Formal class layout	Pupil maintaining high degree of control over learning situation Provided with some structure and assistance
GENERAL CHARACTER AND ATMOSPHERE	Mutual enquiry and study Background guidance Relatively informal	Individual reflection No formal structure	Formal Structured Teacher in control	Individualised learning Background Support
INTENDED OUTCOME OR END PRODUCT	Pupils: develop certain defined enquiry skills and use them to solve a set problem Teacher: maintain formal position in a relatively informal way	Pupil: experience environment in own individual way Teacher: learn more about pupil as an individual?	Pupils: receive and learn a given set of information — develop certain defined skills Teacher: maintain dominance in classroom	Pupil: appreciate aspects of the enquiry process — begin to find answers to own questions Teacher: enhance own enquiry skills through interaction with pupil

Figure 3.2. The four teaching and learning approaches analysed.

PUPIL LEARNING ACTIVITIES

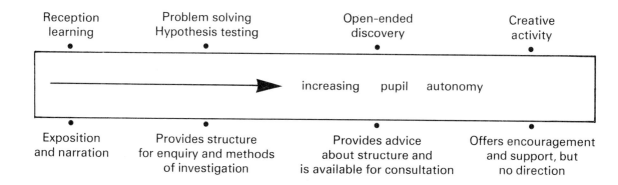

Figure 3.3. The teaching-learning continuum (Adapted from Bartlett and Cox, 1982).

Drama and creative work in geography

Figure 3.4 outlines a series of lessons developed by John Earish, geography teacher at Bishop Kirk Middle School, Oxford, as the geographical component of an Environmental Studies course for 11-12-year-old pupils. The central theme is 'The Formation of Ice and How It Shapes the Landscape', and within this, four units of work draw on a variety of dramatic activities and creative experiences.

If the main objective for these units of work had been to pass on information about glacier ice, and to ensure that pupils memorised it for later recall, then the teacher might have chosen to give an illustrated talk and to ask his pupils to make detailed notes. In this particular instance, however, the class comprised 11-12-year-old pupils of a wide range of ability. Many of the pupils were experiencing difficulties with reading and writing, and as a consequence lacked self-confidence. More able pupils, although competent in reading and written work, were not easily able to appreciate the problems of their slower peers. In the class as a whole, a major difficulty in geography seemed to be that of making the link between everyday environmental experience and the abstract spatial and environmental concepts of an academic subject. Given this situation, the teacher realised that his objectives for the series of lessons were related more to the social and personal development of individual pupils than to the knowledge content of physical geography. Accordingly, his lesson plan drew on drama and creative work. The topic of 'the formation of glacier ice' became a useful medium through which pupils could develop a variety of skills and participate in a range of educational experiences. The teacher listed the following objectives for the lessons:

to encourage pupils to develop imaginative and emotional responses to environmental experiences;

to allow opportunities for the development of communicative skills other than reading and writing;

to assist pupils to link concepts in geography with everyday experiences in their own lives;

to enhance the self-confidence and opportunities for success for pupils who may not excel in more formal geographical work;

THE FORMATION OF ICE AND HOW IT SHAPES THE LANDSCAPE

UNIT OF WORK	IDEAS INTRODUCED	ACTIVITIES UNDERTAKEN
UNIT 1	Snowflakes are made of small ice crystals arranged in random patterns	• Looking at pictures and slides of snowflakes • Making snowflakes using paper cutouts, polystyrene chips, tissue paper, etc. The snowflakes can be made into a mobile
UNIT 2	The formation of glacier ice can be traced through a sequence from snow to névé to glacier ice	• Improvisation — pupils develop a dramatic incident that may have happened to them during snowball fights/sliding/sledging. Pupils work in small groups and then present their ideas to the rest of the class. • Discussion around the idea that snow crystals are crushed together and air is squeezed out. This can be reinforced by arranging the class in the shape of a snowflake, each pupil representing an ice crystal. They then move slowly together until there is very little space between them. • Relating this idea to the pupils' own experiences (eg. crushing snow to make a hard snowball, stamping on the snow to make a slide, etc). • Linking this idea with the processes taking place in formation of a glacier. It is also useful to look at pictures and colour slides to reinforce this idea.
UNIT 3	Rock and debris is frequently frozen into ice sheets	• Improvisation — pupils develop a dramatic story on the theme 'the hidden stone in the snowball'. This can be initiated from a discussion of personal experiences or a story reading. Pupils work in small groups and then present their ideas to the rest of the class. • Discussion — the teacher introduces the idea that rocks and stones can be crushed and frozen into the ice. This can develop from pupils' own experiences of stones and sticks being incorporated into snowballs when they are rolled along the ground.
UNIT 4	Moving ice sheets can shape the environment	• Making sandpaper using various grades of sand and grit. Shaping balsa wood or polystyrene blocks using different grades of sandpaper. Shapes can relate to landforms or animals found in the Arctic. • Looking at a wide selection of pictures and slides and listing the way glaciers smooth and shape the landscape. • Using polystyrene blocks to illustrate how ice is not elastic and will snap and form crevasses and icebergs.

Figure 3.4. Drama and creative work with 11-12-year-old pupils   Source: John Earish.

Figure 3.5. Pupils enacting a snowy day incident.

to enable pupils to work in groups and to appreciate the skills and abilities of others;

to create initial awareness and understanding about the formation of glacier ice upon which more formal geographical teaching can be based;

to stimulate enjoyment and enthusiasm for geography.

Figure 3.5 presents a view of the classroom in which some pupils are enacting their 'snowy day incidents'. It is clear from the photograph and the background provided above, that these lessons incorporate a mix of approaches from the middle and right hand end of the continuum. Within a well organised enquiry structure, the teacher has provided opportunities for open-ended discovery and genuine creative activity as well as for problem solving. Subsequent lessons made use of more directed learning, using maps, slides and simple textbooks. Overall, then, the teacher's choice of teaching-learning approaches for this series of lessons reflected his desire to balance knowledge objectives with broader social objectives.

Enquiry-based learning with 16-19-year-olds

A further example of the link between educational objectives and choice of teaching-learning approach, this time at the broad curriculum planning level for older pupils, is provided by the work of the Geography 16-19 Project. As a result of the Project's early work, it became clear that teachers believed that it should be possible to use 16-19 geography courses to equip students adequately for the real world beyond school. For this to occur, however, it would be essential to stress the development of skills and abilities, and to allow opportunities for values enquiry as well as to provide geographical knowledge. Accordingly, the Project has developed an enquiry-based approach to teaching and learning.

Enquiry-based teaching and learning may be defined as encompassing a range of teaching methods and approaches by which the teacher encourages students to enquire actively into questions, issues and problems rather than merely to accept passively the conclusions and research of others. The main characteristics of an enquiry-based approach with 16-19-year-old students are that it:

identifies questions, issues and problems as the starting points for enquiry;

involves students as active participants in a sequence of meaningful learning through enquiry;

provides opportunities for the development of a wide range of skills and abilities (intellectual, social, practical and communication);

presents opportunities for fieldwork and classroom work to be closely integrated;

provides possibilities for open-ended enquiries in which attitudes and values may be clarified, and an open interchange of ideas and opinions can take place;

provides scope for an effective balance of both teacher-directed work and more independent student enquiry;

assists in the development of political literacy such that students gain understanding of the social environment and how to participate in it.

Reference to Figure 3.3 will show that enquiry-based learning is focused around the 'structured problem solving' and 'open-ended discovery' points on the continuum, although with the facility to draw on expository methods or to expand into creative activity, as appropriate. The character of the approach may be more easily appreciated by reference to an example. Figure 3.6 relates to the set of teaching exercises which comprise the Project booklet *Energy and the Environment* (Geography 16-19 Project, 1984). This booklet was written for students following the Project's Advanced Level syllabus and hence the character of the exercises and activities within it reflect the enquiry-based teaching-learning approach demanded by the course. Figure 3.6 reveals both the broad enquiry sequence followed throughout the booklet and also the wide range of activities incorporated. Study of the last column, the special emphases within each main exercise, will also show that varied teaching-learning approaches can be identified.

Values education — creating the opportunities

One important consideration in choice of teaching approach concerns the opportunities provided for values education. In the past, it was frequently assumed that teaching the right facts about the environment would automatically lead pupils to take up attitudes and values acceptable to society. In any case it was assumed to be the task of parents, the community and society at large to deal with moral and ethical questions in education, not the role of the geography teacher.

Many teachers now believe that this is an unsatisfactory approach, both to geographical education and to values education. Few, if any, topics in geography can be dealt with in a coherent way without reference to the influence of attitudes and values. People's viewpoints, opinions and deeply held values influence the way they react to environmental matters, the way they make decisions about the environment, and the way they take action. Failure to recognise this in geography teaching is to ignore a major environmental factor. Industrial location theories or urban spatial models, for instance, studied without an element of values enquiry, become arid, meaningless and divorced from the real world. More importantly, however, it seems that a failure to provide for the consideration of attitudes and values results in a missed opportunity for geographical education, and a potential narrowing of the curriculum for the individual pupil. Unless issues involving a range of viewpoints and opinions are opened up in the classroom, pupils may find few chances to understand the nature of attitudes and values, to clarify their own values and to develop their own convictions and commitments. Work in geography, focusing as it often does on questions and issues about the environment, can provide a valuable medium for such values education.

One of the main problems for geography teachers is that of choosing an appropriate approach to values education through geography. Many environmental issues are fraught with conflict; it often seems impossible to present a complete picture of all sides of the question. If strategies like role play are used, it is frequently feared that inordinate amounts of time will be spent on a 'mere game' with little to show afterwards in terms of pupil learning.

In any case, there are several different approaches to values education according to whether it is felt desirable to teach some specific values or only to assist pupils in analysing and clarifying values such that they can make up their own minds in every case.

At a practical level, however, a good starting point for the teacher who wishes to consider and develop strategies for values education is to evaluate his or her overall teaching approach. This can be done by referring back to the teaching-learning continuum (Figure 3.3). If formal expository methods or tightly structured problem-solving activities, characteristic of the left hand end of the continuum dominate; if pupils rely solely on the teacher for answers and guidance; if work is always channelled within well established frameworks; if these are accurate descriptions of the teaching approaches most frequently used, then it is unlikely that pupils will be provided with more than a superficial reference to the affective domain of attitudes and values.

In order to develop skills in values enquiry, pupils require experience of analysing and discussing conflict situations, of listening to arguments and opinions, of meeting people with strong viewpoints, of seeing the results of decisions taken on the lives and environments of local people, and of expressing, sometimes in a creative way, their feelings for a situation or environment. It will be apparent that the opportunities for such activities are most likely to derive from approaches from the middle and right hand end of the continuum.

However, it may not be sufficient merely to add strategies like role play and open-ended group discussion to teaching-learning activities and to

HANDBOOK FOR GEOGRAPHY TEACHERS

ENERGY AND THE ENVIRONMENT

Section of booklet	Resources used	Main student activities	Intellectual skills required	Special emphases
EXERCISE 1 'Energy, at what cost?'	impact page: photograph, map, diagrams, news articles.	drawing out ideas from the impact page; writing lists; class discussion.	simple comprehension, interpretation and analysis, initial introduction to values clarification.	student motivation through immediate acquaintance with important and well presented issues.
EXERCISE 2 'North Sea Oil and the Shetland Islands'	pie chart, graphs, simple sketch maps, statistical table, news article, written report.	interpreting charts and graphs; listing ideas; writing paragraphs; producing maps and plotting data; deriving general ideas from a news article; summarising a situation from a variety of data.	simple comprehension, interpretation, analysis, evaluation, synthesis, simple problem solving.	student use of data in a variety of graphic forms (maps, charts, diagrams).
EXERCISE 3 'The Impact of the Sullom Voe Project'	sketch maps, thematic maps, statistical maps, statistical tables, theoretical model, photographs, data sheet, written reports, news articles, and journal articles.	analysing and interpreting maps; report writing; role-playing; completing tables; drawing sketch maps; summarising information; reading articles; creative writing; individual reflection; statistical calculation; applying a model to a new situation; group and class discussion; constructing choropleth maps.	comprehension, interpretation, analysis, evaluation, prediction and theorising, synthesis, decision-making, values analysis.	student involvement in role play/simulation. opportunity for individual reflection and creative writing. student involvement in using and applying a model.
EXERCISE 4 'Decisions and Plan-making for Sullom Voe'	sketch maps, tables, photographs, planning data.	role-playing; interpretation of situation from maps, tables and planning data: writing a brief for a meeting; taking part in group discussion and debate; summarising and report-writing.	comprehension, analysis, interpretation, evaluation, clarifying values, decision-making.	student opportunities to understand alternative viewpoints and to evaluate decision-making about an environmental issue. involvement in group discussion and debate.
EXERCISE 5 'Decision-making and the Environment'	flow diagrams, written report summary.	use and interpretation of flow diagram; writing paragraph; redrawing flow diagram; imaginative reconstruction of a situation; essay writing.	comprehension, analysis, interpretation, generalisation, evaluation, classifying and putting forward views and opinions.	opportunity to identify general ideas reached in the work, and to clarify own values.

Figure 3.6. Analysis of booklet for 16-19-year-old students Source: Geography 16-19 Project, School Curriculum Development Committee, 1985.

assume that in this way effective values education will take place. There is a need to be clear about detailed objectives and about ways of achieving them. Indeed, increasingly, more specific guidelines are appearing for teachers who wish to provide values education through geography. Four different strategies are proposed in an article by Fien and Slater (1981): Values Clarification, Values Analysis, Moral Reasoning and Values Probing. By applying each of these to one issue, the authors show that teachers can use different methods with their pupils. The methods incorporate activities which range from individual reflection and speech writing to group discussion, role playing and structured analysis of conflict situations. An important point which the authors emphasise is that teaching approaches to values education have been designed for different valuing objectives and should be selected to accord with the objectives of lessons and units of work. Only if this is done, Fien and Slater argue, will it be possible to use geography as an effective medium for values education.

The Geography 16-19 Project has also devised strategies for values education as part of its enquiry-based approach to learning. Figure 3.7 shows the Project's route for geographical enquiry. This has been designed to help teachers to structure their teaching approach and their teaching materials and to develop the study methods employed by their students. The route for enquiry has two aspects: a sequence of factual enquiry is set out in the left hand column, and the corresponding stages of values enquiry are itemised in the right hand column. In practice, of course, it is not possible to separate the two aspects of any enquiry. The diagram presents the route in this way in order to assist with lesson planning. Study of the values enquiry procedure in Figure 3.7 will reveal that it seeks to make pupils aware of their own values through the presentation of attitudes and values held by others in the situation under study. In the final stages of the route, they are encouraged to evaluate and judge the situation from their informed personal standpoint, answering questions like: Which alternative would you have chosen? What decision would you have made? Why? If your choice matches the decision made, are your reasons the same? If your choice is different, justify why? Finally, the end stage of this enquiry route is set out as one of personal action. The Project believes that if action of some kind is possible and appropriate, then 16-19 students should be encouraged to commit themselves. 'What will I do?' is an inevitable final question demanding an answer in this approach to values education.

Although the Geography 16-19 route for enquiry has been developed with a specific age group in mind, at a general level it has application to all geography teaching and learning. It emphasises the importance of offering pupils a range of experiences other than formal expository teaching, and it makes clear the direct relationship between specific values education objectives and particular teaching strategies (Naish, Rawling and Hart, forthcoming).

An appropriate approach

It seems, then, that choice of teaching approach is concerned with answering the question 'why?' as much as with responding to 'how?'. It is important to ask 'why are we teaching these pupils?' and 'what are we trying to do?' before selecting appropriate approaches for the classroom.

If the objectives of teaching and learning are predominantly derived from the subject and encompass the desire to put across specific items of knowledge, to develop a certain level of conceptual understanding and to provide for the acquisition of specific defined skills, then it may be possible to select formal exposition-narration methods and structured problem solving approaches to carry this out effectively.

If, however, the objectives of teaching-learning derive predominantly from the needs of pupils, and in addition to knowledge and intellectual skills incorporate the development of a wide range of social, practical and communication skills, and the provision of opportunities for clarifying and developing values, then a wider range of approaches selected from across the full continuum will be necessary.

The idea of developing an approach to teaching and learning which encourages the teacher to draw on a variety of methods in a balanced way is one which finds favour amongst many educationalists (see Lawton, 1981). Too often, however, the demands of examination syllabuses, and the difficulties of classroom management and resource provision, inhibit the teacher from selecting the classroom approaches most appropriate to the needs of the pupils. Exposition and narration or closely structured hypothesis testing are often the easiest ways to impart the content required for the

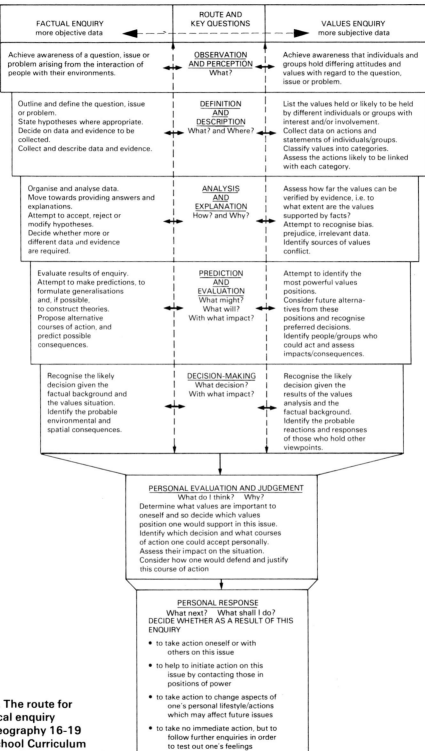

Figure 3.7. The route for geographical enquiry
Source: Geography 16-19 Project, School Curriculum Development Committee, 1985.

examination, or to manage the class as a uniform group when resources are short. On their own, however, these are not the best methods to ensure that pupils develop a wide range of skills and abilities, enhance their self-confidence and achieve the enjoyment and stimulation to carry their learning into the outside world.

Equally, however, a total commitment to open-ended discovery and creative activities is unlikely to be advantageous to the pupil. As Lawton states 'Clearly it is very important not to go to extremes in this ... No child left to his own process of "discovery" can produce a structure in science or social studies which will have taken hundreds or thousands of years' expert work to evolve; the teacher's job is to provide short cuts to significant learning; guided discovery, not blind groping in the dark, is the correct teaching formula. The art of teaching is to enable children to make discoveries they would not encounter on their own' (Lawton, 1981).

Lawton refers to the 'art of teaching'. Indeed, a final important comment on teaching approaches must inevitably return to the role of the teacher. There is no easy formula or short cut to managing the teaching-learning situation. Ultimately when objectives have been set, methods selected and individual strategies planned, it remains for the individual teacher to make them effective. The most formal lecture can become an exciting and exhilarating experience — a stimulus to real enquiry; the most innovative and open-ended game can invoke apathy and boredom. The teacher is the key to a successful and effective teaching approach.

Given the wide range of pupil needs and the fact that transmission of knowledge is only one small part of the teacher's role, it has been suggested that the teacher needs to make available all kinds of other opportunities and activities for pupils. More than this, however, each teacher is an individual person and as such has an immeasurable potential for enhancing his or her pupils' education.

Each person has individual attitudes, viewpoints and feelings; each has experience of the wider world, of relationships with other people and of taking part in other activities; each has potential for continued learning. In all these aspects, teachers, as people, can assist the development of their pupils.

If, for instance, pupils are to learn to discuss and debate retaining tolerance, sympathy and good humour, then the teacher must display these characteristics as well. If it is thought desirable to reveal that environmental issues create conflict and that people hold different views, then this point should become clear in debating the issues with the pupils in the classroom. If it is important that pupils develop a realistic view of the subject's potential and of its contribution to real world problems, then a realistic view must be reflected in the attitude and approach of the teacher. If it is believed that geography is worth teaching and can provide a basis for action, then the teacher in the classroom must be prepared to encourage pupils towards commitment and action, and indeed, reflect this belief in the way that he or she lives.

Such a personalised 'teaching approach' may sound frightening — in fact it need not be so, because it is essentially a matter of revaluing oneself as a person and by so doing assisting pupils to revalue themselves as people.

'I am a teacher more than a geographer and a person more than a teacher' writes Robin Richardson (1983). It might be suggested that this extract makes an appropriate concluding comment to this section on teaching and learning in geography, by returning to the sentiments of the quotation which headed it.

Having analysed 'how we teach' and 'why we teach' and then having identified appropriate strategies, the most effective approach is still a matter for each one of us to resolve with our pupils, '... the truth is that the best resource is a sensitive teacher'.

References

Bartlett, L. and Cox, B. (1982) *Learning to Teach Geography*, John Wiley, Australia.
Brough, E. (1983) 'Geography through Art', in Huckle, J. (ed) *Geographical Education: Reflection and Action*, Oxford University Press.
Fien, J. and Slater F. (1981) 'Four strategies for values education in geography', *Geographical Education*, vol. 4, pp. 39-52, reprinted in Boardman, D. (ed) (1985) *New Directions in Geographical Education*, Falmer Press.
Geography 16-19 Project (1984) *Energy and the Environment*, Longman.
Lawton, D. (1981) *An Introduction to Teaching and Learning*, Hodder and Stoughton.
Naish, M., Rawling E. and Hart, C. (forthcoming) *Geography 16-19: The Contribution of a Curriculum Project to 16-19 Education*, Longman.
Richardson, R. (1983) 'Daring to be a teacher', in Huckle, J. (ed) *Geographical Education: Reflection and Action*, Oxford University Press.

3.2. Talking, Reading and Writing

Margaret Roberts

Introduction

Language plays a vital role in every geography classroom. Pupils fill pages of exercise books and files with written work. They read textbooks, resource sheets, library books and worksheets. Teachers describe, explain, instruct, ask questions, encourage, reprimand and listen. Pupils listen, answer questions, discuss and sometimes ask questions.

Since the publication of the Bullock Report *A Language for Life* (DES, 1975) teachers and researchers have given increasing attention to all this writing, reading and talking which is taking place in our classrooms. The work which has been done so far is valuable to geography teachers for three reasons. Firstly, it provides examples of ways in which teachers can begin to investigate what is going on in their own classrooms. Secondly, the work suggests strategies for developing pupils' language skills. Thirdly, the research gives us insights into the way language enables us to learn. All pupils learning geography already have some sort of world picture, however inaccurate or incomplete. They make sense of new experience and new knowledge in terms that they already know. They have to reconstruct new information in their minds to give it meaning. Most research suggests that the role of language is very important in the learning process. Geography teachers can help this process of reconstruction by careful planning of teaching and learning or they can prevent or hinder it by excluding some uses of language from their lessons.

This section draws upon the work which has been done on talking, reading and writing in the classroom to suggest investigations which teachers might carry out, strategies to use to develop pupils' language abilities, and types of activities which will help pupils to make sense of geography for themselves.

Talking

Geography teachers who want to be more aware of the spoken language of their lessons could start by making some tape recordings. These will show that a lot more is going on than the teaching and learning of geography. Teachers decide the kinds of talk which are permissible by pupils, who can talk, when and for how long. They are establishing the social context within which learning takes place. Underlying the surface meaning of classroom talk are messages conveying to pupils what teachers think is important about geography and what pupils' part in the learning process should be. Often the meaning of a tape can be understood only with knowledge of previous pupil-teacher interactions. Also some messages are conveyed in gestures, others are conveyed more by intonation and pause. So listening to recordings needs to be done with caution.

A first step in analysing the tape could be to measure the amount of pupil and teacher talk. Is the balance consistent with our aims? Most teachers talk too much. Listening to teachers is a relatively passive activity.

'The thing I hated was listening to you talking ... just standing in front of us all and going on. That's why people get bored.'
— *Comment by 14-year-old boy on student teacher's lesson.*

Yet some teacher talk is highly desirable. It is a means of conveying excitement in the subject, of motivating pupils and of introducing pupils to the specialist language of geography which is best done in talk where meanings can be fully explored. Research has shown that teachers are more conscious of their use of specialist vocabulary than of the general level of vocabulary and sentence structure.

'The teacher explained using all long words and none of us understood. When a boy said he didn't understand the teacher said the work was easy and he ought to understand it.'

— *Comment by 13-year-old girl on student teacher's lesson.*

Often the work is easy but teachers can make understanding difficult by using unfamiliar phrases such as 'limited by', 'the distribution of', 'in excess of', 'giving rise to', assuming that pupils understand. Listening to tapes of themselves can make teachers aware of the language they use and how much they make its meaning clear.

Most teachers break up their talk with question and answer sessions. There are many analytical frameworks to describe teachers' questions. They need to be used carefully. Often the meaning of a particular question can be understood only in the context of several exchanges or in the context of particular teacher-pupil relationships. Some sort of analysis, however, can indicate what we are up to when we ask questions.

It is worthwhile to consider two dimensions of questioning. Firstly, what type of *thinking* does a question encourage? Secondly, is the question *open* or *closed*? These two dimensions are shown in Figure 3.8 with some examples.

The first dimension (shown on the vertical axis) ranges from factual recall to hypothetical thinking. Research indicates that the majority of questions asked require factual recall or limited comprehension. Modern syllabus objectives emphasise comprehension, analysis and developing hypotheses rather than memory. Yet the message that we convey in our questioning is that memorising knowledge is still of overriding importance.

The second dimension of questioning (shown on the horizontal axis) gives an indication of how much scope we are giving pupils to develop their own meanings. A closed question is one which has only one acceptable answer. An open question is one in which a range of answers is possible.

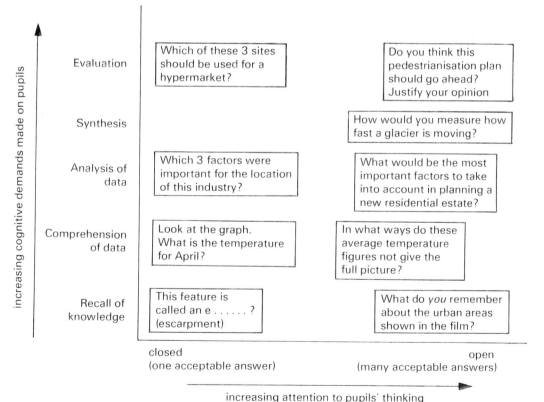

Figure 3.8. Two dimensions of questioning.

Most questions asked in geography lessons are closed. They are asking pupils to tell teachers what teachers already know. The teacher's purpose is to take pupils through a particular line of reasoning, in this way keeping control of the way geographical knowledge is structured. As pupils get older they become reluctant to take part in this curious procedure of telling the teacher what is already known. Teachers complain that they cannot get pupils to talk in class:

'It is infuriating. I ask questions and I know they know the answers but they sit there silently.'
— *Comment by student teacher.*

Open questions allow pupils to put into words what is in their own minds rather than guess what is in their teacher's mind. They enable pupils to make their own sense of new knowledge and to interpret it in the light of what they already know. Open questioning assumes that what pupils have to say, what they understand or misunderstand is important even if it does not fit the teacher's line of thinking. It leads to exploratory talk, to much better class discussion and to greater willingness on the part of pupils to participate. The difference between an open and closed question is often not evident in the words. For example a teacher might ask, 'What do you remember about the shanty towns you saw on the film?' This could mean 'I want you to tell me the things *I* think are important which I remember' (closed question). Or it could mean, 'I'm interested to know what kinds of things *you* remembered'. This is open because all recollections are valid. The essential difference is the intention of the teacher. Is it to understand the pupils' thinking or is it to limit the pupils to the teacher's thinking?

Similarly, 'Why do you think people moved to the shanty towns?' might invite the reproduction of information given in the film or it might invite the pupils' opinions, speculations, hypotheses. The differences are subtle, but if teachers listen to the way they question pupils, they can consider how much opportunity pupils have to express their own thinking and communicate their own ideas. Consideration of these two dimensions of questioning is also valid for teachers talking to individuals and small groups.

Class teaching, however well organised, gives limited opportunities for each individual to talk. Small group discussion is an alternative which has a lot to offer. Each pupil in a group of 4 or 5 has a chance to participate and to control the discussion and this can be motivating and stimulating. The pace of the talk can be related to individual needs. Transcripts of small group talk might appear initially very disappointing because the talk is hesitant, disjointed and tentative. Analysis by Barnes and others has shown that, in spite of this, pupils talking in small groups demonstrate high levels of thinking, of generalisation and of speculation which are rare in class talk. Children are more likely to ask questions of themselves and each other. Group work also encourages social skills (Figure 3.9.)

The following transcript is of four first form (11-12-year-old) girls discussing their next move in the computer program *Treasure Island*.

Pupil A: If the wind doesn't change and we go along to 100, no to 270, and then if it takes us to 15, um, 9 — then if it takes us to 14.9, no — 15.9, sorry, then we go down here in this direction and that wind — and that will take us into port cos then we'd be blowing against that, wouldn't we? We'd be blowing against that. Get what I mean?

Pupil B: No Karen — it's north east now, not north west.

Pupil A: I know, I told you.

Pupil C: If we go down here.

Pupil B: Excuse me.

Pupil A: Then we're going to be shipwrecked cos it's blowing . . .

Pupil B: I don't . . .

Pupil C: Can I speak?

Pupil A: Wait a second. I'm just trying to tell her something. She can't seem to pick it up. We're here — right — 17.9. We go across . . . 270 degrees which should take us over to here somewhere for one hour. I don't know if it will take here, here or here, but it's worth a try to get there.

(Cummings, 1984)

This brief extract from a lengthy discussion illustrates the ability of young pupils to make deductions from information given, to consider hypothetical situations (*if* the wind doesn't change), to correct themselves, to ensure that everyone understands and to correct each other. There are characteristics of this situation which encourage purposeful activity and which can be applied to any group activity. Firstly, there is a specific task, in this case to collect treasure. Secondly, the task is relatively open-ended, ie. has many possible outcomes. Thirdly, there are materials and data to support the talk. The following tasks all have these features and involve group work.

1. Selecting or ranking exercises

Several possibilities are presented to the group. The task is either to select one or to place them in rank order.

(a) Allocating scarce resources, eg. financing one of several schemes; or allocating a council house to one of several applicants.

(b) Choosing a representative sample, eg. slides or photographs can be chosen to illustrate particular aspects of an area.

2. Making a planning decision

(a) Choosing a route for a new routeway.

(b) Planning part of an urban area, eg. an area of land for redevelopment, using cardboard cut-outs or wooden models which can be moved about on a base map. Manipulation of materials reveals more possibilities than drawing plans on paper and encourages a tentative approach where each individual can explore implications of decisions. Attention becomes focused on the resource materials which provide an alternative means of communication to support talk.

3. Devising a Questionnaire

This can be a preparation for fieldwork. It is useful for groups to exchange draft questionnaires for criticism and revision.

4. Devising a Method

Pupils are asked to devise a method for measuring eg. the flow of a stream or the movement of a glacier. This task prompts questions, arouses curiosity and may produce some ingenious methods.

5. Guesswork based on existing knowledge

Pupils can categorise a list of countries eg. according to specific criteria such as density of population, or proportion of people living in towns. Eavesdropping on these discussions gives insights into pupils' existing world picture and reasoning. It also makes the actual figures far more interesting.

6. Designing something

Pupils are asked to draw a sequence of plans eg. for draining an area of sea for reclamation. The role of the teacher is to assess the plans, feed in information when necessary and offer advice. This discussion precedes work on Dutch reclamation.

7. Preparing for a role-play exercise

Pupils are given roles to play and are provided with background information for each role.

8. Devising a marking scheme

This helps pupils clarify what is important and why. It is useful preparation for external examinations.

Teachers who have not used group work worry about the upheaval, the noise and the time it takes. If it is used frequently classes will organise themselves quickly and make it possible to have flexible group sessions lasting from as little as five minutes to a whole afternoon. Teachers who use group work value the greater pupil participation, the enthusiasm it generates and often comment that the discussion continues after the lesson, an indication of involvement (Figure 3.9).

Reading

The ability to read well has become increasingly important with the development of an enquiry approach to learning geography. Pupils need to follow written instructions, to find out information, to skim through reference books, to read intensively and to evaluate what they read. What actually happens in most classrooms does not encourage the development of these skills.

The Schools Council Project on the Effective Use of Reading found that the time spent reading in social studies lessons, which included geography, was only 15 per cent of each lesson. Most of this time was spent reading textbooks, worksheets and

Figure 3.9. Second year pupils in small group discussion.

the pupils' own writing. The most significant finding was that over 80 per cent of the reading was in short bursts of less than 30 seconds. The authors of the report wrote, 'It is unlikely that "short-burst" reading provides pupils with an adequate means for developing a critical or evaluative approach' (Lunzer and Gardner, 1979). Continuous reading was, however, frequently required for homework when pupils might not be able to get the help they need.

Geography teachers can help the development of reading skills firstly by being aware of the difficulties pupils have, secondly by providing a variety of reading materials, and thirdly by devising activities which enable pupils to read intensively and grapple with the meaning of what they read.

1. The difficulties of reading

Reading a text is more difficult than listening to spoken language because there are no additional clues to its meaning in intonation and gesture. Teachers can help by reading extracts out loud and by recording extracts on tape for poor readers in mixed ability classes.

The actual process of decoding words and sentences is difficult for many. Geographical texts and worksheets contain many new words, both technical and general. For example, all the words in italics in this extract from a geography textbook might need to be explained:

'*Irrigation* can *render* deserts green and *fertile*, as in the Nile Valley, and it can be used to *supplement* rainfall in the dry seasons in India'.

Most geography teachers would ensure that children understood 'irrigation' and 'fertile' but what about 'render' and 'supplement'?

The length of some sentences and their unfamiliar structures make them difficult to read, as in this extract from another geography textbook:

'Finally, in stage five, production is at such a high level and goods are produced so fast that, instead of concentrating on increasing production, industry concentrates on encouraging customers or consumers to buy more and more goods to use them up.'

I am not arguing against the use of texts containing new vocabulary and long sentences, but I would suggest that teachers study texts to find out how difficult they are. One way of doing this is to apply a readability formula such as the Flesch Formula which is based on the number of syllables in words and the length of sentences. This gives a rough guide to the *minimum* reading age necessary to read the text. The use of readability formulae is explained in *Readability in the Classroom* (Harrison, 1980).

Some children can understand the words and sentences, yet still make no sense of the overall meaning. The structure of non-fiction depends on linking words such as likewise, moreover, consequently, nevertheless, conversely and so on. If the significance of these words is not recognised, then comprehension is difficult. Comprehension questions tend to focus on the meaning of particular words or sentences. They rarely expect pupils to get to grips with the meaning of the text as a whole. The reading activities suggested in section 3 below involve the pupils in a close study of the structure and are therefore preferable to the random piecemeal sampling of texts characteristic of comprehension questions.

2. Variety of reading materials

The use of reading materials other than textbooks can bring pupils closer to first hand sources. Newspaper reports and letters, planning enquiry reports, travel brochures, foreign embassy and commercial handouts, autobiographical travel, writing and fiction can all contain information relevant to geographical enquiry (Figure 3.10). Some would appear initially to be more attractive than textbooks for pupils to use. They do, however, present two problems. Firstly, most have been written for adults and demand high reading ages, so can be used only with considerable support.

Figure 3.10. Different types of reading material.

Secondly, many are written from a particular viewpoint, which is evident in the selection of information and the underlying values. Teachers need to help pupils detect bias and to evaluate the text in the light of this. These particular skills should also be encouraged when pupils use textbooks which inevitably present a particular perspective on geographical information.

The use of fiction in geography is particularly helpful in deepening children's grasp of remote experiences and difficult concepts. For example, Anita Desai's *Village by the Sea* conveys powerfully the experience of rural poverty in India and the search by a teenager for work in Bombay. Extracts could be read to illustrate aspects of the experience (eg. the arrival of a newcomer to a large city) or of the Indian environment (eg. the start of the Monsoon). The intention, however, is to explore the total experience so that pupils are encouraged to read the whole book.

The abilities to use the contents page and index of a book, and to skim through reference books for information, do not come naturally. They are best taught in context. If a geography teacher sets up an investigation or a project which demands these skills, then that is a good opportunity to teach and practise them.

3. Reading activities

The Schools Council Reading for Learning Project devised a series of activities which enabled pupils to study a text closely and to make sense of it as a whole. The activities were called DARTS: Directed Activities Related to Texts.

The texts used by the project were chosen by subject teachers to suit their own subject aims and objectives, while at the same time developing reading skills. The activities are done by pupils working in pairs so that they can reach agreement about the meaning of a text through discussion.

DARTS have two stages.
(i) The pupils analyse the information given in the text according to categories chosen by the teacher. The pupils underline categories and in this way discover what information is in the text and how it is put together. Some texts are better analysed by dividing into segments.
(ii) When the analysis is complete the pupils reorganise the information into another form such as a diagram or a table. The reconstructed information should be in a simpler more memorable form than the original text.

Several types of geographical texts are suitable for DARTS.

A. *Description of a mechanism or structure:* eg. a volcano; a power station; an HEP scheme; a fishing trawler.

(i) Pupils start by underlining the parts of the structure. A different type of underlining can be used to label the characteristics of the parts eg. the function.
(ii) The information is re-assembled on a diagram, either drawn and labelled by the pupil or given to the pupil with incomplete labelling. The task of labelling entails close reading of the text.

B. *Information about a classification:* eg. rock types; types of agriculture; uses of coal.

(i) Pupils underline different categories of information. Sub-categories can be distinguished separately with different types of underlining.
(ii) If the text is mainly a classification, a hierarchical tree-diagram can be made. If there is additional information about each category, a table is more suitable. If pupils are likely to need help, parts of the tree diagram or table can be completed for them.

C. *Texts describing a process or a sequence:* eg. making steel; a farmer's year; an ecological cycle.

(i) The stages in the process are analysed by underlining or segmenting putting marks in the text dividing the stages.
(ii) The information is re-organised into a flow line diagram — drawn by pupils or a given diagram is labelled from the text.

D. *Texts developing an argument:* eg. advantages/disadvantages; problems/solutions.

(i) Different types of underlining are used to pick out the categories given.
(ii) The information is re-organised into tables or lists.

Geography teachers can invent their own DARTS, consisting of analysis of text followed by reorganisation of information, for any text.

An example of a DART is shown in Figure 3.11. It was used with mixed ability 2nd form (12-13-year-old) pupils. All were able to do the first stage — analyse the text. The second stage was very demanding — far more stretching than copying a blackboard or textbook diagram of the water-cycle. The task required careful re-reading and linking of parts of the text. Weaker pupils needed help starting their diagrams.

Writing

Writing requires the same mental processes as those involved in all learning; selection; sequencing: generalisation; making connections; and making meanings clear. Writing can help children to learn.

The Schools Council Writing Across the Curriculum Project found that the kinds of writing done in geography classrooms were very limited. 88 per cent of the writing was transactional, that is, factual, impersonal writing to convey information and ideas. Expressive writing, which is more personal, allowing expression of feelings, personal responses and opinions, was not widely encouraged. Given the nature of geography, the dominance of transactional writing is not surprising. Yet most children find it extremely difficult and cannot produce what is expected. Some teachers by-pass the problem by dictating notes or by reducing written work to copying and gap filling. This achieves the purpose of filling exercise books with transactional writing. It does not help pupils to acquire writing skills, nor does it allow them to use writing as a means of learning.

The Writing Across the Curriculum Project also analysed the types of audience children were writing for. In geography 81 per cent of written work was done for teacher as examiner. This means it was to be measured and tested against what the teacher already knew.

The prevalence of transactional writing, produced for the 'teacher-examiner', discourages skills which many geography teachers want pupils to acquire: the abilities to question, to speculate, to

---

THE WATER CYCLE

Read the passage below carefully. *Underline in pencil* all the times that water is mentioned in any form (include rain, snow, hail and water vapour).
Underline in biro each time something happens to the water.
Work in *pairs*.

Water falls from clouds in the sky as rain, snow or hail. Which you get depends on how cold it is. Some of this water soaks into the ground and some of it stays on the surface — like puddles in the playground. Puddles like these usually evaporate into the air in a few hours.

Some of the water that soaks into the ground is sucked up by the roots of plants and trees. This water evaporates through the leaves. When plants do this it is called 'transpiration'. Water also evaporates into the air from the sea, from lakes and rivers and from the ground itself. Water evaporates more quickly when it is warm. Have you noticed things drying more quickly on a warm day?

Most of the water that falls on the ground flows across the surface to make streams and rivers. Some water seeps through the ground into the rivers, too. Rivers flow down into the sea.

The water that evaporates into the air becomes invisible. It is called water vapour. If the air cools the water vapour might condense into water droplets which you can see as clouds.

Make a list of all the things that happen to the water (underlined in biro). Try to draw a picture showing these things happening. Put on your picture as many of the things that you underlined as you can.

Figure 3.11. Example of DART for 12-13-year-old pupils.

form hypotheses, and to make judgements. There are, however, many ways in which we can help pupils to develop these skills and also improve their written language.

1. Preparing for written work

Class or group discussion before written work can help to sort out difficulties at three levels: word, sentence, and overall structure. Brainstorming (jotting down random ideas, words and phrases) is motivating and can extend vocabulary. Words suggested can be put on the blackboard for reference. Pupils can be asked to write out sentences in draft form, eg. sentences giving a definition, or an explanation, or a reason. The aim would be to explore the possibilities in discussion rather than to produce an ideal sentence. Discussing possible definitions, for instance, would give children insight into the process of defining as well as providing them with a definition. Dictating definitions leaves out the defining process. At the structural level pupils can be helped by having examples of written work read out to demonstrate what is possible.

2. Shifting the emphasis away from transactional writing

Pupils can use their writing to make sense of geography more easily if they can include their own reflections, attitudes, comments and questions. The enquiry approach of the Geography 14-18 Project and the Geography for the Young School Leaver (GYSL) Project permits reflective writing in the course studies. What emerges in the writing is an awareness of the process of making geographical knowledge as well as the knowledge itself.

> 'Of the eight ideas that we set out to test, only one turned up results that were debatable — idea 4. 'The particles of sediment on the bed of a stream are larger in the areas of fast flowing water.' However, if the stream had not been disturbed at all, I believe this test would have given us expected results. Therefore I suggest that next time only a certain number of people are allowed in the stream, and only to measure the channel cross sections'.

— From a pupil's GYSL course study

Course studies are usually based on first-hand experience and often produce good, coherent writing. In writing based on secondary sources, we can still allow freedom to explore a topic. Pupils can be encouraged to reconstruct knowledge rather than reproduce it. They can ask their own questions of it and reshape it instead of merely answering the teachers' questions. The following piece was written by a second year pupil during an examination.

> Lake Maracaibo
> In the past: a group of dirty mud huts on the edge of a stinking slimy lake.
>
> In the present: A South American Geneva or a booming Manchester. What has happened to the lake dwellers? Gurgling oil pipe lines under their huts. Telegraph wires overhead. Their stilted, primitive homes being destroyed by oil and spluttering tankers.
>
> The new lake dweller: Mr. Smith with a job (good prospects travel passes, three weeks holiday a year) at the oil refinery. Mrs. Smith, belonging to the Women's Institute; Friday, shop at the supermarket, Monday bridge, and Wednesday visit mother, Smith Jnr. attending a state school perhaps, and, if there is one, going to Sunday school.
>
> The architects, a dreamland, space as much as desired. New, clean, prosperous. The builders' paradise, houses to be built.
>
> Who makes a profit? The owner of the oil company, the ship-owner, the builders, the shop-keepers.
>
> What would happen if some egg-head of the future invents something easier than oil, cheaper than oil? Would the Smiths build a house on stilts? Would Mr. Smith fish in the muddy waters? Would Mrs. Smith grow potatoes in the mud? Who would make a loss? The oil company owners, the ship-owners, the builders, the shopkeepers? Would the Smiths go north to rich North America? South to Brazil?

— From a script by a 13-year-old boy

This pupil's ability to question, to speculate, and to suggest alternatives demonstrated more geographical awareness than could ever be shown by simply reproducing the information on which this writing was based.

3. Widening the audience

It is difficult to know how much detail to write, how much to explain and what to take for granted if the writing is done for somebody who knows it all already. It is easier to write for a specific audience. Possible audiences are:

(a) *The whole class.* Extracts from written work can be made into a booklet so that the whole class can read them. Or extracts can be combined with maps, graphs and pictures to form a display. Flexibility in using different lengths of extract, ie. words, phrases, sentences and paragraphs, make it possible for everyone to contribute. Written work can provide the starting point for the next lesson. Extracts can be copied out and duplicated for group or class discussion.
(b) *Imaginary readers.* Pupils can write letters, newspaper reports, etc. Or they can be asked to write for a younger child who knows nothing about a topic.
(c) *For a public audience.* The following piece was written by a fourth year girl as a speech to be made at a simulated public meeting. The imagined audience helped the girl sort out her arguments and suggested how they should be written.

Against the construction of Killingholme Refinery.

1. The refinery will be built in an area which is popular with tourists. It's very unpleasant when you go to the coast and find out there is an oil refinery next to you, blowing pollution towards you. Most families go to the coast to get away from the city and the noise. They don't want to sit staring at a refinery all day.
2. People think that it will be much cheaper to build a refinery because it will put more men in jobs. But this is wrong. Money will have to be paid to buy and build the refinery and to buy the land to build it on. This money would be better spent on education and activities.
3. Oil slicks often occur and they cause many deaths of sea creatures. Just think back to 1982 — all the dead birds, fish and a few mammals on the beach.
4. The land is beautiful and if men build a refinery they will ruin the landscape.

Conclusion.
   The Humber Estuary and areas around should be thought of more than the oil industry.

— *From a fourth year girl's report*

4. Encouraging first draft writing

Children can be helped with the mechanics of writing if they start by making a rough draft. Words and phrases can then be changed, and meanings can be made clearer. The revision can be done in pairs or groups with teacher guidance.

5. Marking

Teachers' marking falls into three categories:

(a) *Evaluation.* Work is judged as 'superb', 'excellent', 'disappointing', 'messy', 'careless' or 'untidy'. This type of marking emphasises the teacher as examiner, especially if it is accompanied by a mark or grade. Pupils try to produce what the teacher values rather than explore what is in their own minds. They conceal lack of knowledge and misunderstanding.
(b) *Correction.* Errors of grammar, spelling and geography are corrected. This is helpful but if there are many corrections it can be discouraging. It is preferable to correct a few errors and to monitor improvement. Use of first drafts and preparation of work for a wider audience provides motivation for improving accuracy.
(c) *Comment.* Comments can focus on geographical thinking, on use of data, on reasoning, on value judgements and on misunderstandings. Teachers can question pupils on their writing and invite replies so that marking becomes a dialogue aimed at increasing understanding. Comments emphasise the importance of writing as a means of learning.

Should writing be seen as an end product which takes place after learning, something to be assessed and corrected? Or should it be part of the learning process? How we mark children's written work indicates why we think writing is important.

Conclusion.

Many geography teachers want to help their pupils' language development yet are uncertain how to do it. This brief resumé suggests three approaches which are common to talking, writing and reading.

1. Investigate what is going on at present: what talk is taking place, what questions are asked, what written work is done, what reading ages the texts require.

2. Encourage uses of language which enable pupils to make sense of geography for themselves.

3. Use the talking, writing and reading activities suggested to develop language skills at the same time as geographical skills.

References

Barnes, D. Britton, J. and Rosen, H. (1970) *Language, the Learner and the School*, Penguin.

Barnes, D. (1976) *From Communication to Curriculum*, Penguin.

Barnes, D. and Todd, F. (1977) *Communication and Learning in Small Groups*, Routledge and Kegan Paul.

Britton, J. Burgess, A. Martin, N. McLeod, A. and Rosen, H. (1975) *The Development of Writing Abilities 11-18*, Macmillan Research Series.

Casholan, A. (1979) *Language, Reading and Learning*. Blackwell.

Cummings, R. (1984) 'Pupil talk in groups during a CAL simulation game'. Unpublished M.A. dissertation, University of London.

D'Arcy, P. (1978) *The Examination Years. Writing in History, Geography and Social Studies*, Ward Lock.

Department of Education and Science (1975) *A Language for Life* (The Bullock Report) HMSO.

Harrison, C. (1980) *Readability in the Classroom*. Cambridge University Press.

Lunzer, E. and Gardner, K. (1979) *The Effective Use of Reading*, Heinemann.

Lunzer, E. Gardner, K. Davies, F. and Greene, T. (1983) *Learning from the Written Word*, Oliver and Boyd.

Marland, M. (1977) *Language Across the Curriculum*, Heinemann.

Martin, N. D'Arcy, P. Newton, B. and Parker, R. (1976) *Writing and Learning Across the Curriculum 11-16*, Ward Lock Educational.

Mercer, N. (ed) (1981) *Language in School and Community*, Arnold.

Richmond, J. (1982) *The Resources of Classroom Language*, Arnold.

Spencer, E. (1979) *Writing Matters: Across the Curriculum*, Hodder and Stoughton.

3.3. Games and Simulations
Rex Walford

Introduction

Games and simulations vary from simple to quite elaborate structured activities but they all have a common element — they invite pupils to imaginatively 'put themselves in other people's shoes' and exercise thought and reflection in making a decision of some kind.

The essence of simulation is to provide learning through experience (and through subsequent reflection on experience) rather than by the processing of information through more didactic means. Pupils involved in a game or role play are likely to come to grapple with ideas based on their own experience and on discussion with their peers, rather than being *told* about the ideas. For instance the *Oxfam Farming Game* involves pupils actually grappling with the experience of being in a poverty spiral, as they make decisions about which crops to plant year by year; the consequences of their choice and of unexpected chance factors create a changing situation with which they have to cope.

Although the experience of classroom simulation is avowedly partial (it is not possible to properly simulate the *morale* or cultural mores of a Third World farmer) there is evidence to suggest that such activities do help pupils to empathise more effectively with people of other lands, cultures and occupations. It also notably helps them to understand the dynamics of a rapidly changing world more effectively, since circumstance does not allow a 'blueprint for success' to be successful, as events and factors change during a simulation. Both empathy and the understanding of process are key objectives of much modern geography teaching in schools.

Beyond that, and irrespective of content, simulation activities encourage much talk, discussion, negotiation and peer-group learning in the classroom. General social and communication skills are well developed by such techniques. In geography, the effect of students talking each other into an understanding of (say) the best location for a park or the fairest way of allocating a budget for a Government, are often spectacularly more effective than 'chalk and talk' from the teacher.

One other attribute of simulation is acknowledged almost universally: its motivating effect on classes — as long as it is not over-used. Pupils who are often quite lukewarm about other class activities enjoy being given some chance to discuss and exercise judgement on a set of decisions if these are presented to them effectively and at an appropriate level of comprehension. If anything, the disadvantage of simulation is over-motivation; sometimes the power of roles may carry pupils on into de-briefing discussions when they need to stand back and be more reflective about their own actions.

Types of simulation

Simulations can be divided into four main types:

1. *Role Play*. In role play students take on the role of another person (either consciously or unconsciously) and take part in a simulated meeting or set of negotiations of some kind (eg. a Cabinet meeting, a public enquiry, a parish council) where a particular decision has to be made. Such meetings are often helpfully aided by the preliminary distribution of documents (eg. a report from an expert, a simulated front page of the local paper) which can be studied. Some documents can be common to all participants, some given only to certain people.

Spring Green Motorway (Community Service Volunteers) is a role play about the effects of a proposed new road on a small village community. All the participants are given a hypothetical front page of the 'Ildridge Gazette', a map of the area and a short paragraph of *individual* role description. They meet with their local councillor (the teacher) who chairs the meeting to advise him on how to vote at a forthcoming County Council debate on the proposed motorway.

Some role plays do not need to have a conscious ascription and adoption of roles. In *The Aid Committee Game* (Oxfam) pupils assume the roles of members of an aid-giving agency as themselves. They then bring their own judgements to bear on the allocation of a finite sum of money to various projects in a particular Third World country.

If role cards are written, it is important to write them in direct style, to keep them succinct, and to avoid the gratuitous ascription of attitudes. It may be helpful to provide a starting point for thought by suggesting an initial question on which to ponder. Compare, for instance, the two examples in Figure 3.12: role B is likely to be more effective than role A in persuading a pupil to be imaginatively empathetic to the role.

2. *Games*. These simulations have sets of rules and procedures which help to define and limit the scope of decision making and enable it to be followed through a number of stages (rounds). Thus a topic like 'farming' is a common theme with participants measuring the success of their policy over a period of simulated time. Chance factors are often introduced either by the throw of a dice (eg. to simulate weather probabilities) or by cards drawn randomly from a pack (often based on the real life chance events which occur to complicate everyday decision making in the particular context under scrutiny).

One example of this kind of simulation is *Caribbean Fishermen* (Longman) where pupils are invited (either singly or in pairs) to become fishermen on a small 'model' Caribbean island. They can fish for shell-fish each day either in-shore (where the fish are small) or off-shore (where they are larger). They sell their fish at the fish quay each morning. The initial pay-off matrix for their catch is shown in Figure 3.13. This varies during the game, however, according to random chance factors. Though the fishermen know that there is a

---

Role A

JANE BROWN

She is a secretary who has recently come to live in rented accommodation in a dilapidated semi-detached house in one of the roads on the eastern outskirts of Melbury.

Despite the fact that she enjoys the gardens of Western Park and has taken out membership of a tennis club which is based there she favours a western by-pass route which cuts through the park, since she does not want noisy traffic to disturb her reading and TV watching on the evenings when she is at home.

A few months ago a friend was knocked down and slightly injured by a slow-moving lorry in the town centre and so she has been convinced about the need for some kind of by-pass road for the town ever since the incident happened.

Role B

JANE BROWN

You are a secretary and have recently come to live in a rented flat in an old part of eastern Melbury.
You often go to Western Park to enjoy the gardens and to play tennis there.
The town centre is so crowded with shoppers that traffic is usually at a walking pace on busy days; but a few months ago a friend of yours was injured when she stepped off the pavement in front of a lorry.

Questions to consider

1 Will the town centre be safer for shoppers if they build a by-pass for through traffic?

2 Should personal concerns influence your view about whether the western or eastern route is best?

3 If so, which ones are most important to you?

Figure 3.12. Specimen roles from a simulation: building a by-pass.

	Inshore	Offshore
Good weather	2	6
Bad weather	4	0

Figure 3.13. Pay-off matrix for fishing game.

1 in 6 chance of strong currents removing their off-shore pots, they cannot predict the current. They must therefore work out a strategy for placing their pots each day. The teacher rolls a dice to see whether the current runs and then the pupils work out their takings. As the game progresses the more complex ideas of weekly basic needs, higher standards of living for successful fishermen, changes in the price of fish, etc, are introduced.

Another style of game is concerned with the exploration of attitudes and viewpoints about particular aspects of society. The classic game *Starpower* (Christian Aid) sets up a society in which three groups of pupils exchange tokens in a trading game. Belatedly the 'poor' group come to realise that the 'rich' group are able to change the rules of the game to maintain their own advantage.

Another game of this kind is *Ra' Fa Ra' Fa* or, as it is known in Britain, *Outsider* (Christian Aid). In this game two separate cultures are set up in adjoining rooms. One has the injunction 'keep smiling' and the other 'keep trying' and they follow differing sets of rules. Visitors from one society to another are encouraged and they report back to their own room on what they see. The de-briefing session after this game usually reveals many misunderstandings and misconceptions by the 'visitors' and is a valuable aid to pointing out how easily a little knowledge produces cultural misunderstanding in real life.

3. *Hardware Simulations.* Some geography teachers have used simulations of the 'hardware' kind for many years; the wave tank and the sand box have an honourable history in teaching generations of pupils about coastal and river processes. More recently there have been extensions of this technique in conjunction with role play. A number of planning games have sought to simulate city growth with lego-type building blocks. The seminal model *Portsville* in the American High School Geography Project involves pupils in role-playing urban developers working with a disguised model of Seattle.

Other kinds of hardware simulations have sought to develop theoretical models or replicate industrial processes. *The Paper Bag Game* (Christian Aid), for instance, has pupils involved in the potentially monotonous task of actually making paper bags as children do in the bustees of Calcutta. Doing it over a limited time produces reaction to the difficulty and the monotony of the task; the de-briefing materials in the game emphasise how ill-rewarded the task is in real life and how many paper bags need to be made and sold in order to earn enough to buy even a single meal.

4. *Mathematical Simulations.* Mathematical simulations are often useful to replicate those processes which are not directly related to human decision making and thus are notably used in physical geography. In this style the use of random numbers or cards simulates the chance events in processes such as the development of an off-shore spit, or of the stages in river capture. Monte Carlo simulations are used to simulate the long-term large processes of urban growth.

One American game, *Grasslands* simulates the degradation of a grassland environment by overgrazing. It takes the form of a card game in which numbers of cards are marked to represent meadowlarks, grasshoppers, cattle and good weather. The cards are played against an environment which begins as long grass, but which can turn to short grass or even to desert by the combination of particular cards. The key to the game is the number of cattle cards in the pack; beyond a certain number, the 'long grass' environment can rarely survive in the long term however much 'good weather' cards sustain it in the short term.

Within these types of simulation, there is considerable scope for variety. Some simulations can be developed informally with a class in a spare five or ten minutes at the start or end of a lesson; for example, 'Imagine we were an expedition and we came to a land we did not know; here's a rough map of what we can see (draw on blackboard) — where would it be best to settle, do you think? . . .' Others may involve considerable equipment and be planned as an activity which would last through two or three double lessons; for example, the use of a game like *North Sea Exploration* (Longman) in which pupils study geological evidence and decide what kind of rigs they will use and where they will drill in order to find oil.

Simulations can be given to pupils as individual problem solving exercises within a formally set out classroom; they can even be used as examination questions. If the exercises are used with pairs of groups, however, there is the added advantage that particular ideas are expressed and considered by another person.

Some simulations and games need the teacher to act as 'manager' of the game, setting the class off on different projects and then comparing and communicating work done en route to a corporate decision. There may well be the hum of discussion from such a classroom — it is difficult to simulate in silence. But, as most teachers well know, the lively babble of 'work noise' is different in kind from the unlicensed chatter of boredom. Often the need to throw a dice, or announce a decision, or move to a subsequent stage in the game, allows the teacher to gather the class's attention and in doing so to insert a rhythm of pauses into the animated discussions.

Some of the most regularly used themes for simulations in geography teaching are:

Location (of settlement, industry, parks, reservoirs, hypermarkets, etc.).

Route Building (of new roads, railways, pipelines, etc.).

Search and Exploration of Mineral Resources (both on land and under the sea).

Development of a Land Surface (farming, regional planning, urban development, etc.).

Primary Activities in the Environment (farming, fishing, herding).

Issues of Environmental Conflict (competition for use of land, effect of a proposed location on an existing community, clash of cultures, etc.).

Some important points to consider

1. *Adapting Material.* Many teachers begin their use of simulation in the classroom by using a well known and well tried existing model. But it is often useful, in tailoring the technique to the needs of a particular syllabus, to subsequently adapt the model and make up a variant of the original. There are a few robust structures and styles of simulation which bear any amount of adjustment; sometimes it is necessary simply to insert one set of data (or one map) for another. On other occasions, a teacher may want to make an existing simulation more complex or more simple than the published version. Such adaptation is to be encouraged; simulations are ideas rather than rigid and unchanged structures.

2. *Simplicity.* Some early simulations were notable for their copious documentation and their myriads of equipment. Time has shown that models of elegant simplicity are more useful than the elaborate kits. Games which can be used within the confines of one 40-minute period are especially useful.

The essence of simulation is to provide a simplified model which reveals a structure of the real world with clarity. It is usually true that complex models run the risk of being unusable. The temptation to keep adding material to a simulation in order to 'bring it nearer to all the facts of the real world' may be considerable, but it inevitably reduces 'playability'.

3. *Sensitive Management.* It is unwise to use simulations with classes who are not well known to the teacher. The activity demands attention to individuals and group activity and decision making may put particular stresses on individuals as well as provide much enjoyment and stimulation. There needs to be keen observation and sensitivity to what is happening to individuals in the class as the simulation proceeds.

The teacher should not attempt to run a game in a 'sergeant-majorish' way, giving interminable interpretations on every point and ploughing on to some pre-determined end. Sometimes players suggest changes and improvements to the activity and it is wise for the teacher to consider them and adopt them — the comparison of the model with known events is a most educational activity in itself. And it is best to make no pre-judgements about how long a simulation activity should run; some of the most fruitful geography may emerge when animated (and perhaps unexpected) discussion engulfs the activity at a particular point.

It is also possible to find classrooms in which simulations are used but in a very 'directed' way ('No, no, don't put your airport there — can't you see that is clearly not flat enough land . . .'). Pupils need a chance to explore the consequences of their own decisions and poor decision-making can usually be self-monitoring. If the teacher has provided an effective initial stimulus, he should be relatively silent for some of the time — observing and listening to what is going on as the pupils engage in activity.

4. *Integration.* If role play or a game is reserved only for afternoons at the end of term, or for some unexpected single lesson on emergency cover, its value is largely nullified. The strength of simulation is as a keystone experience within a topic, from which increased motivation and desire to work is a spin-off result leading to more interest and learning.

A teacher working within a theme (say Transport or Industry) for half-a-term will want to do some preliminary work on the topic in order to provide some background for the exercise of decision making; the simulation will then become more animated and more informed.

But following the simulation there also needs to be 'de-briefing' and follow-up work. De-briefing should include:

(a) an action-replay of some of the key moments of the simulation;

(b) a communication of viewpoints between different groups and/or individuals (since they don't all see the same picture);

(c) a consideration of the basic themes and ideas involved in the exercise;

(d) a comparison with some convenient case study examples (especially the one on which the simulation may be based);

(e) a discussion of how the simulation may differ from real life (eg. simplification).

Most simulations are best used in the middle of units of work, though there are occasions when teachers use them to spark off an unmotivated class at the start of a topic, or as a synthesising conclusion to a piece of work.

Possible difficulties and disadvantages

1. *Producing Evidence.* "We all know that simulation is the answer," said Alice Kaplan Gordon, "but what is the question?" Her tongue-in-cheek comment eloquently outlined both the promise and the problem associated with the technique. Simulation users often speak of 'success' in widely differing terms. They value the intertwining of cognitive with affective learning and the elements of surprise and unplanned learning which develop, and point to the value of the experience as expressed by those who take part. Orthodox evaluative procedures do not fit easily with these characteristics and neither do they easily capture the 'flavour' of simulation experiences. It is still true that you are most likely to use simulations if you have been involved as a participant in one; and that the 'hunch' of sensitive and informed observers produces more useful evidence than batteries of questionnaires and pre-and post-tests.

2. *The Illusion of Open-endedness.* Simulation may seem open-ended and democratic as a way of learning but it is important also to remember that the model *itself* needs to be put under scrutiny. If a pupil finds a simulation powerful and educationally attractive, that does not, of itself, make the model truthful. It is important that both students and teachers subject the models to critical scrutiny and examine the assumptions and philosophies which may underpin their structure, the selection of evidence and balance of characters.

3. *Replicating Reality.* It is a mistake to see simulation as a replica of reality. It can never go that far. It is better seen as a vehicle for teaching about reality and a help in bringing pupils into an empathetic frame of mind in order to better understand it. What pupils may receive is a kind of 'shadow-reality' in which they (say) enter into the decision making situation of a farmer *to some degree*. This weakness is transformed to strength as it allows dispassionate self-analysis of decisions and observation of total events whilst working from a particular standpoint.

Conclusion

Games and simulations have clearly taken root in the secondary school classroom. There has been time for ideas to be thoroughly tested and a generation of surviving and frequently-used ideas bears ample testimony to the durability and usability of good simulations.

For instance, Brian Fitzgerald's original *Iron and Steel Game* (Fitzgerald, 1973) has gone through many adaptations in other hands. Some teachers have simplified it, others made it more complex; some have used the idea with different locational maps, or with different mineral resources and products. Neville Grenyer has adapted the exercise to take account of the criticism that it sees the world only from the point of view of 'economics'; his variation includes the revelation of environmental consequences of an industrial location decision (Grenyer, 1981).

Such activities have become part of the teaching beyond the lower school and are included in examination courses based on the Geography for the Young School Leaver (GYSL) Project and the Geography 14-18 (Bristol) Project. At Advanced Level the examination for the Geography 16-19 Project has an examination paper solely devoted to a decision-making exercise.

If suspicion or scepticism about simulation remains it is usually for one of three reasons.

1. A teacher may feel that simulation approaches do not teach factual information effectively or quickly enough. This view is not sustained by research; as many studies show simulations *superior* in this respect as not. But, as Jacquetta Megarry (1978) points out, the attempt to compare on a simple input-output basis may be quite mistaken; the role of simulation and its effects may alter many other variables in the classroom context. Pate and Mateja (1979) found simulation more effective in aiding *retention* of information when reviewing a number of research studies.

2. A teacher may believe that it is better to teach about the 'real thing' rather than an invented model. But this is to ignore the fact that many simulations are based closely on real life evidence; sometimes only a thin veneer of name changes and re-orientation of maps disguise the reality of the material. Few simulations survive if they are not representative of reality, since they cannot effectively be corroborated in follow-up work.

3. A teacher may doubt whether it is possible to manage simulations successfully in the classroom. There is an understandable concern about liberating pupils to discuss and perhaps persuade and negotiate, but few teachers trained in contemporary courses will be unskilled in the management of such activities. Most simulations can be used in a variety of ways, ranging from the individual and/or formal environment to very informal contexts.

Though simulations emerged into geographical education at a time when positivist and quantitative geography was dominant, their emphasis on empathy and understanding human decision making make them appropriate methods as more 'humanistic' geography develops. Their effectiveness and practicability as a classroom technique is now a matter of common observation.

References

Fitzgerald, B. P. (1973) 'The iron and steel game', in Walford, R. (ed) *New Directions in Geography Teaching*, Longman.

Grenyer, N. (1981) 'The impact of the Industrial Revolution on an English city — a location exercise', in Walford, R. (ed) *Signposts for Geography Teaching*, Longman.

Megarry, J. (1978) 'Retrospect and prospect' in McAleese, R. (ed) *Perspectives in Academic Gaming and Simulation: 3*, Kogan Page.

Pate, G. S. and Mateja, J. A. (1979) 'Retention: the real power of simulation gaming?' *Journal of Experiential Learning*, vol 1. pp. 195-202.

Walford, R. (1980) 'Caribbean Fishermen: some reflections about the development of a simulation game', *Simulation Games for Learning*, vol 10, no 2, pp. 75-85.

Walford, R. (1983) 'Spring Green Motorway: a question of reconstruction', *Simulation Games for Learning*, vol 13, no 4, pp. 156-165.

Addresses

Community Service Volunteers, 237 Pentonville Road, London N1 9NJ.

Christian Aid, PO Box 1, London SW9 8BH.
Oxfam, 274 Banbury Road, Oxford OX2 7DZ.

4. Resources for Learning

4.1. Managing Resources
Patrick Bailey

Working out a policy for resources

The physical resources available to a school geography department are always limited. The severity of this limitation becomes evident when one notes that the total amount of money which a maintained British secondary school has to spend on books, stationery and equipment each year is of the order of one-twentieth of the sum paid out in staff salaries; and that it can be substantially smaller than either the bill for examination entries or for heating. From this initially small amount of money the geography department has to secure its fair share. Because money is in short supply it has to be used in close support of the school's and the department's educational policies and with careful regard to priorities. Once obtained, resources have to be conserved; and whenever an opportunity occurs of tapping extra resources, such as those contributed by parent-teacher associations and the community, it has to be taken with enthusiasm.

A department will try to build up its resource stock in support of its educational policies; these are policies about what should be taught and how learning should be brought about. In practice, ideal policies have to be modified to take account of various constraints such as the experience, expertise and ideology of a department's members and the money available to buy new resources.

Teachers are the most important, most expensive and potentially the most versatile resource possessed by a department. It is easy to overlook this fundamental point and to concentrate one's energies on developing policies for the use of physical resources. In fact, all such resources are of secondary importance; they depend for their educational effectiveness upon how well they are used by teachers. The teaching staff of a department are a resource to be systematically developed like any other. This is done by making certain that all members know what each has to offer in the way of special knowledge and teaching expertise; and then by finding ways of sharing this knowledge and expertise so that all members learn from each other and continue to develop professionally as a result of working in the department. Good departments 'grow' their teachers, to the obvious benefit of their pupils.

This section is concerned with the management of resources other than staff. To obtain and use such resources to maximum effect, a resources policy has to be worked out. When doing this, the following considerations have to be borne in mind.

1. *How are the resources going to be used?* This is always the very first question to consider, because no resource, be it book, worksheet or colour transparency, has a fixed educational value. Its value depends upon how it is used by teachers and pupils, working in classrooms which can be organised in various ways. Thus teachers whose classes are relatively homogeneous in ability, who believe that their principal aim is to transmit knowledge, who see themselves as the principal learning resource and who adopt a mainly didactic teaching style, will use books and visual aids in a quite different way from teachers whose classes are of mixed ability, whose main aim is to develop the pupils' capacity to learn, who see the pupils themselves as a learning resource, and whose style is that of a manager of learning.

In working out its resources policy, a department will therefore first try to agree on the approaches it will adopt to teaching and learning. Its members may, of course, agree to disagree and use a variety of approaches. Whatever decision they make, it will carry clear implications for the resources the department needs. Inevitably, their decision may also be influenced by whole-school policies about teaching and learning methods and the organisation of teaching groups.

2. *Public examination syllabus requirements.* These affect the whole of a school geography course and therefore its resource requirements. Public examination demands can sometimes be used to strengthen a department's claim on resources, for example, money and equipment for necessary fieldwork. Some Local Education Authorities make grants available *only* for fieldwork which is an integral part of a course leading to a public examination.

3. *General LEA and school arrangements for ordering stock.* It is often obligatory to order from suppliers nominated by the Local Education Authority. This may limit a department's freedom to order exactly what it wants.

4. *School policies for linking subjects and for team teaching.* Such arrangements have resource implications for the years directly affected and 'knock-on' effects elsewhere.

5. *The characteristics of the learners.* The resources appropriate to pupils in one school may not be at all appropriate to those of similar age in another because of variations in their home life, the environment in which they live, their different experiences in feeder schools and other variables.

6. *The experience and levels of expertise of department members.* Generally speaking, teachers with long experience will be more versatile in their use of resources than those with less experience.

7. *The number of non-specialists who help with the department's teaching.* These may need the support of resources prepared by the specialists and of textbooks which are relatively 'teacher-proof'.

8. *School responsibilities held by members of the department.* Members who are also heads, deputy heads, heads of house and year or careers tutors may find it difficult to participate in certain kinds of time-consuming, innovative geography teaching and therefore certain kinds of resource use.

9. *The time available to teach geography.* Total allocations of time per year and the way it is subdivided always affect teaching approaches and therefore the kinds of resources required.

10. *Teaching rooms.* Specialist, fully-equipped geography rooms allow a far wider range of resources to be used than do general teaching rooms or mobiles. The availability or otherwise of storage space and security for display work are further considerations when deciding what kinds of stock to buy.

11. *The availability of central reprographic and other facilities.* A department will not spend its money on replicating facilities available elsewhere in the school; and it will plan to make full use of all centralised facilities, such as reprographic unit, library and resources centre.

12. *Stock in hand.* The usefulness of existing stock always affects one's policy for ordering new stock. If most of a department's book stock is usable, money can be directed towards bringing about development in one part of the course in one year. If trickle replacement of existing stock is unavoidable, it becomes difficult to achieve a break-through in any one area. It is therefore very important to have a policy for conserving resources as well as for using them.

All these items are related. Their order of importance varies from school to school and also from time to time. Any department engaged in reviewing its resources policy will probably find it useful to re-write this list or one like it in the order of importance which seems right for its members, adding and subtracting items as appropriate. The resource implications of each item can then be considered.

The rest of this section considers the characteristics and management of the most frequently used resources in a school geography department: the blackboard, overhead projector, book stock, map collection, and slides and filmstrips. Subsequent sections of this chapter are concerned with evaluating textbooks, producing resource sheets, discussing photographs and using videotapes.

Blackboards — and whiteboards

Almost all classrooms have blackboards. In non-specialist teaching rooms and mobile classrooms a box of chalks, a blackboard and a blackboard rubber comprise the only form of visual aid available; this is also the case in most Third World schools. Therefore it is important to make the most of what this long-established visual aid has to offer.

Blackboards can be used in three main ways:

(a) to put up headings or make jottings, in written and diagrammatic form, as a lesson proceeds;

(b) to prepare more elaborate presentations in advance of a lesson;

(c) to complement other forms of visual aid, such as a projected colour slide.

Ideally, therefore, a geography room needs to have both fixed blackboards for use during lessons and at least one portable board for prepared maps and other presentations. If this can be kept out of harm's way in the stock room when not in use, so much the better.

When showing a slide or similar picture, it can be very helpful to draw attention to some aspect of it or bring out some relationship, by drawing a sketch map on the blackboard. The geography room therefore needs to be arranged so that at least part of the board space and a projection screen can be used simultaneously. Screens which pull down to cover the whole blackboard are to be avoided.

Whiteboards are sometimes regarded as alternatives to blackboards, but they suffer from two major disadvantages. They begin to discolour in an unpleasing way as soon as they are used; and it is excessively difficult to write neatly or to draw well on their slippery surfaces. They are also more expensive to operate than blackboards, needing special felt-tip pens and cleaning fluid. The writer recommends that manufacturers' publicity which favours 'modern' whiteboards in place of 'old fashioned' blackboards should be firmly resisted by geography departments.

The overhead projector

This is an extremely useful item of equipment and has the merit of not needing full blackout. There are two main ways of using it: either as an alternative to a blackboard on which notes and headings can be written during a lesson; or as a means of projecting prepared work, either the teacher's own or a commercially produced item.

The overhead projector offers several possibilities.

(a) A base map or diagram may be drawn on a sheet of acetate in spirit-based ink. Details can then be added during the lesson in water-based ink, and washed off afterwards. This is an effective way of building up annotated maps, weather charts, three-dimensional diagrams and so on.

(b) Sequential drawings can be built up layer by layer, to show for instance stages in the recession of a waterfall, a plant succession, or the growth of a town.

(c) Items shown on the overhead projector can be linked with slides or filmstrip pictures, simultaneously projected. A field sketch can be built up and annotated alongside slides of the same view. Map symbols on the overhead projector can be matched with slides of the same features and so on.

Although acetate rolls are useful, especially when the overhead projector is used mainly as a blackboard, single sheets of acetate are better. What in effect are lesson plans can then be built up as a series of transparencies. If they are drawn or written in water-based ink they can readily be modified. It is as well to photocopy elaborate drawings on acetate in case they are lost or accidentally wiped clean. Acetates may be stored in teaching files; but if they are to be developed as a departmental rather than a personal resource, photocopies of all acetates should be kept in a central file and the acetates themselves stored in a filing cabinet or trays.

The book stock

Books and atlases are extensively used resources in all geography departments. Two main considerations apply to their management: getting as many books as possible into pupils' hands for as much of the year as possible; and keeping books and atlases in good condition.

Books are requisitioned to be used, not to be stored. Storage space for the whole departmental stock is therefore not necessarily needed throughout the year and the design of the stockroom can take account of this fact. The convenience of shelf arrangements found in a library is not essential, especially as there is a chronic shortage of storage space in most schools.

Building up the book stock is an essential part of a department's in-service training function. All publishers offer an inspection service and the examination and reviewing of a regular flow of new books should be shared out among all the department's members. Reviews of new books are published regularly in *Teaching Geography, Geography* and *The Times Educational Supplement*, especially in the latter's Geography Extras. Arrangements need to be made for these reviews to be read regularly and filed if they look useful.

However, the best way of seeing all the available books, new and established, is to visit the Publishers' Exhibition at the Geographical Association's Annual Conference. All the major publishers and producers of educational materials have stands there and these provide a unique opportunity to compare offerings.

Once a book has been purchased, the first thing to do is to number it and record it in the department's stock book. Unless books are numbered, it may be impossible to know which pupil has, or has had, which book. In the last resort this information is needed so that parents can be sent a bill for its replacement.

Stock control is especially necessary when resource-based teaching is in progress. A large number of books and other resources may have to be out with individuals and groups simultaneously, exchanges being made at intervals. Generally speaking, the more the pupils themselves can be encouraged to take responsibility for signing out and signing in resources, on a simple class master-sheet, for example, the more smoothly this form of organisation is likely to work. Whatever the system adopted, the teachers in charge of the class need to know which books and other resources are with which pupils at any time in the lesson. If they do not know this, then they will be unable to ensure that each individual pupil makes progress and follows a properly balanced programme.

The high price of books makes their conservation a crucial matter. If books can be backed by the pupils themselves with brown paper or something more decorative but equally robust, such as wallpaper, they will last longer than if they are left unprotected.

In addition to the book stock for pupils, a department should build up a reference library for staff. This can include:

(a) a book or two and a file of cuttings on aspects of departmental management and administration;

(b) books and articles on teaching methodology, including fieldwork;

(c) topical reference books, such as *The State of the World Atlas,* annual publications such as Philip's *Geographical Digest,* and *The Statesman's Yearbook;*

(d) geographical texts written mainly for a university readership, parts of which can be used to prepare sixth form materials;

(e) current and back numbers of journals such as *Geography, Teaching Geography, The Geographical Magazine* and *Bulletin of Environmental Education;*

(f) collections of cuttings and articles on selected geographical topics taken from newspapers and magazines.

The map collection

A map collection is a necessary part of every geography department's equipment. When building up a map collection, two questions have again to be considered: What will the maps be used for? What maps are therefore needed? Maps can be used in two ways: for display (wall maps) and for detailed study and interpretation (Ordnance Survey and other large-scale maps).

Maps for display should always include a selection of world maps. Geography is a global study, therefore frequent references to worldwide distributions, patterns and relationships are necessary. A straightforward world relief map, drawn on almost any projection except Mercator, is the most versatile. To support this, a selection of thematic maps may be useful: world population, political units, climate and so on. Every teacher will have particular preferences for such maps, related to what other resources are available.

Wall maps other than world maps may be useful but their precise manner and frequency of use needs to be questioned. A large display map of the local area may seem to be an attractive idea, but unless the pupils actively work on it, it will be no more effective as a teaching aid than patterns on wallpaper.

A globe complements flat maps and may be used to remind pupils that the world is round and that therefore all maps give a distorted impression of reality. A globe is particularly useful when demonstrating the different 'world views' one has from, say, Tokyo, Sydney, Capetown and Buenos Aires and for explaining great circle routes, the true nature of latitude and longitude, the changing seasons and other aspects of earth-as-a-planet geography.

Rolled-up wall maps store well in the horizontal cupboards provided behind some fixed blackboards, or in frames of the umbrella-rack variety. They can also be suspended from hooks at picture-rail height. Mounted and folded wall maps store well on shelves or in large drawers.

Maps for close study and interpretation mainly consist of Ordnance Survey Maps. 1:50,000 map sheets are large and cumbersome to handle. They are most conveniently used on tables of the variety which fit closely together. They are particularly inconvenient on any kind of school desk.

The safe storage of such maps requires the purchase of a map chest with drawers not less than one metre square. Smaller drawers mean that maps have to be folded, which always leads to damage; it also makes the identification of sheets difficult.

It is cheaper to order Ordnance Survey maps in paper-flat format. However, maps for fieldwork, whether local or residential, are more durable if bought in the more expensive folded format. Folded maps are most conveniently stored in labelled boxes or on bookshelves.

A substantial discount is available to schools on certain Ordnance Survey maps when these are ordered directly from the main agents in London, Edinburgh or Belfast. Foreign maps are obtainable from Stanford's International Map Centre, London. The Ordnance Survey issues comprehensive catalogues annually; Stanford's respond best to specific questions about areas covered, scales and costs.

The best map teaching is usually done directly from maps, but certain books containing map extracts help pupils to develop the skills of map interpretation. They also help non-geographers to teach mapwork.

It can be helpful to make colour transparencies of sections of Ordnance Survey and other maps in order to draw the attention of a whole class to a particular feature, or to compare a map with a photograph. Such transparencies can be made with any 35 mm camera able to focus down to one metre or preferably less. Best results are obtained by photographing the map on daylight colour film out-of-doors.

Copying at once raises the thorny question of copyright. No copying of Ordnance Survey maps may be done without the written permission of the Director-General. Most Local Education

Figure 4.1. Slide storage in boxes and plastic envelopes.

Authorities pay an annual fee to the Ordnance Survey which allows limited copying for strictly educational purposes. Teachers should always ascertain the position in their own school or LEA.

Colour slides and filmstrips

Colour slides and filmstrips and tape-slide sequences such as the BBC Radiovision series are an essential part of a geography department's stock-in-trade; but their management always presents something of a problem.

The first thing to do with a slide, whether one's own or a commercial product, is to label it, then to link this labelling to some form of master list. Labelling is most easily done on card mounts; plastic mounts require the added labour of sticking on adhesive labels.

Slide boxes offer a practical solution to the storage problem. They stack on shelves like books and contain index sheets (Figure 4.1). The precise details of a cataloguing system do not matter as long as all members of the department understand it; simplicity is always to be recommended.

It is a good idea to use a card index or a list held on computer disc to link individual slides with teaching topics. For example, a slide of sixth formers levelling a beach profile at Durdle Door on the Dorset coast might be indexed under the headings of coastal features, chalk topography, techniques for beach study and public pressure on favourite coastal locations. Whenever a slide sequence is made up, its details should be noted; it will almost always be needed again.

Commercial sets of slides come in small boxes of the 'Kodak' type or in various forms of pack or plastic envelope. Sometimes it is convenient to break up the sets and integrate them with the departmental collection, but generally it is easier to keep them together and catalogue them as units. Perhaps the most useful way of storing commercial sets of slides is to hold them in the specially made plastic envelopes which fit into a filing cabinet (Figure 4.1).

Filmstrips are most conveniently stored in holes punched into a vertical sheet of hardboard. Accompanying booklets should be filed and indexed for ready reference. Most filmstrips are of little use without their booklets.

Colour transparencies (slides) are preferable to filmstrips because they can be shown in small numbers and in any order. It is inconvenient to show a filmstrip in any other than the order of the strip itself. However, filmstrips are generally cheaper than slides and may be the only form of illustration available. Given time and preferably some pupil help, strips can be cut up and mounted as slides, which makes them far more useful.

If they are to achieve their maximum educational effect, projected pictures must be technically acceptable. This means that they must be sharp and render colours accurately. Television pictures do this, and they are now the pupils' standard of comparison. Some filmstrip pictures fail this test, but any medium-priced camera will produce slides of good quality. If possible, pictures intended for teaching should be taken by someone who understands the problems which pupils have in seeing reality through flat, still pictures and who has considered the technical devices which help to overcome these problems.

So many geographers teach from their own pictures that some hints on effective picture making may be appropriate. The basic rules are:

1. First decide precisely what it is you wish to photograph, then make sure it fills most of the picture. Pictures of 'everything' are rarely satisfactory, photographically or pedagogically.

2. Teaching pictures must be sharp enough to allow for detailed examination. Therefore, always use the fastest shutter speed possible while maintaining a lens aperture which allows the lens to perform at its best. When shutter speeds slower than 1/125th second are used, the camera should be rested on a rigid object.

3. A good teaching picture is first of all a well-composed picture. This means that large, boring foregrounds such as tarmac have to be avoided; that the pictures should not be cut in half by a strong horizon or by strong vertical lines; and that the main point of interest should be placed almost anywhere except in the centre of the composition.

4. Where possible, some immediately recognisable indication of scale should appear in the picture. It helps to have people in most pictures; then the viewer can mentally step into the picture and stand beside them.

The ways in which projected pictures are used depend upon the teaching style employed; projection arrangements in the geography room will vary accordingly. If pictures are to be used with a whole class, or as part of a lead lesson with several classes, blackout is essential. For small-group and individual picture study, a back-projection screen or mains slide viewers may be better.

If possible, there should be two screens in the main geography room, so that a slide and overhead projector or two slide projectors can be used together. It may be convenient to have one screen permanently in position across one corner of the room, tilted for the overhead projector; and another painted on a section of roller blackboard for slides. Pull-down projection screens are to be recommended; those of the tripod variety are easily damaged and cumbersome to store.

Projection facilities should be designed to produce the largest possible picture and to make the use of projection equipment as simple as possible. A projector trolley is to be recommended. This allows projectors to be wheeled in and out of position and materials for a lesson to be assembled on the trolley beforehand. Projectors should never be switched on and off at short intervals, nor moved while they are hot. Both practices shorten or end the lives of their very expensive bulbs.

4.2. Evaluating Textbooks

David Wright

Textbooks are the most widely used resources in most school geography courses. Teachers have the freedom to select the textbooks which they consider to be most suitable for their classes. Publishers respond by producing a great variety of books. It is important, therefore, to consider carefully the attributes of these books before they are used in the classroom. This section suggests ten questions to ask when evaluating a book before using it with a particular class.

1. What is the true cost of the book?

Price is all-important nowadays, but the price in the publisher's catalogue is only one factor. Much more important is an approximate calculation of 'pence per lesson'. A £2 paperback book on farming may be useful for a month's work — at a cost of 50p per week, while a bound £5 book on developing countries may be the basis of a year's work — at only 12p per week. Furthermore, the £2 paperback might fall to bits after a few weeks' use, while the bound £5 book may be usable for many years. The cheaper book can thus represent a real cost per lesson of ten times the cost of the apparently expensive book. On this type of calculation, the strongly-bound course book for the year always comes out as the best buy. But a leavening with shorter topic books is always welcome, if funds permit.

2. Is the language appropriate?

The clearer and the simpler the English, the better is the book likely to be for all pupils. Many books use unnecessarily complex language, too many pseudo-technical terms, and sentence-structures which are far too difficult. Whilst clear, simple English is desirable for *all* pupils, it is vital in mixed-ability and lower-ability classes.

'Readability' exercises are useful but not essential: pupils will willingly list words that they find difficult, and a picture of the level of difficulty will quickly emerge.

An interesting style of writing is equally important. A good geography text should be able to grip pupils' interest. This cannot be measured, but it is vital. Sadly, few textbooks seem to qualify for good 'readability' in terms of an interesting style of writing.

The points made about clear, simple language apply with even greater force to clear, simple, mathematical 'language', except perhaps for the most able pupils. Extremes of quantification are comparatively rare, but there is still the danger of a class being alienated from geography by encountering more complex mathematics in geography lessons than in mathematics lessons.

3. Is the book up-to-date?

The date of first publication is a vital clue. Books which do not indicate the date of the first edition are immediately suspect. In many cases, revision is only superficial, and numerous out-of-date facts, statistics and concepts may be found in the book, since full revision is an expensive process. A book that is over ten years old is unlikely to be a good buy: it will still be in use fifteen or even twenty years after it was written, by pupils who were not even born when the author put pen to paper!

However, a recent date of 'first' publication is no guarantee in itself that the book is up-to-date: closer study of one chapter that tackles a local theme is often very revealing. Herrings were still being landed at Great Yarmouth in geography books for many years after they had deserted the North Sea!

There is also the danger of being so up-to-date that the plans appear as reality in a textbook when they have been abandoned in the real world. The London-Glasgow Advanced Passenger Train was running a regular service in several geography books whilst in reality it was still alternating between breakdowns and withdrawal from service.

4. Are the themes and topics interesting and relevant?

It is arguable that pupils are ready for higher concepts at younger ages than they have been offered in many geography lessons in the past. Some children entering secondary schools may be ready to discuss pollution and acid rain, desertification and the destruction of tropical forests. They have grown up with increasingly sophisticated children's television, and welcome books that tackle real issues. Provided that the *language* is simple, the topics can be more complex than many traditional authors have thought appropriate. Publishers who describe a book in all-inclusive terms are unhelpful: 'suited to mixed ability classes aged 11 to 14 and to less able 14- to 16-year-olds' may in fact mean a book which does not suit anyone very well.

5. Are the pupil activities varied and appropriate?

A distinguishing feature of a textbook, as opposed to an information book, is the wide variety of pupil exercises. There is a major contrast between the book where text and exercises are closely integrated and interdependent, and the book where the exercises are 'tacked on' at the end of each section, appearing almost as an afterthought. The former type of book is preferable, provided that the exercises are interesting, workable, at the right level and not too time-consuming. This ideal is rarely achieved in practice — a distant author cannot tailor-make his exercises for a specific class. Thus many teachers prefer to use the latter type of book, sometimes developing their own exercises to accompany the text. For either approach, a wide variety of tasks and ideas is preferable, giving plenty of choice to the teacher.

6. Does the book have full colour printing?

Good full colour, not merely full colour, is highly desirable for geography textbooks. The world is colourful; our role is to motivate and educate pupils, but they will not be motivated by a monochrome world. A book with some pages in full colour and others in monochrome is not fully satisfactory: the monochrome pages can look very dull when they are in such close proximity to colour. A checklist for evaluating photographs is given in Figure 4.2.

There is no clear correlation between the quality of colour printing and the price of a book; so the common argument that economic factors prevent full colour printing does not seem to be valid. A comparison of pence per square centimetre between two recent textbook series from well-known publishers produced the surprising result that the black-and-white book cost over twice as much per unit area as the excellently produced full-colour book.

7. Is the design and printing clear and attractive?

Design and printing affect more than the photographs and diagrams. An overcrowded page; too few subheadings; print that is too small; a double-page spread without visual input — any of these faults can destroy fragile motivation, and hence destroy pupils' willingness to work and to participate constructively in lessons. At the other extreme, textbooks which look like comics are unacceptable, too, as pupils may find them patronising. A book that looks cheerful and interesting, without descending to a 'pop' style, has the greatest chance of success.

8. Are the maps and diagrams comprehensible and interesting?

One of the main functions of maps and diagrams in a geography textbook is to clarify and simplify complex concepts. Unnecessarily complicated diagrams and maps defeat the main object — as well as destroying motivation. Diagrams should be clearly labelled and the names on maps should be easy to read. Many publishing houses have neither a specialist geographer nor a specialist cartographer, so 'howlers' occur: brown sea, green mountains, grey for both the sea and the high land have all appeared in otherwise good geography textbooks. The remedy is in the hands of geography teachers, who can help to bring about a general improvement if they stop buying such books.

1. *Are the photographs big enough?*

    'Postage-stamp' size photographs do not interest pupils, nor do they provide sufficient detail for deductions to be drawn from them.

2. *Are the photographs sharp enough?*

    New methods of printing have greatly improved the clarity of most photographs, and some publishers have set new high standards of clarity, attractiveness and correct colour balance.

3. *Is the tone satisfactory?*

    Some photographs suffer from underexposure (too dark) or overexposure (too pale); the strong light in the tropics can make different parts of a photograph unsatisfactory on both counts.

4. *Are the views hackneyed?*

    There is a big contrast between books that use well-known photographs and books which emphasise unfamiliar scenes.

5. *Are the photographs merely free propaganda?*

    It is easy for publishers to get free photographs of high technical quality from embassies and multinational corporations because they are good advertisements; it is worth checking the list of sources of photographs.

6. *Do most of the photographs show relationships?*

    The study of causal relationships in geography is very important; photographs should make the relationships sufficiently concrete for all pupils to attempt reasoned explanations.

7. *Do most photographs show people?*

    Pupils relate much better to a picture which includes people: they can envisage being there, and imagine asking the people about their life and landscape.

8. *Is there a good variety of types of photographs?*

    So many books only show landscapes and townscapes, yet the potential of other types of photographs is vast, for example, oblique air photos, vertical air photos, satellite photos, close-up photos and wide-angle photos.

9. *Do the photographs encourage varied work?*

    Photographs can be used for analysis, for explanation or for creative responses.

10. *Are the photographs up-to-date?*

    Pupils are quick to reject out-of-date photographs: they can seem irrelevant and even boring, and can also cast doubt on the validity of the text.

Figure 4.2. Checklist for evaluating photographs in textbooks.

9. Is the value-system of the book acceptable?

This is perhaps the most complex issue, as well as one of the most interesting. It is also perhaps an insoluble issue, at least in part: if the book reflects my prejudices and ignorance, I may well consider it ideologically sound, and up-to-date too!

Racism and ethnocentrism need to be identified and rejected; a simple test might be: 'If I was a black pupil in a class using this book, would I feel offended or embarrassed?' There is some evidence that sexism is widespread in geography texts, too: most contain far more photographs of men than women. Analysis by pupils of alleged racism and sexism in textbooks can be an effective way of discovering and criticising racist and sexist attitudes in society.

There are many other types of bias. There can be bias towards middle-class assumptions: for example, a lack of empathy with the inner city, seeing it only as a 'problem'. In books on tropical countries, the bias towards high-technology 'solutions' is widespread.

In all these controversial areas, the need is for books which look at both sides of an issue, and which raise questions. This approach both helps to resolve the problem of bias and produces more interesting lessons — an ideal answer to a problem which can seem intractable.

10. Does the book suit me and my pupils?

Happily, this is still the key question. It is salutary to reflect that the word 'me' would be missing from this question in many countries. Even in a country such as the USA which prides itself on 'freedom', many teachers do not have the freedom to select texts. The 'approved' list of textbooks is all that is available to them. But the freedom we enjoy needs to be balanced with a greater professionalism in assessing and selecting texts. The subjective 'feel' of a book is still important — but many elements in the selection process can be handled more objectively than they have been in the past.

4.3. Producing Resource Sheets
Fred Martin

Introduction

The quality of school textbooks has greatly improved in recent years. But no matter how good textbooks become, the need to produce resources within the school will continue. While some textbooks do aim for specific age and ability ranges, others aim more broadly in order to appeal to the widest market. It is rare to find one that caters exactly for your own department's specific syllabus and pupil needs. When the content is right, the reading age may be inappropriate. The pupil activities, when present, are often a particular source of discontent. The cost of textbooks which may only partly be used is another important consideration in the light of departmental economies.

Duplicated resources for the classroom, whether as single worksheets or more lengthy booklets, are produced to meet a number of needs.

(a) To provide material to meet a specific need, such as for a locally based topic, a topic of current interest, or additional case study material to supplement textbook generalisations. Integrated and combined studies courses are particularly badly served by existing texts.

(b) As back-up to textbooks in the form of alternative pupil activities, often where the textbook is being used as a reference source with suitable illustrations or sections of text. Textbooks retain an advantage in the use of colour, especially for photographs.

(c) To provide homework resources when pupils cannot take textbooks home for whatever reason.

(d) To give pupils guidance and activities to go with audio-visual aids, such as filmstrips, videotapes or computer programs.

(e) For examination and test papers.

A set of duplicated resources is a great convenience for the teacher. They save the need to write work on the blackboard, perhaps the same work several times a week. Besides, the blackboard may well be a completely inappropriate technique to cater for a mixed ability class, even during a relatively short lesson.

But there are other valuable by-products to producing your own materials. In order to produce resources of high quality, the workload must be shared. When the workload is shared, then the resources themselves will be shared. Teachers are notoriously reticent to share duplicated resources, partly because of a circular process of lack of time, rushed production, poor quality, then a natural dislike of wider circulation. The 'Banda morning' after the night before is all too common.

Breaking the cycle needs a planned approach with work being allocated to all members of a department. This allows time to prepare materials properly, materials which can be used by others while they in turn are producing whatever else is needed. The 'blackmail factor' is important in that work which is to be shared is more likely to be produced to higher standards. It is also likely that staff will become more critical of their own standards, and in the process become capable of doing better. Time is certainly needed to experiment with some of the techniques described below.

The end result is a set of well produced materials which can be used 'off the shelf' with confidence. Revisions and new materials can be produced on a planned and long-term basis. Shared resources give the additional advantage of ensuring that a reasonable uniformity of syllabus is achieved.

Designing resource sheets

Producing resources involves both design and technology. *Design* refers to the way in which the material is arranged and presented on the page. This will normally be a combination of text, illustrations and pupil activities. *Technology* refers to the machinery available in the school to prepare, print and complete a resource. Unfortunately, the technology varies to a remarkable degree from school to school, making production easy for some and acting as a definite barrier to others.

Principles of design described in this section and illustrated in Figure 4.3 can be appreciated irrespective of the levels of technology that are available. There are, however, severe limitations to some aspects of design if there is a lack of equipment which can perform tasks such as enlarging and reduction, and printing by offset litho.

Other aspects of writing resources also need to be considered such as appropriate reading ages and the content itself. These topics are not included here but they do of course play an equally important part in making sure that the resource is the teaching aid it is intended to be.

Choosing the style

An early decision must be taken over the way the final product is to be laid out. Resources can be in single sheet or booklet form. The booklet allows a longer topic to be tackled with adequate material to cater for a wide ability range. It also avoids adverse reaction to yet another loose sheet of paper and it is certainly easier to store. Balance this against the pupil's satisfaction in actually finishing work on a single sheet.

Most work is produced on A4 size paper, but there are at least three styles of use, each with their own advantages and disadvantages.

(a) Vertical (portrait), using the longest length along the side. This allows a simple design to read down the page, but makes for problems if a booklet is to be stapled.

(b) Horizontal (landscape), using the longest length along the top. This allows more interesting designs and possibly has a longer life when stapled along the side to form a booklet.

(c) A4 folded in half from the horizontal. This form of booklet is complex to organise, but the small size makes it ideal as a homework resource, a simple field study booklet, or a back-up resource to a textbook. Page design can be very straightforward as line length is already largely prescribed by page width. When stapled in the middle, it makes a handy and strong resource with a good life span.

There is much to be learnt by looking critically at professionally produced materials. This may be a geography textbook, a comic or daily newspaper. Immediate reaction is usually to know what we like, but not to know why. The principles, however, are easy to work out, and applying these principles to our own resources is quite possible.

The grid

The page of any book is based on a *grid*. This gives the page an overall structure. It divides the page into column width and depth. The grid may be simple, allowing only two columns. More complex grids allow several different column widths to be used on the same page. The latter is especially useful to allow for variations in the size of illustrations.

Using a grid does have some constraints, but the advantages of structure and clarity generally outweigh the problems.

Text

It is common to see resource sheets which contain *text* written continuously from one side to another. At worst, the text continues on for line after line, forming a visually boring and impenetrable block to all but the most able and motivated pupils. But textbooks do not do this and for good reason. It is easier to read text in a more compact block than stretched out in long lines. As a general guide, a typewriter setting of about 50 characters should probably be a maximum. A setting of slightly less would allow a simple two column layout on vertical A4 to be produced. Obviously a page design must be worked out at least roughly before anything is typed or drawn.

HANDBOOK FOR GEOGRAPHY TEACHERS

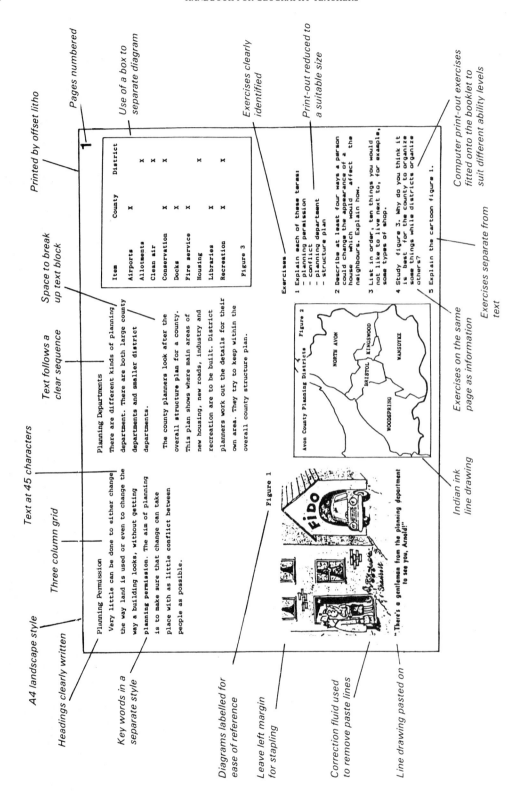

Figure 4.3. Extract from an upper school booklet.

Layout

The reader should be able to follow the *sequence* of text and illustrations easily, without being unsure of where to look next. Forcing the reader to jump from side to side and top to bottom creates unnecessary confusion.

Illustrations, as well as containing information, can add visual variety and interest. *Location* should be thought out with regard to any related text or activities. Illustrations often determine the layout of the whole page as, for example, a map which takes up more horizontal than vertical space.

A related aspect involves making sure that pupils do not have to keep turning over pages to find either questions or information. A page or double spread should be *self-contained*. This may constrain the organisation of material or reduce the amount of material to be included, but the price is small compared with the confusion that will otherwise ensue.

The balance of space

The amount of *space* devoted to text, illustrations and pupil activities (the latter usually also in the form of writing) depends on who the resource is intended for. Large blocks of closely written text must be avoided for average and below average pupils. Leaving some 'white space' on the page may not be economical, but it does avoid cluttering the page and makes it easier to follow — where to look and what to do.

Careful interspersing of text and illustrations can also break up text blocks. Locating illustrations combines the technique of making them fit and the art of making the page 'look right'.

Writing resource sheets

A range of techniques can be used to make the resource as 'workable' as possible with pupils. This involves making sure that the finished product is visually clear enough to read and easy for pupils to follow. They need to know where to look for text, illustrations and exercises in the sequence. It is also possible to make the page visually interesting by using the techniques described below and illustrated in Figure 4.3.

Writing and typing

Handwritten resource sheets can be produced effectively by a rare minority of teachers. Handwriting involves not only having a style that is without 'quirks' and is neatly written, but also requires extreme care to avoid reprographic problems. Writing for spirit duplication involves writing with an even pressure throughout. This is something which nobody does naturally. Mistakes on a spirit master cannot be easily corrected.

Printing by either ink duplicator or offset litho requires writing with as dark an ink as possible for best results. An Indian ink is ideal, though several makes of cheap fibre-tip pen are quite effective.

Using a typewriter removes these problems, especially a good quality electric typewriter with a *carbon* ribbon. Nylon ribbons tend to fade quickly with use. Carbon ribbons are only used once and give a much sharper and darker image.

A computer *wordprocessor* can be used in exactly the same way as a typewriter, though with the advantages of being able to correct errors cleanly, alter wording until it says what you intend, and experiment with line length. Wordprocessor chips can be fitted into most microcomputers at tolerable cost (such as Wordwise, Edword and View for the BBC B). Disc systems are also available which do the same thing (such as Txed and Wordstar for the RML 380Z).

The finished text can be turned into hard copy on a printer, but care must be taken over the quality of print-outs. Dot printing made up to a double density gives a very acceptable result, though more expensive printers are available which give a quality no different from an electric typewriter. Again, check whether a nylon or carbon ribbon is fitted to the printer. A bonus on the wordprocessor is that if the resource is a disaster, alterations can be made to the original with minimal retyping effort.

Typefaces

The *typeface* used on typewriters varies from machine to machine. The most versatile typewriters in this context are those with either a 'golfball' fitting or a 'daisywheel' fitting. These allow the typeface to be changed making letters fat, thin, straight, sloping and in different sizes. This facility is useful as one of a range of techniques which solves the pupil's problem as to exactly which part of the page needs to be 'done'. One style used for text and another for exercises, goes some way towards solving this problem. The golfball or daisywheel also allows key vocabulary words to be picked out in a style chosen to stand out more clearly than the rest, such as bold or italic. These words can be typed separately, then cut out and pasted into the appropriate space.

A range of typeface styles can also be obtained using the combination of wordprocessor and printer. Capabilities differ from machine to machine, but a good range of styles and sizes is usually possible.

The most suitable letter size for different age and ability groups is a matter for judgement. 'Jumbo' size for the less able is sometimes recommended, though the effect is rather condescending with pupils of secondary school age. It is doubtful if the end result is any clearer than normal typewriter size on a well designed and clearly printed page. Avoiding clutter on the page is probably far more important. Typing on a spirit master is likely to be clearer than handwriting, but corrections are almost impossible.

Illustrations

The kind of illustrations which can be included depends on the reprographic equipment available. Although spirit duplicating does have the advantage of colour, it is in many ways the most limiting method of duplicating resources. Use of a heat copier to produce a spirit master can extend the types of diagram that can be included. But the spirit duplicator has no advantages over either the ink duplicator or an offset litho printer.

Line drawings are the easiest to reproduce on offset litho or ink duplicator. The quality of diagrams or photographs in colour depends on the colours used because stencils or plates are made as variations of black and white. A platemaker or photocopier being used for making offset plates, for example, will not distinguish between white and yellow. Photographs can be enhanced by hand with additional shading or by making outlines clearer. Unless a dot screen is available, however, their inclusion probably creates more problems than they are worth. School reprographic equipment tends to be inadequate to cope with large areas of dark or solid shading.

The microcomputer has opened up possibilities for including illustrations which have been 'dumped' directly from the screen to a printer. Some programs can be altered so that this can be done and home-produced programs can build in this facility from the start. Use of a graphics tablet enables illustrations to be drawn directly on a screen, then dumped.

Remember that using items from books and tampering with computer programs can cause a breach of *copyright* laws.

Headings

It is very helpful, as well as being informative, to provide *headings*. A short heading answers the pupil's question as to what to use as a title. It also serves to break up blocks of text, provided that a suitable space is left between paragraphs.

Headings which say 'Things to do', 'Work', 'Tasks' or whatever other term is used, help pupils to locate where the exercises are. A clear and standardised method of exercise numbering will also be an advantage, both for pupils who have to do the work, and for teachers who have to answer questions about them. A large number of sub-divisions should be avoided when numbering questions, such as 2(b)iv.

Using headings should also encourage pupils to divide their own work into sections and give an instant guide as to what the work is about.

Graphics

Simple techniques can be used to make a sheet more interesting and easier to use. Ruled boxes or 'cloud' enclosures can highlight areas of essential text or pupil activities. Use of arrows, pointing fingers and more elaborate flow lines can also help, though there is a danger of overdoing it and creating too many different styles. Rub-on graphics are commercially available, though they tend to be rather expensive. Making a collection of items cut from magazines is a cheaper source.

Plastic stencil strips can be bought which contain a wide range of letters, figures, technical drawings and abstracts. A special mapping nib or 'Rotring' style pen is needed to use these.

It is a good idea to include page numbers in a prominent place when producing a booklet. It is easier to tell pupils to 'turn to page 3' than to try to hold up a page which at long range looks indistinguishable from all the others. In this context, it is also useful to make an overhead transparency from the printed sheet. Any good photocopier or heat copier has the ability to do this.

Covers

A well designed cover to a booklet may be an additional cost, but it reinforces what the topic is about and adds a professional touch to an item that will have taken many hours to produce. A bold heading with design, cartoon, or some other illustration can be simple and effective.

Cut and Paste

Bringing text, graphics and illustrations together on one original involves an extensive paste-up procedure (unless the item is being produced on a spirit duplicator). Items need to be cut out and glued onto a blank sheet, so be prepared in advance with sharp scissors, glue and tweezers. Glue lightly as heavy lumps of glue will cause problems when making photographic plates. At this stage, problems of space become obvious and emergency action may be needed, usually to shorten what is there.

Before a stencil or plate for printing is made, the sheet needs to be as clean as possible, and of course free from errors. Lines often appear between different parts of a paste-up when a stencil (for ink duplicator) or plate (for offset litho) is made. These can be removed by making a clear photocopy, then painting out the lines, or simply painting around the cut-out edges. A small bottle of correction fluid is invaluable at this stage.

Originals need to be stored safely. Unlike a spirit master, they can be used many times, though even the best original fades in time.

Assembly and Protection

Methods of glue binding are available in some schools, but they are too expensive for everyday use. Ring binding or slide grips are effective with large booklets, but again, the cost makes them prohibitive for general use.

A collator saves considerable time if a booklet is being produced. Booklets can be straightened out on a jogger, then stapled, preferably with an electric stapler. A strip of sellotape along the binding edge gives added protection, though sellotape yellows with age and begins to peel. By then, your booklet is probably in no fit state to be reused anyway!

Storage and Indexing

Careful thought should be given to the system of storing and indexing of printed resources. Class sets of resource sheets are conveniently stored in strong A4 size cardboard boxes or in tiered trays of the 'Lawcobox' type. Alternatively they can be placed in card folders, document wallets or envelopes and stored in filing cabinets. Whichever method is adopted, all materials should be systematically labelled and indexed. A departmental reference collection of single copies of each sheet should be compiled, indexed and reviewed annually.

Conclusion

Even a limited selection of the techniques described above will go some way towards the production of presentable resources. It should also help to remove at least some of the sources of confusion for the pupils.

It is recognised that many teachers do not have access to the kind of technology which may be required. Where this is the case, priority ought to be placed on obtaining the equipment. The value of good reprographic equipment has to be appreciated. Once installed, it opens up new resource possibilities.

Duplicated materials are not the panacea to all problems. They have a place alongside other resources such as textbooks, filmstrips and the rest. A box of chalks will still be useful.

Glossary of terms used

Banda: a trade name now commonly used to describe any spirit duplicator.
Carbon ribbon: a typewriter ribbon which gives a sharp and dark image in contrast to a nylon ribbon which fades and is less clear.
Collator: a machine which compiles sheets into booklets.
Copyright: a complex set of laws restricting the use of anything contained in published materials.
Design: the process of laying out a page and making it both workable and interesting.
Dot screen: a machine which breaks a photograph into dots so that it can be reproduced more easily.
Dumping: obtaining a print-out from a computer program.
Electric stapler: a machine which takes the effort out of stapling.
Enlarging: making text or illustrations larger on a photocopier (or by other photographic processes).
Golfball: a typewriter fitting which enables the typeface to be changed to give print of differing size and style.
Graphics: drawings used as design features.
Grid: the structure of a page in terms of column width and lines.

Hard copy: a computer print-out.
Heat copier: a copier capable of a wide range of tasks including single copies, spirit and ink masters, and overhead transparencies.
Ink duplicator: a duplicating machine which gives adequate results, but with limitations in terms of special preparation and storage.
Jogger: a machine which shakes booklets into straight edges prior to stapling.
Landscape: a style of booklet using the longest axis as the top.
Line-drawing: a drawing produced using outlines and minimal shading.
Master: the sheet used in the duplicating process (see also stencil and plate).
Offset litho: the most versatile and high quality type of duplicator available in schools, invariably operated by a technician.
Original: the final paste-up sheet which is converted into a plate or stencil; the sheet that must be stored for future reproduction.
Paste-up: the process of assembling items of text and illustrations on to a master prior to duplication.
Photocopier: a machine to produce single copies; longer runs are usually too expensive.
Plate: the sheet used on an offset litho when printing; plates can be produced either on platemakers or on adapted photocopiers.
Portrait: a style of booklet using the longest axis along the side.
Printer: a machine linked to a microprocessor to give hard copy.
Reprographics: the process of producing printed materials.
Reduction: making text or illustrations smaller.
Resources: any form of sheet used by pupils, whether for information, as a worksheet, in single sheet or booklet form.
Rub-on (graphics): individual figures, letters or designs which are rubbed directly onto paper (as in Letraset or other brand names).
Scanning: a method of producing a stencil for ink duplication which will work from a composite paste-up of text and illustrations; useful in the absence of offset litho equipment.
Spirit duplicator: commonly known as 'the Banda'; a hand operated or electric duplicator which is useful where more advanced printing technology is not available.
Stencil: the sheet used when printing on an ink duplicator, reusable in theory, but in practice messy to store; stencils can be produced by scanning, by heat copying, or may be typed on directly.
Stencil (plastic): a strip of plastic with figures, letters or designs which can be used to draw neat standardized results.
Typeface: the figure and letter images on a typewriter or printer.
Wordprocessor: a method of obtaining perfect text using a microcomputer, with the ability to store text on cassette or disc for future alteration.

4.4. Discussing Photographs
Roger Robinson

A major part of school geography is about what can be seen in the world, and geography teachers rely heavily on visual material to bring some reality into their classrooms.

Most trained geographers are skilled in the analysis of photographs and guide pupils to 'see' certain patterns or activities in them. Much successful work is based on this conventional interpretation of photographs, typified in public examination questions, and it is an important part of school geography. There are, however, other approaches to learning through photographs that relate especially to two important concerns for geography teachers.

The first concern is that relatively little attention is given to the subtleties of image building and to the generalisations and stereotypes that are created or reinforced partly through visual images. The second concern is that many geography teachers use closed learning strategies that do not make it easy to explore issues involving values and attitudes.

In chapter 3 of this volume strong arguments are made for loosening the control and teacher-monopoly of knowledge in the classroom, and encouraging open learning atmospheres where pupil contribution is highly valued. Such classroom ethos, involving learning through discussion and honest response, needs teaching skills unfamiliar to many geography teachers.

This section suggests ways that photographs can be used to provide a resource both for building images that involve alternative viewpoints and develop empathy, and for encouraging discursive learning based in the 'real' world of the pupils and involving their own attitudes and values.

The types of activities outlined below are numbered for convenience but they are not sequential, and complete flexibility is envisaged in the use or modification of the techniques. Some of the activities are possible with a slide projected in a semi-darkened room; others need individual photos (or slides with slide viewer) for each pupil, pair or group of pupils; others require a set of photos. Two black and white photos (Figures 4.4 and 4.5) from published photopacks are used as examples, when appropriate. The questions labelled a, b, c, d, appear in the form in which they are addressed to the pupils. Examples of pupils' answers are shown in italics.

1. A photo is just one image. It is not *the* truth, for there are many images. Pupils can attempt the following activities when using photos to build an image of an unfamiliar place.

    a. In a group or pair discuss and decide what you would photograph in your own area to show someone what it is like.
    As a class look at some of the lists and discuss how and why they were chosen.

    b. Look at a selection of photos of the other place (displayed on wall or laid on central table).
    Discuss in pairs what you think the place is like to look at, to visit, to live in.
    How do you think the pictures were chosen, by whom, and why?
    What do you like/dislike about the place from these pictures?
    What do you think is missing?

    c. Which photos in the selection do you react strongly to, like or are pleased by, dislike or are annoyed by, find interesting, familiar, strange?
    Why? Share your reactions with the class.

    d. Choose one or two photos from the selection on the basis of like/dislike, interest, irritation, agreement, familiarity, etc.
    Show them to your partner. Why did you choose them?

Figure 4.4. Birmingham: a street in Balsall Heath. Photo: DEC, Birmingham.

What questions would you like answered?
Share these with a larger group or the class.

Eg. Figure 4.4. We chose it because it's like our street and there are people like us in it. We'd like to know if the street is closed off at the end to make it safer.

2. A photo must be studied to be fully understood. When decoding a photo for information, use 'buzz' groups for slides (ie. give a few minutes for discussion with neighbour before getting class response), or work in pairs with photos.

   a. Study photo, turn it over, and then write down as many things as you can that are in the photo. Compare your list with your partner. Why did you remember certain things? Are they the important things?

   b. Imagine you are a detective. What things are in the photo? What is going on? Where is it? When is it? What is the evidence?

   c. News flash — search the picture for a news item.

   Eg. Figure 4.4. Missing boy last seen strolling in Balsall Heath.
   Dealers are trying to purchase Victorian chimney pots cheaply from unsuspecting householders.

3. Understanding of, and reaction to, a photo depends not only on the photo but also on the viewers' experience, attitudes, perceptions, prejudices and values. The photo can help the viewers to express their own ideas.

Figure 4.5. Ghana: a local chief receives visitors. Photo: Maggie Murray.

a. Before studying photos of a new place: Make a list of what you expect to see, explain it to your partner and make an agreed list. Then compare the photos with what you expected to see.

b. Clip a partial 'mask' of card or paper to cover all but part of the photograph. Pass it on to someone else for them to explain what they can see and what they think the rest of the photo will show. Remember that the whole photo is only 'part' of the real scene!

c. Respond to the photo:
Are there things, people, activities in the photo that seem unreal or unlikely to you? Why?
Are there very familiar things in the photo?
How would you feel if you were in the photo?

Eg. Figure 4.5. I would feel strange and wouldn't know what to say if I were sitting on the step with these villagers.

Is there anything in the photo that surprises you? Why?
What does it suggest that you hadn't expected?

Eg. Figure 4.5. The villager has a wristwatch. I didn't think he would have enough money, visit shops that sell watches, live by 'clock' time.

4. Learning by discussion is particularly effective for understanding concerned with feelings, attitudes and alternative interpretation.

   a. Discuss with your partner or group the

relationships between the photographer, the people in the photo and others present.
Were the people co-operating with the photographer?
Was the photographer 'unseen' — perhaps using a telescopic lens?
Was the photographer intruding?

b. Make captions or titles for the photo:
Say what you think the photographer wanted to show.

> Eg. Figure 4.5. *A relaxed meeting between the chief and the villagers.*

Suggest something quite different.

> Eg. Figure 4.5. *Poor villagers have to laugh at the chief's jokes.*

Show your own feelings about the photo. Share them with the class.

c. With a partner produce arguments or evidence either as the photographer to defend the photo — the way it was taken, the image it gives, or as a person who thinks the photo is biased, showing people in a certain way.
Join with another pair and present your arguments. Come to an agreement about the photo. What is good about it? How could it be improved? What other view, scene, people should be in it?

5. Trying to think the pupils 'into' the photo can develop an appreciation of the people's situation, and the possible reality of the scene.

   a. With your partner or group list and discuss:
   What is each person doing?
   What do you imagine each is thinking?
   Why are they doing and thinking this?
   What sounds do you imagine there are?
   What languages are the people speaking?

   b. Make up a 'story' for the photo.
   What happened before?
   What will happen next?
   Share it with the class.

6. All photos leave a lot of questions unanswered — but thinking and discussing these leads to an appreciation of alternative possibilities.

   a. What would the scene be like at other times?
   In the night, early morning, rush hour, or in other weather conditions?
   Choose a picture from several to describe under other conditions.
   Let others guess which one you have described.

   b. Questions on photos — in pairs:
   Place the photo (blue-tack) on a large blank piece of paper. Write as many questions as you can around the photo.
   The questions may have a 'factual' answer:

   > Eg. Figure 4.5. *What is special about the design of the chair?*
   > *What is the window shutter made of?*

   An answer that can be deduced from evidence:

   > Eg. Figure 4.4. *What are the two girls doing?*

   An answer that may be a matter of opinion:

   > Eg. Figure 4.4. *Would you like to live in this street? Why?*

   An answer that may require imagination:

   > Eg. Figure 4.5. *What is the man on the right thinking about?*
   > *Why isn't he laughing?*

   Remember all the 'question' words — who, when, why, what, how, where, which.
   Pass your photo on to another pair for answering.
   Compare your questions with some prepared by the teacher.

7. A photograph can stimulate a desire for further information and possible explanation.

   a. Read the caption and photographer's information about the photo.
   Does it confirm your own ideas?
   Are there some surprises?

   b. Search for other information to explain, or to help you to understand, what is going on in the photo and why?

The emphasis in this section on photographs is on open response and discovery, so that pupils have some chance of internalising their learning. One important problem does arise in such activities — how much and when should the photo be put into its context?

Lack of context and inevitably ethnocentric responses to a photograph can lead to misinterpretation — and certainly at some stage background information is needed. The photo can stimulate interest in what may otherwise be 'boring' description, and really the integration of 'context' — as case study work, simple background, role-play simulation, readings etc. — depends on the photo, the theme, the teacher and the pupils.

Common open responses to photos include emphasis on differences between 'them' and 'us' and sometimes a resort to crude stereotypes and generalisations. The geography teacher should hope that over the long term exposure to alternative views and open discussion will reduce these tendencies.

Sets of photographs are not easy to obtain, and so often the teacher has little choice. However, photos that involve people, preferably doing something, and with facial expressions and body posture visible, are ones that most pupils easily relate to. A good variety of photos of the same area, people or theme helps to break down simplistic generalisation and crude stereotyping.

Note

Most of the activities here and many others are discussed in detail in the following publications:

DEC (Birmingham) (1982) *The World in Birmingham: Development as a Local Case Study.*

DEC (Birmingham) (1985) *People before Places: Development Education as an Approach to Geography.*

Both of the above are available from the Development Education Centre, Gillett Centre, Selly Oak Colleges, Bristol Road, Birmingham, B29 6LE.

Richardson, R. et al, (1976) *Choices in Development* (Kenya and Tanzania photopack), VCOAD.

Taylor, N. and Richardson, R. (eds) (1979) *Living with the Land* (Ghana photopack), CWDE.

4.5. Using Videotapes
Jeff Serf

The potential of videotapes

At one time a geography teacher's 'resource bank' consisted of five sets of textbooks (one for each year), a few filmstrips and the slides from family holidays. Of the many additional resources to appear in schools, the video cassette recorder is one with very great potential. Geography teachers, with their tradition of using visual aids, ought to be able to use this piece of equipment to good effect.

Video cassette recorder (VCR) is really a misnomer. It certainly does record sound and moving pictures from a television set on to a cassette tape. However, that on its own is of limited use. The machine also plays back the recording time and time again, and therein lies its true value. The recording may be stored until its use becomes appropriate, allowing it to be slotted into the curriculum at the point where the teacher deems it to be of most value. No longer need a teacher look at the list of schools broadcasts and lament that the ideal programme to use with the third year in the summer term is being transmitted two weeks before Christmas during morning break. The BBC and ITV programme lists that arrive in schools annually can be studied and programmes can be selected for recording on the VCR at any time.

The mechanics of using VCRs are relatively simple. The instructions for recording and playing back can be reduced to two sides of A4 and then circulated to staff. Obviously, each of the many models on the market is different, but there are certain points common to all.

(a) School VCRs tend to be one of three types: VHS, Betamax or U-Matic. The VHS is probably the most widely used, is portable and is fairly easy to set up. All machines are small enough, however, to be carried by one adult.

(b) The cassettes for each system are not interchangeable. Thus, ensure when purchasing, or borrowing cassettes from other schools, that you obtain the correct one. Cassettes vary in length of tape time from 1 to 4 hours. Cost naturally varies with tape time, but the retail price of tapes is not fixed, so it is worth 'shopping around'.

(c) To record, a VCR needs to be linked to a television set, but as the VCR only uses the set's aerial, the set does not have to be switched on. To play back, both video and set need to be switched on. Recording is made possible at any time because the VCR can be 'programmed' to record in advance of transmission. Again, this facility varies with the individual model, but most can be programmed to record one programme up to 7 days before it is transmitted; some videos can be set to tape up to 3 programmes. Programmes may be recorded outside school hours, therefore, if the VCR is set up in advance.

(d) Many teachers argue that it is not the mechanics of recording or playing back that stop them utilising the VCR, but the organisational problems of getting the pupils and the equipment together. There seem to be two possible solutions to this problem. Either a room is put aside in the school to act as a TV/video 'theatre', where staff may take classes when they wish to use the video, or provision is made for transporting the equipment from classroom to classroom. One possibility is the use of a mobile stand or trolley which carries the VCR on a shelf below the TV which is set at head height, thus ensuring that pupils can see the screen.

Assuming that neither the mechanics of the equipment, nor its location with regard to the class, hinder staff using the VCR, then they are faced with the all important question: 'What do pupils do other than watch the programme?'. My answer to that is 'Hopefully the same sorts of exercises that they do when the film projector is used'.

In designing work units involving the VCR, it must be appreciated that there are certain traps into which it is easy to fall. It is essential that the VCR should be seen as a teaching resource and recognised as such by the pupils. It must not be seen as a reward for good behaviour, when they will be allowed to sit passively and let the images 'wash over them'. The other extreme must also be avoided; it is such a pity if they spend more time looking at a question paper or recording sheet than at the screen. A balance must be struck.

Examples of videotapes in use

The examples which follow are offered as illustrations of exercises that have been used in lessons with different age groups. However, one must bear in mind that each has been plucked out of its context. Although some attempt is made to explain what each particular context is, the exercise is a small part of a much longer unit of work.

Example 1

Class: Third year mixed ability.

Video: A Story From Ghana (BBC)

This tape is used as part of an eight-week unit of work which examines the contrasts that exist within different societies. Although the pupils study Ghana and Britain, the main emphasis is on contrasts within each country rather than between the two countries. One of the objectives of the unit as a whole is to develop some appreciation of such contrasts, and to challenge the stereotype of a developing country held by many pupils.

The programme attempts to present the viewer with the differences in life style found in three settlements in Ghana — Gowrie, a small village; Bolgatanga, Gowrie's local market town; and Accra, the capital city. Although this programme may not be ideal, it does help the pupils to focus upon the contrasts within one country. To this end they are asked to fill in a simple recording sheet (Figure 4.6). This enables the pupils to store

	GOWRIE	BOLGATANGA	ACCRA
How do people dress?			
How do people travel?			
How are goods transported?			
What jobs could you get?			
What could you buy?			

Figure 4.6. Recording sheet for 'A Story from Ghana'.

information that they can use later in either discussion or written work.

The programme lends itself to this exercise as it deals with each settlement in turn. The procedure in the classroom is fairly straightforward. The pupils can either fill in the sheet as they watch the tape, or watch the entire programme and then complete the sheet. With some classes it is preferable to view the section on Gowrie, stop the tape, allow the pupils to fill in the first column of the sheet, and then move on to Bolgatanga and subsequently Accra. In many ways the programme provides the link between the two countries being studied. The contrasts between Gowrie and Accra are obvious to all, and besides providing a conclusion to the work on Ghana, the programme also acts as the initial stimulus for pupils to examine the magnitude of contrasts in Britain.

Example 2

Class: Second year mixed ability.

Video: Brazil — Skyscrapers and Slums (BBC).

The use of this tape helps to achieve similar objectives to the previous example. However, attempting to illustrate the differences that exist in modern Brazil is not the prime aim of the six-week unit of work, which is directed towards the links between Latin America and Western Europe.

The programme looks at the lifestyles of two families — one from a shanty area of Sao Paulo, the other from an upper middle-class area. The first 15 minutes of the tape deal with the poorer family, concentrating on one boy. He is shown at home and at work (part-time shoe-shining in the city centre), and we are given information about him, his family, the area in which he lives and some of the issues concerning squatter settlements. The pupils are asked to watch and listen. After watching this section of the tape a brief discussion takes place on what they have seen. The pupils are then told that the next section of the tape deals with a different family. They view the next two minutes of the tape in silence — without the sound commentary.

The tape is replayed, again in silence, and the pupils are asked to write the commentary that they think fits the pictures they see. There is usually no shortage of volunteers to read their 'voice over' to the rest of the group as an accompaniment to the silent video, and of course all are eager to compare their efforts with the real commentary. The exercise provides a wealth of information for further discussion, verbal or written; for example, a comparison of lifestyles, what each family might think of the other, the influence of North American and West European cultures on both families, the 'fairness' of the unequal distribution of wealth and opportunity.

This exercise goes some way towards countering the stereotype held by many pupils. If they simply watched the tape through, the pupils might remember something about the boy who lives in a shanty and cleans shoes. By concentrating on the other family as well, they may remember that there is another side to Brazilian city life. Perhaps another exercise of this kind could be of use to highlight other generalisations.

Example 3

Class: Fourth year (Geography 14-18 Project).

Video: Geography Casebook: Britain — The Inner City I and II (Glasgow) (BBC)

The following activities are taken from a twelve-week unit on urban geography. The pupils spend some time looking at the urban structure of Birmingham and the changes that have occurred since 1945. The first of these two videos expands their experience, illustrating that similar conditions existed outside Birmingham. It provides the pupils with a second example of council action taken to alleviate inner city problems and the results of such action. The programme raises several issues that it would be a mistake not to at least discuss with the group. Such points include the lack of job opportunities in the 1930s compared with today, and the value system that allows an expression of public affluence in the form of many public, and one assumes private, buildings, whilst condoning the squalid housing conditions of the majority of Glaswegians.

Much of the first programme is presented in the form of personal reminiscences, and this has provided the stimulus for pupils to interview grandparents and family friends about what they recall of pre-war Birmingham. Many of the pupils live, or know people who live, on one of the large housing estates built on the edge of the city to house people displaced by redevelopment. Thus some of the pupils can contribute with first-hand accounts of living in such areas, and so provide the link with the second programme. This looks at such problems, and the swing towards urban renewal as a possible solution. Again calling on personal recollections, the programme helps the pupils to develop an empathy with the residents of Glasgow. This process is helped by the fact that many of the pupils are aware of Birmingham's urban renewal policy. For a second time interviews with local residents have been arranged to see if the problems of Glasgow are shared by Birmingham, and it is a relatively simple task to arrange visits to both inner city areas and edge of city housing estates.

Here then are three examples of how the use of the VCR has made a vital contribution to three units of work. Each of these programmes probably has greater potential than has been discussed here, but only individual teachers can view a programme and assess its value for use in their classroom. As with any resource, clear thought and adequate preparation must go into its use. Even with previewing, the use of the VCR does not, of course, guarantee a successful lesson. Any programme will present the pupils with a great deal of information, and it is the task of the teacher to help them to extract that which will be of most use to them. I recognise that whatever I direct them towards is based on my own value judgements and that another teacher could see a different use for each of the programmes.

There are two main obstacles to the greater use of the VCR. The first is the cost of purchasing the equipment and the time involved in mastering the mechanics of using it. Every school should make the purchase of the equipment a priority and every geography teacher should learn how to operate it, as it really is straightforward.

The second obstacle is the copyright law. At present teachers may record any schools programme provided that their school or LEA has a licence. A glance at the annual BBC and ITV lists of schools programmes will show the great variety that is produced. Nevertheless I must admit to watching a number of other television programmes with a warped sense of envy. Although they contain a great deal of potentially valuable educational material, the copyright law prohibits their recording for use in the classroom.

A licensing scheme, however, does enable educational institutions to register to record designated television programmes transmitted by the Open University and by Channel 4 and use them for educational purposes. Only designated titles may be recorded by those licensed to record and annual fees are payable for each programme. Advance information of Channel 4 programmes available for recording are listed in *The Times Educational Supplement*. If this scheme were to be enlarged then we, as teachers, would be offered the greatest possibility of using the VCR to its full potential.

Note

Addresses for details of the licensed off air recording of television programmes.

1. BBC Schools Broadcasts: The School Broadcasting Council, The Langham, Portland Place, London W1A 1AA.
   Tel. 01-580 4468.
2. ITV Schools Broadcasts: The Education Officer, ITCA, 56 Mortimer Street, London W1N 8AN.
   Tel. 01-636 6866.
3. Open University Programmes: The Guild Organisation, Guild House, Peterborough PE2 9PZ.
   Tel. 0733 63122.
4. Channel 4 Programmes: The Guild Organisation, Guild House, Peterborough PE2 9PZ.
   Tel. 0733 63122.

5. Microcomputers in the Classroom

Howard Midgley and Peter Fox

Introduction

"Microcomputers can't be any good for education — the children like them too much." "Just think how my lesson preparation and marking is going to be cut down once we have full use of computing facilities." "I don't think it is worth trying to use the computer in my lessons — it won't achieve anything that I can't do already."

We are living in a time when the technology of education has been altered, upset, even invaded, by the microcomputer, with consequences which have yet to be fully appreciated and come to terms with. The above statements are typical of those which have been made in staff-rooms up and down the country over the past few years. Since microcomputers became available at a price many schools could afford, they have been coming into schools in increasing numbers.

This chapter sets out to look at the *reality* of the computer in the geography classroom — a reality, because the computer has entered the geography room in many cases already — and to illustrate some ways in which computers can aid the learning of geographical concepts.

The Government, since 1980, has been encouraging the growth of computing in education in two main ways. First, through funding from the Department of Industry, it has enabled microcomputers to be purchased by schools at low cost. Second, through the Microelectronics Education Programme (MEP) the Government is helping schools to use microcomputers and educate children for life in the microelectronics age. This programme is reaching schools in three main ways, all of which are relevant to geography.

(a) *Curriculum Development*. MEP is sponsoring research and development into new subject areas, and into the development of materials which enable the new technology to contribute to existing syllabuses. As well as classroom materials, MEP has developed a range of packages for use in teacher training, and in-service computer familiarisation courses.

(b) *In-service Training (INSET)*. MEP has created fourteen Regional Information Centres, covering England, Wales and Northern Ireland. It is hoped that, with all schools being within reach of one of these centres, the dissemination of knowledge and materials relevant to the new technology will be more effective and more rapid. One of the functions of each centre is to provide in-service training for existing teachers. Indeed, for schools to take advantage of the Department of Industry scheme for 'microcomputers in schools', two of their staff have to attend computer familiarisation courses. Generally speaking, INSET courses have absorbed much of MEP's energy and a great many teachers have attended courses relevant to their subjects.

(c) *Information Service*. Each Regional Information Centre collects and catalogues software and some even distribute (usually through LEA advisers) commercial and MEP sponsored software on a loan basis. It also compiles and distributes information and advice about the uses of microcomputers and software in the classroom.

Figure 5.1
Microcomputer
and peripherals
on trolley.

Hardware

A selection of hardware is shown, mounted on a trolley, in the photograph (Figure 5.1). The term 'hardware' refers to the *Computer* and its *Peripherals* — the various electrical gadgets which enable programs to be loaded and run.

(a) *The Microcomputer*. This is the heart of the system, containing the memory and processors which enable computing to take place, and the keyboard for the input of information by the user. With some microcomputers, the keyboard is separate from the processing unit.

(b) *Disc Drive*. This, when connected to the computer, enables programs and data to be stored on a floppy disc. Disc drives are expensive, but have the great advantage of reliable and rapid information storage and retrieval.

(c) *Cassette Recorder* (not illustrated). A much cheaper means of information storage, which is rapidly being superseded by disc drives in schools. That part of the computer's memory which enables programs to be stored and run only works when the computer is switched on. Thus discs or cassettes are vital for long-term storage of programs.

(d) *Visual Display Unit* (VDU). This may be a television, connected to the television output socket of the microcomputer, or, perhaps more usually in schools, a monitor which enables the computer to display its output with greater clarity. Some computers do not work satisfactorily with television sets.

(e) *Printer*. This is a great boon, for example, if pupils need paper copies of computer calculations. However, for most computer programs designed for geography, a printer is not necessary, and it is thus likely to be of low priority on the school's hardware 'shopping list'.

(f) *Other Peripherals*. A range of other peripherals is available, and may prove useful to the geographer, notably the graphics tablet — a device which enables maps to be copied by the computer from paper, areas and distances to be calculated, and even topological transformations to be made. Graphics tablets will probably find their way into classrooms in increasing numbers in the future.

The photograph (Figure 5.1) illustrates a particular make and model of microcomputer — the BBC Model B. It is very important to be aware of the make and model available to you in the classroom. Differences between them are important for several reasons. Different makes vary in the programming languages used. Even though they may use BASIC as their principal language, there are significant variations in each manufacturer's version of BASIC. This is especially true where the programs use the *graphics* functions of the computer — its ability to display maps and diagrams on the VDU screen.

Each make of microcomputer uses a different means of recording its information on disc or cassette. Thus a disc or cassette prepared for a particular model cannot be used with a different make of computer.

For several reasons, then, programs written for a particular microcomputer cannot be used on any other. This is further complicated by the memory size of the machine available to you. It is important to ensure that your machine has sufficient memory to enable the program in question to be loaded and run. Memory is measured in kilobytes, and a typical figure is 32k. for schools computers.

Thus it is important to know what hardware you have available, especially the make, model and memory size of microcomputer, so that this can be borne in mind when selecting and purchasing software. Most software is available in versions to suit the more common schools microcomputers — particularly the BBC Model B, RML 380Z, SPECTRUM, and APPLE.

Software

Software, the complement of hardware, consists of programs and documentation. The program, when loaded into the computer, enables it to perform the required tasks. A computer without software is about as useful as a camera without a film! The documentation supports the program. This is as important as the program itself, because it includes all the necessary background information, together with full instructions about how to use the program.

The growth in the availability of software in geography has increased rapidly within the last few years. The material at first was produced mainly by interested geographers in universities, colleges and schools that were lucky enough to have a microcomputer. It was then distributed by the Geographical Association Package Exchange (GAPE).

The growing availability of computers in homes and schools, the encouragement by the Microelectronics Education Programme and the development of the software departments of major publishing houses, together with some individual enterprises, have provided the subject with an ever increasing amount of software.

About 200 programs are now available in geography. They vary widely in variety, price and suitability. Some of these programs have been produced for the home computer/leisure market and tend to be quizzes or atlas tests often using the computer graphics and a means of scoring. Many of these programs test factual knowledge only and have limited use for a geography teacher. Of the rest there is great variety in price, quality and suitability for school use — although most of the programs are now available for a variety of computers in a variety of formats (disc/cassette).

A large proportion of programs have evolved through long stages of testing and evaluation in schools, others have been developed almost overnight. The programs which have a clear design, good documentation and clear aims are probably best and may not be the most expensive.

Since most computer software cannot be obtained on inspection from the publisher it is important before buying to view the program. This can be done at MEP Regional Centres and LEA Teachers' Centres. Reviews of software are to be found in many computer magazines and on the Computer page of *Teaching Geography*. A list of microcomputer software is available from the Geographical Association.

Some lesson outlines using microcomputers

The increased availability of computers and computer software has meant that geography teachers can be selective in the type and level of program they use. Rather like sets of textbooks, individual teachers will find the sort of software they are happiest with in the classroom with their pupils.

Many computer programs are now very complicated and have been designed for older students. There are too few good simple programs for the younger, less able groups in the secondary school. How best to use the different sorts of software available, and which types of software are of most use to the geography teachers, are questions to which there are no simple answers.

The lesson outlines below illustrate four uses of the computer in the classroom:

1. Drill Practice.

2. Role Play Game.

3. Simulation.

4. Graphic Display.

In each of these examples the pupils are organised in a different way. The examples are not necessarily the best possible examples of the classroom use of computers, for that depends on the class, the situation, the software and most of all the teacher and his or her relationship with the class.

1. Drill Practice

Class. Second Year Mixed Ability.

Time. One hour.

Program. SLOPES, GAPE/MEP (GA/Hutchinson).

Context. As part of a unit of work on the use of maps — and man and the landscape.

Aim. To illustrate the way in which transport and slope interact.

Equipment. BBC 'B' Micro. One Monitor.

Layout. Class in rows, twin tables. Monitor at front for initial demonstration of program and then moved to one corner for pupils to work with.

Details of the Program. Examines the relationship between slopes of varying gradients and different modes of transport (walking, cycling, lorry, tractor, train). The program gives an animated display.

How the Program is Used

(a) For those who need the recap, one or two points about the use of the computer are made. The program is demonstrated to the whole class so that they can see what it does and how it works.

(b) The class are then divided into five groups, one for each form of transport, and asked to think about the different types of problem illustrated on a worksheet (Figure 5.2).

(c) Whilst the class are working in groups on this, each group is invited to work at the computer for about 10 minutes using a recording sheet (Figure 5.3).

(d) Before the lesson ends the class talks about the results of the tasks, and works out whether the computer is correct!

2. Role Play Game

Class. Third Year Mixed Ability.

Time. One hour.

Program. SETTLEMENTS, GAPE/MEP (GA/Hutchinson).

Context. As part of a unit on settlement, why and where people settled in Britain in Anglo-Saxon times.

Aim. To examine the factors which may have influenced settlers in Anglo-Saxon Britain.

Equipment. BBC 'B' Micro. One Monitor.

Layout. Class divided into groups around tables. Monitor at side of class (not facing class) with computer. Later turned round for pupils to see.

Details of Program. A decision making exercise based on siting and location of village. Resources are given different weightings. A map is displayed — pupils have to locate villages. An effort factor is printed for each.

How the Program is Used

(a) An introduction to settlement and the sort of factors which are important are completed in the lesson prior to this one.

(b) Each small group is given a map (Figure 5.4).

(c) Each group is asked to mark out of 10 each of the following factors thought important for settlement and to note any other factors. (10 = most important, 1 = least important.).

**BUILDING MATERIAL
WATER
ARABLE (CROP LAND)
PASTURE (GRAZING LAND)
FOREST**

The pupils think of the *three* best locations for their village on the map — based on these ratings. The pupils then use *one* of these.

(d) Two members of each group (in the order in which they finish) input their information into the computer (with some help from the teacher) and the location of one of their villages (one of their three best locations).

Look at the illustrations below and answer the questions under each.

A — 100m — **B**

How would you get from A to B?

Any disadvantages?

How would you reduce this slope to make it easy to climb?

A — 50m — **B**

How would you get from A to B?

Any disadvantages?

How could you reduce this slope to make it easy to climb?

A — 50m — **B**

How would you get from A to B?

Any disadvantages?

How does man cross a valley such as this?

Man reduces slopes so that they are easy to climb. In the boxes below draw some examples of how he does this. Title your drawings, eg. cutting.

Complete the table below:

Method of transport	How that method overcomes steep slopes
Walking	
Cycling	
Lorry	
Tractor	
Train	

Figure 5.2. Worksheet on slopes.

What form of transport is your group?_____

Try your form of transport on these slopes and find out if your type of transport finds this type of slope

        EASY

        HARD

        IMPOSSIBLE

SLOPE %	EASY	HARD	IMPOSSIBLE
5%			
10%			
15%			
25%			
35%			
45%			

Then find out up to which % your type of transport finds

        EASY _____ %

        HARD _____ %

        IMPOSSIBLE _____ %

Figure 5.3. Recording sheet for use with computer program.

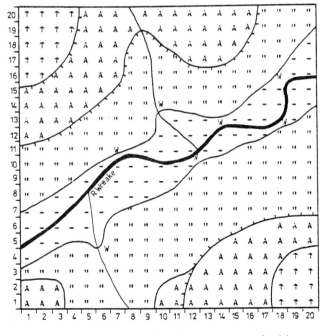

↑ Woodland   " Grazing   ⋏ Arable

w Water point   - Marshland   ⤴ Slope

Figure 5.4. Map of Wreake Valley.
Source: MEP/Hutchinson Software.

(e) This continues until the computer has all the groups' information. In the meantime the group are asked to think and note on the back of their map any disadvantages they could think of for their site, in relation to the computer's factors and other factors.

(f) At this point the computer is turned round so that all can see it. The computer then completes all its calculations and shading.

(g) Each group then shades on their map their 'parish' and notes down the results (these can be printed out if a printer is available).

(h) The winning group is asked to say why they won.

(i) The next geography lesson is used for another 'round' but this time the locations are in the end compared with actual locations on the map of the Wreake Valley.

3. Simulation

Class. Fourth Year Mixed Ability.

Time. One hour.

Program. WATER DROPLET, GAPE/MEP (GA/Hutchinson).

Context. As part of the Hydrology Unit based on the Geography 14-18 Project.

Aim. To improve understanding of flows and factors which change flows in this system.

Equipment. BBC 'B' Micro. One Monitor.

Layout. Class in rows, twin tables. Monitor in a corner visible for everyone but then turned round and used by small groups of five pupils.

Details of Program. Program displays simplified diagram of hydrological cycle. Pupils can vary temperature, rock type, ground cover and soil cover. User can follow the path of a water droplet. Animated display of hydrological cycle.

How the Program is Used.

(a) Some general material on the hydrological cycle has already been completed by the class before this program is used. Words such as evaporation, throughflow, percolation, groundwater flow have been introduced.

(b) A demonstration of the program is given to the whole class so that in small groups they understand what to do — the program itself is quite self-explanatory.

(c) Each group is given copies of the flow diagram (Figure 5.5) under which they enter their selected temperature, rock type, ground cover and soil cover.

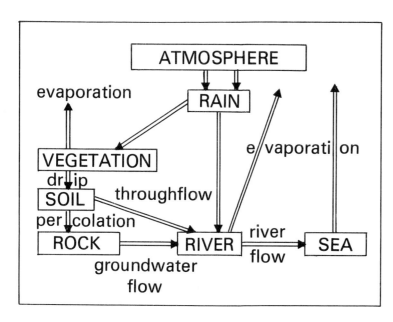

Figure 5.5. Flow diagram for use with computer program.
Source: MEP/Hutchinson Software.

What will happen to a droplet of water if

1. The temperature is   High   _____
                                 Low   _____

2. The rock type is   Clay   _____
                                 Sandstone _____
                                 Chalk   _____

3. The ground cover is   Forest   _____
                                 Ploughland _____
                                 Scrub   _____
                                 Crop   _____

4. The soil cover is   Thin   _____
                                 Deep   _____

5. The land is built over   _____

Figure 5.6. Table for the path of a water droplet.

(d) The groups each use the computer in turn for about 8-10 minutes, running the program three times with three showers of droplets, and changing at least one of the conditions each time they run the program.

(e) Groups not working at the computer are asked to complete a table (Figure 5.6).

4. Graphic Display

Class. Fourth Year Mixed Ability.

Time. One hour.

Program. URBAN LAND USE. GAPE/MEP (GA/Hutchinson).

Context. As a follow-up to local fieldwork in Beeston in the Urban Unit of the Geography 14-18 course.
The class has already undertaken a fair amount of work on land use.

Aim. To show distribution of various land uses, ages, types of building in Beeston.

Equipment. BBC 'B' Micro. Disc. Large TV. Slide projector and screen.

Layout. Class surrounding TV.

Details of Program. The program simulates by coloured map the land use of Leicester. It enables the user to select different combinations of land use, building date, type of dwelling and display them on the screen.

How the Program is Used.

The program data for Leicester is replaced by data on Beeston. The program is used to illustrate differences in land use, ages and building types in Beeston.

(a) Housing pre-1870 is selected and illustrated by map on the TV. At the same time a slide of early housing is shown. This type of housing is now only found in one square of the map of Beeston. The pupils are asked why this is so.

(b) Housing 1870-1918 is added to this map — again illustrated.

(c) The screen is cleared and the pupils asked where most of the post-war council houses might be found — as illustrated by slide. These are then displayed.

(d) Detached houses are added — and illustrated.

(e) The pupils suggest the locations of different types of land use, buildings of different ages, and types of building for housing and industry. These are illustrated on the screen in a variety of combinations.

Classroom management

A glance at the previous section will make it obvious that the microcomputer can be used in a variety of classroom situations. It has the advantage of being easily carried, so that it can be used equally well by small pupil groups and by the teacher in the same way that he would use the blackboard or film projector. As with other teaching aids, there are one or two practical aspects which the user must be aware of before the lesson starts.

(a) *Hardware*. The chances are that the computer equipment will not reside permanently in the geography classroom, though it would be nice to think that, in years to come, all classrooms will be equipped with computers. This necessitates a certain amount of forward planning so that you can be sure that there will be a computer available when you want it. Your school may have a booking system, either formal or informal. It is important to ensure that you can have the equipment set up before the start of the relevant lesson.

The hardware will require at least two — probably three — plug sockets. The position of the computer, monitor, and disc drive or cassette recorder may need to compromise between the layout of the classroom, and the position of a suitable socket. It must be stressed, however, that electricity should *never* be compromised with anything. Do not contemplate extension leads — many local education authorities bar these anyway because they are a risk to both pupils and hardware.

When using the computer for a large class of pupils, it is important to ensure that the monitor or television is large enough to be seen by all. It is frustrating and boring for the pupils at the back of the room if they cannot see what is going on. Partial blackouts will make the screen even more visible, especially on a sunny day.

Hardware, then, should be appropriate to the particular teaching situation, and needs to be set up and tested before the lesson begins.

(b) *Software*. Perhaps the biggest myth, which has yet to be exploded in many cases, is that a computer program removes the burden of lesson preparation and classroom control from the teacher. Far from it — the microcomputer is one of the best teaching aids available, but it is no more than an aid. Software involves a great deal of lesson preparation. First, the chosen program(s) and documentation needs to be previewed, so that the teacher knows exactly how it runs and what responses are needed from the user. Computer programs inherently prompt much useful class discussion, and so it is important for the teacher to know their limitations as well as their facilities. Second, the teacher should know precisely how to load the program into the computer, and, if it is to be loaded from a cassette recorder, how long it takes to load — it can take several minutes. In such cases, it is best to load the program before the start of the lesson.

The safest rule to remember is 'be prepared'. Not only should the pupils have adequate work to do, before, during and after the program is run (the essence of thorough lesson preparation), but the teacher should be prepared for unforeseen problems. A faulty disc, or cassette, or momentary power cut, or loose connections in the various hardware connecting leads, will leave the unprepared teacher stranded. Believe us, these things do happen! So, it is best to be prepared by having alternative, or back-up, work available — Murphy's law being that, as long as you have back-up work available, nothing will go wrong!

When not to use microcomputers

We believe the microcomputer to be one of the most significant teaching aids available. Its versatility, calculating ability, information storage capacity and graphics display, make it useful in a wide variety of contexts. It is *not* however, the universal panacea for all teaching and learning situations. It has not rendered any other teaching aids obsolete, and we believe it never will. It is as important to be aware of the computer's limitations as it is to be aware of its capabilities.

One of the great advantages of using a microcomputer is that it removes the tedium of lengthy calculations and the drawing of maps and graphs. Particularly when the end result is more important than the process of calculating or drawing, this tedium can be of little educational value, and hence the computer is performing a useful service. On the other hand, there are many occasions where it is important for the pupils to understand *how* to perform the calculations or draw the graphs. The only way to learn these skills is through practice. In these circumstances the computer fulfils no useful purpose.

The microcomputer is an extremely versatile calculator, and can be programmed to perform all sorts of statistical tests, for example. Nevertheless the simple pocket calculator should not be underestimated, and, before going to the trouble of using the computer, it is worth asking oneself whether the pocket calculator might not do the job just as well. This is particularly relevant to pupil project work, where each pupil might well be working with a different set of data. Moreover, a great many pupils nowadays possess their own calculators.

One of the most useful capabilities of the microcomputer as far as the geographer is concerned is its ability to display graphics. It is worth considering what other graphics facilities are available to the teacher — not such high technology maybe, but equally effective educationally.

The blackboard, for example, enables maps, graphs and diagrams to be drawn quickly, modified, erased, and annotated. It is large enough to be seen clearly by all the pupils in the room, and it needs no elaborate setting up. If the classroom is not being used beforehand, then quite elaborate maps or diagrams can be prepared before the beginning of the lesson.

Posters have a greater degree of permanence than blackboard material. They can be prepared in advance by the teacher or pupils or purchased from commercial publishers, and be displayed in the classroom for as long as they are relevant. They can easily be made large enough to be seen by all, and can be tailor-made to suit the individual teacher's requirements.

More flexible, and potentially more useful than the blackboard or poster, is the overhead projector. Acetate sheets can be prepared beforehand and can be stored for future use. Transparencies can be prepared accurately, either by tracing an original map or diagram, drawing freehand, or even photocopying from an original. The use of overlays allows the teacher to demonstrate geographical processes (such as marine deposition), sequential developments (such as the growth of urban areas), and spatial distributions. The transparencies can be annotated in the classroom, and modified during the course of class discussion.

Whilst we believe that the microcomputer can significantly aid the teaching and learning of a great many concepts, we would urge the teacher to always ask the following question before using the computer — 'Does the computer perform the task better than any other means at my disposal?' The answer will often be 'yes', in which case it would be silly not to use the computer; but there are many occasions when the answer will be 'no'.

Do it yourself computing

Software is rather like a filmstrip — it can be invaluable as an aid to teaching, but it does not *quite* suit personal requirements. These requirements relate to the way you want to teach a particular topic, or the way your syllabus suggests it. In such cases, it is worthwhile modifying the software — a relatively easy task if it is written in BASIC – or even creating your own.

A BASIC course on computing is beyond the scope of this chapter, but we do make one or two points for those teachers who feel confident enough to dabble with programming. First, programming takes hours and today's teacher often just does not have that sort of time, especially if he or she does not have frequent access to a microcomputer. It is not so much the programming which is time-consuming, it is more the detection and correction of errors, the testing of the program in a classroom situation, the modifications that may result from this, and the preparation of a straightforward set of instructions about how to run the program. Even the simplest of programs requires a minimum of documentation, especially if colleagues within the school or at neighbouring schools are also going to use it.

The modification of existing software is perhaps a more practical proposition, especially where it consists of straightforward programs clearly written in BASIC. One often finds listings of programs in journals or books, and it is these which are the easiest to modify. At the other extreme, some very polished complex programs have been written in *machine code* — the computer's own internal language, and modification of these is out of the question to all but highly knowledgeable programmers.

With the rapid growth of computer literacy — how many of your pupils have computers at home? — the need and ability to create custom-built software will no doubt increase. This is to be welcomed, but it must be approached with care. Commercial software has generally taken a long time to produce, by specialist teams of authors, and the problem for most teachers is that they are already stretched to their limits.

Where to get help and information

There are several sources where the teacher can get help and information about using the microcomputer in the geography classroom. Some advice should be available from within the school. There are teachers in almost every school who have attended in-service courses on computer familiarisation. They may not be geographers, but they can at least help you with the equipment.

Outside the school there is the local education authority advisory service, both in the fields of computer education and geography. There is also the network of MEP Regional Information Centres. They produce a wealth of information by way of newsheets, journals and information leaflets, and are very willing for teachers to pay them a visit. Communication should in the first instance be made via the head teacher or LEA adviser.

References

Fox, P. et al, (1984) *List of Geography Microcomputer Software,* Geographical Association.
Kent, A. (ed) (1983) *Geography Teaching and the Micro,* Longman Resources Unit.
Nash, A. and Ball, D. (1983) *An Introduction to Microcomputers in Teaching,* Hutchinson.
Watson D. (ed) (1984) *Exploring Geography with Microcomputers,* Council for Educational Technology.

6. Maps and Mapwork

6.1. Map Reading Skills

David Boardman

Learning to read and use maps is an important aspect of the development of graphicacy in children. It is in geography lessons that children learn the essential map skills and responsibility for teaching these skills rests firmly with the geography teacher. During their secondary school geography courses pupils should encounter a wide range of maps drawn on different scales and for different purposes. Map reading is not an end in itself, however, so it should not be limited to exercises in the classroom. A map is a means to an end: for showing the location of places; for finding the way from one place to another; for indicating what places might look like. In other words, maps are not just for reading — they are for *using*.

Mental maps

Mental maps form a useful starting point for a course of mapwork for pupils in their first term at secondary school. A mental or cognitive map is one that is carried around in the mind of the individual person. It is a freehand memory map which is based upon the recall and reconstruction of the individual's experience. This mental map may be drawn as a sketch on a piece of paper and shows the individual's subjective personal view of some part of the spatial environment.

Children can be asked to draw from memory on a piece of scrap paper a freehand map of a walk, cycle ride or bus journey with which they are familiar, such as that from home to school. If pupils attempt this in their early days at secondary school they will have followed the route for a sufficient number of times to enable them to represent it in some way as a sketch on paper. After they have drawn the map from memory, they can be encouraged to think carefully about the way they came from home to school and discuss it with one another.

The pupils should take their maps home and on the following day should be asked to notice more closely the objects they pass on the way to school and note changes in direction. When they look again at their original maps they can ask themselves whether they are accurate and if appropriate

Figure 6.1. A mental map is unique. © 1984 by United Feature Syndicate Inc.

they can alter or redraw them. Several pupils who follow the same route to school can compare one another's maps and look for similarities and differences. Pupils who take different routes to school can look at one another's maps to see whether they recognise the routes and features shown on them.

The mental maps drawn by pupils may not be very accurate. The less familiar the pupils are with their spatial environment, or the longer the distance they travel to school, the less accurate will be their maps. Nevertheless the maps should show in elementary form the concepts of abstraction, orientation and scale. Abstraction has to be attempted because the pupils are trying to represent in two-dimensional form part of their environment which they will normally have seen only from ground level. The pupils also have to try to orientate their maps and show direction on them, and to show distances on the map in approximate proportion to the corresponding distances on the ground.

Mental maps depict the images of the pupils who draw them, and these images are the result of their own experiences of the environment. The pupils' freehand maps show their images of the environment that they cannot see as a whole from their earthbound viewpoint. Drawing mental maps is a creative activity which gives pupils the satisfaction of producing maps that are personal to them. A mental map is unique to the individual: no one else would draw quite the same map (Figure 6.1).

The pupils' perception of features, the nature of their mental maps and the kind of detail shown on them, all indicate to the teacher the pupils' ability to represent in graphic form their environment as they experience it. Mental maps provide the teacher with insights into the pupils' perceptions and representations of space. In this respect the drawing of mental maps forms a useful starting point for teaching the conventions used in maps and for subsequent work on direction, location, symbols and scale.

Direction

The relative positions of features shown on a map have to correspond exactly to those in the real world. Only in this way is it possible for map readers to orientate a map, relate their own position to features shown on it, and thus describe direction. When pupils are learning to describe direction by using the points of the compass they should thoroughly master the four cardinal points. It is not uncommon for pupils to confuse left and right, and if they do so they are also likely to reverse east and west. Errors with these two cardinal points may tend to persist as a result of the way in which routes are usually described as 'east-west' (right-left) rather than as 'west-east' (left-right). The word 'we' is a simple but effective way of helping pupils to remember west and east correctly.

When learning the four intermediate points of the compass pupils should be taught to remember that the north and south points always precede the east and west points. Thus the order is north-east, south-east (not east-south), south-west and north-west (not west-north). Similarly, when pupils are learning the complete sixteen points of the compass, they only have to remember the rule that the four cardinal points are stated first and to these are added the adjacent intermediate points. The remaining points are thus north-north-east, east-north-east, and so on (Figure 6.2). By convention the top of a map is north, but it is still good practice to insert the north point on it.

Learning the points of the compass should go beyond drawing the compass rose on paper. Pupils should be able to point correctly to each of the various directions in the classroom and state the direction in which they are moving when they walk round the school building or follow a route marked out in the playground. They can note which of the exterior walls are south-facing or north-facing, linking this observation to the comparative warmth of south-facing walls and rooms when the sun is shining.

It should be emphasised to pupils that the direction from one place to another is always the direction *from* which they are standing *towards* which they are looking. Confusion may occur because wind direction is stated as the direction from which, and not towards which, the wind is blowing.

The pupils should compare the mental or cognitive maps of the route to school which they drew earlier with an A to Z or Geographia street map. They will then be able to see how accurate or inaccurate their mental maps are. It is particularly valuable to introduce pupils to large scale street maps at an early stage because they are designed for a specific purpose — route finding — and do not contain other distracting information. They are also maps which are commonly used in everyday life.

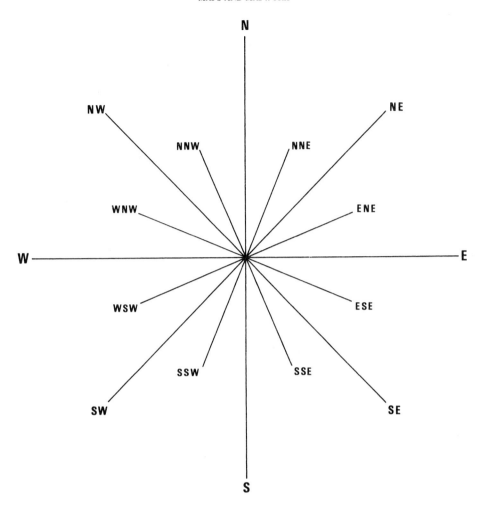

Figure 6.2. Learning the points of the compass.

Location

Every feature on the surface of the earth occupies a place on it, either permanent or temporary. When represented on a map that place becomes a fixed position. The main function of a map is to enable objects and events to be located in space. If a feature cannot be located it cannot be mapped, so the concept of location is fundamental to all mapping.

Pupils who have drawn their own mental maps will have shown a rudimentary appreciation of the relative locations of objects and the relationships between them. This can be used as an introduction to ways of stating precise location on a map. An A to Z or Geographia street map again forms a good basis for teaching the concept of location to younger pupils, because the squares along one edge are lettered A, B, C etc and those along the other edge are numbered 1, 2, 3 etc. The pupils can use the squares for their intended purpose, that of finding the location of roads after looking them up in the index. Pupils can play games in pairs, one asking the best way of getting from one place to another, and the other working out the shortest route on the map.

When they have spent time searching for the names of roads within squares, the pupils will appreciate the advantages of using a more precise method of locating features on a map. When younger pupils learn to use grid lines as reference systems, a natural transition from the A to Z system is provided by the Ordnance Survey map of the area in which the school is situated on the largest published scale, either 1:1,250 (50 inches to one mile) or 1:2,500 (25 inches to one mile). Small sections of these maps can be pasted on to workcards, the grid lines being numbered along the top, bottom and sides.

Provided that the pupils have learnt to draw simple linear graphs in mathematics, the geography teacher should explain that, just as the starting point for numbering a graph is the lower left-hand corner, so the starting point on a map is the south-west corner. The numbering on the map moves from this point to the right and upwards in exactly the same way as the axes of graphs in mathematics. On a map, however, the lines are called eastings and northings. The vertical lines are drawn east of the point of origin and so are called eastings; their numbers are printed at the ends of the lines along the lower and upper margins of the map. The horizontal lines are drawn north of the same point and so are called northings; their numbers are printed along both sides of the map.

The intersection of an easting and northing gives the four figure grid reference of the square to the east and north of it. It is important to emphasise that eastings are always given before northings; pupils frequently give them the wrong way round. They can be reminded that, just as in mathematics the value of the x axis is always given before the value of the y axis, so eastings are always given before northings. A useful memory aid is to 'walk along the corridor before you climb the stairs' so that the number along the lower margin of the map is always read before that along the side margin.

Once this has been grasped, pupils can learn to locate points precisely within squares by means of six figure grid references. Initially younger pupils may find it helpful to draw a 10 by 10 grid of small squares on tracing paper to superimpose on the large squares of the map. The method of giving grid references which was previously printed on all Ordnance Survey maps, including the first edition of the 1:50,000 series, was confusing because the instructions for reading eastings included references to west, north and south, and those for northings included references to south, west and east. Teachers are recommended to teach their pupils the much simpler and more easily remembered method adopted for the *second* edition of the 1:50,000 series (Figure 6.3).

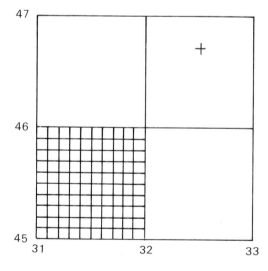

FIRST QUOTE EASTINGS	
Locate first VERTICAL grid line to LEFT of point and read LARGE figures labelling the line either in the top or bottom margin or on the line itself	32
Estimate tenths from grid line to point	5
	325
AND THEN QUOTE NORTHINGS	
Locate first HORIZONTAL grid line BELOW point and read LARGE figures labelling the line either in the left or right margin or on the line itself	46
Estimate tenths from grid line to point	7
	467
SAMPLE REFERENCE	325 467

Figure 6.3. Giving a six figure grid reference.

Symbols

Maps contain point, line and patch symbols to depict objects in the real world or to show relationships that may not be readily discernible. Point symbols include those which represent individual buildings such as churches and public houses as well as spot heights. Line symbols include roads, railways and rivers, and also grid lines and contour lines which, although they do not exist in reality, are added to the map to assist the map reader. Patch symbols include fields, factory sites and the areas covered by settlement and woodland. In addition most maps contain letters, words, abbreviations and numbers which are just as much symbols as the graphic ones.

It is important to recognise that, when pupils are examining symbols on maps, they are performing two distinct tasks. Firstly, they have to *perceive* the symbol on the map, and secondly they have to understand the *concept* for which it stands. A pupil could, for example, perceive a thick, curved blue line without realising that it represents a motorway and not a river, or perceive a green patch without appreciating that it represents an area of woodland and not grassland. It should be remembered that areas left white are the most common of all on Ordnance Survey maps, and few clues are provided to indicate what these areas represent!

The meaning of some symbols on 1:25,000 and 1:50,000 maps is obvious because they depict in pictorial form easily recognised features. Other symbols require reference to the map key for their meaning until the pupils are familiar with them. The teacher should explain that some point symbols show buildings in plan form, such as a square for an isolated farm; others show buildings in side elevation, such as a lighthouse or windmill; others use a combination of both, such as a church with a tower or spire; yet others are purely symbolic and do not resemble the actual buildings either in plan or side elevation, such as the red dot representing a railway station or the red triangle indicating a youth hostel. It should be pointed out that the amount of space taken up by each symbol on the map is proportionately much larger than the area which the building occupies on the ground.

In the same way pupils should be taught that line symbols representing roads on medium and small scale maps exaggerate their true width. Road widths are drawn to scale on 1:1,250 and 1:2,500 maps, but this is not possible for minor roads on 1:10,000 maps. On 1:25,000 and 1:50,000 maps colours are used to differentiate types of roads and their width is exaggerated. Thus an A-class road with one lane in each direction is shown on a 1:50,000 map by means of a red line $\frac{1}{2}$mm in width. This is equivalent to 25 metres on the ground, the width of a six-lane motorway!

When pupils have become familiar with the more common symbols, they can use them in conjunction with a route finding exercise. They can be given a starting point and a destination which they have to reach. They can then decide on the best route and describe what they would pass or see during the journey. The exercise can be made more difficult if the pupils are given a particular route to follow and asked to describe what they would find at specified grid references. The use of symbols also provides opportunities for creative map drawing, pupils constructing their own maps of imaginary areas and showing the location of a variety of features on it by means of symbols and a key.

Scale

A map is a scaled-down representation of reality; it shows a large surface on a small surface. A statement of scale indicates the extent to which a portion of the real world has been reduced to fit on to a piece of paper. The concept of scale as consistent proportional representation is one which younger pupils may take considerable time to master. When attempting to draw a scale plan, pupils have to work out not only position and distance, using a system of co-ordinates, but also perspective and proportion, using a system of measurement conversion. The concept of a diagrammatic layout incorporating all of these properties, with accurate measurement of distance and proportional reduction to scale as well as the correct positioning of objects in relation to one another, requires careful teaching. Pupils should be taught scale in mathematics lessons before they use it in geography lessons.

The reduction in size of a desk or table top so that it fits on to a piece of paper forms a useful introduction to scale drawing. The pupils can then attempt to draw a plan of the classroom and selected objects within it. The scale chosen should be one which is easily calculated from measurements taken with a 10 metre tape. Thus if the classroom is 10 metres long and 8 metres wide an appropriate scale might be 2 cm to represent 1 metre. After each measurement has been taken in metres, therefore, the pupils simply have to double the reading and change it to centimetres. Measurements and scale conversions may usefully be listed in tabular form for reference purposes before they are used to draw the plan.

The teacher can build up a sketch plan on the blackboard whilst pairs of pupils take turns in measuring with the tape. It is advisable to draw the blackboard sketch so that the front of the room is at the top and the back of the room at the bottom. This means that when the pupils look at the blackboard sketch it is the same way up as their own plans. After the pupils have drawn the rectangle to represent the room and inserted the door, the pieces of furniture such as blackboard, bookshelf, cupboard and cabinet can be shown in their correct positions. The teacher's and pupils' desks are then added to complete the plan and the pupils can identify and mark each other's desks. The scale may be given as a ratio, such as 1:50, or as a statement, such as 1 cm represents 50 cm; it is mathematically incorrect to say that 1 cm equals 50 cm. A scale line should appear on the plan, graduated in units from zero to the required length. This can also be extended to the left below zero, the resulting unit being divided into tenths, so that smaller lengths can be read off.

When pupils have successfully drawn a plan of a room to scale they should have an opportunity of using a scale plan which has been prepared for them. A scale plan of the interior of the school, or of the school and playground, is useful for this purpose. Pupils can make measurements from the plan and calculate the size of the features represented, such as the actual length of a corridor, the width of the hall, or the perimeter of the school grounds. This provides practice in the use of scale lines extended to the left beyond the zero mark and subdivided into smaller units.

Scale measurements on published maps can initially be made from the 1:1,250 or 1:2,500 plans of the area in which the school is situated. The pupils can attempt a variety of measuring exercises on these maps; for example, they can calculate the distances along roads and the lengths of rows of houses or their gardens. Subsequently maps on smaller scales can be used as the basis for measuring exercises, for example, straight line distances on 1:25,000 and 1:50,000 maps.

Pupils should have an opportunity to compare maps on different scales from 1:1,250 or 1:2,500 to 1:50,000 or 1:250,000 (Figure 6.4). For this purpose a set of maps of the local area on ever decreasing scales should be displayed on the wall.

Scale	Equivalent	Map size	Ground area
1:1,250	1 cm to 12.5 m or 50 in to 1 mile	40 cm x 40 cm	500 m x 500 m
1:2,500	1 cm to 25 m or 25 in to 1 mile	80 cm x 40 cm	2 km x 1 km
1:10,000	1 cm to 100 m or 6 in to 1 mile	50 cm x 50 cm	5 km x 5 km
1:25,000	4 cm to 1 km or $2\frac{1}{2}$ in to 1 mile	80 cm x 40 cm	20 km x 10 km
1:50,000	2 cm to 1 km or $1\frac{1}{4}$ in to 1 mile	80 cm x 80 cm	40 km x 40 km

Note. The table does not include some other Ordnance Survey topographical maps, such as the 1:63,360 (1 in to 1 mile) Tourist maps and the 1:25,000 ($2\frac{1}{2}$ in to 1 mile) Outdoor Leisure maps, which are available for certain parts of the country.

Figure 6.4. Comparison of Ordnance Survey maps on different scales.

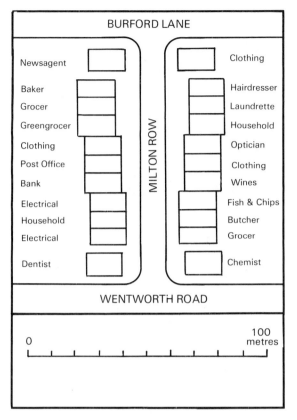

Figure 6.5. Part of outline map after fieldwork.

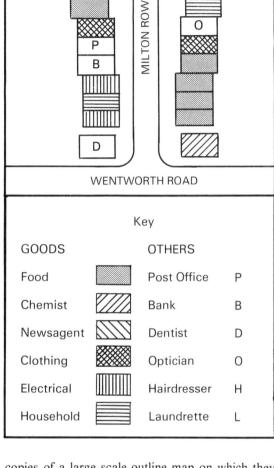

Figure 6.6. Part of completed map of local shopping centre.

Pupils should compare, for example, the detail shown in a kilometre square on the 1:10,000 map with that in the same square on the 1:25,000 and 1:50,000 maps.

Conclusion

It is important that all pupils should attempt at least one piece of fieldwork which enables them to practise their mapping skills and use their knowledge of direction, location, scale and symbols. An example of a fieldwork exercise which can be carried out in any urban area is a study of a local shopping centre. The pupils are provided with copies of a large scale outline map on which they record the different types of shops as they walk past them. In order to complete a map such as that shown in Figure 6.5 the pupils have to relate each building on the ground to its representation on the map. Upon returning to the classroom the pupils can make a fair copy of the map showing categories of shops in different colours or types of shading, as in Figure 6.6. This can then form the basis of a discussion on, for example, the proximity of shops in the same category and the adequacy of the shopping centre in meeting the needs of the community.

6.2. Relief Interpretation
David Boardman

Maps are two-dimensional representations of the three-dimensional earth's surface. The problem of showing the third dimension on maps is one which has always exercised map makers. Generally the problem is tackled by using symbols to depict relief features in the landscape, the most commonly used being the contour. The third dimension also presents problems for map readers who have to develop the ability to visualise the landscape from the map. The representation of height, slope and relief on a map is often a source of difficulty for pupils of all ages. Teaching this aspect of map reading requires the use of visual aids and other materials.

Simple models

Some simple aids for teaching the concept of contours can be made quickly. A piece of card can be cut and rolled into the shape of a cone and lines drawn round it at regular intervals. A potato cut in half can be further cut horizontally into sections which can be dismantled and reassembled. A piece of clay or plasticine can be moulded into the shape of a hill and similarly cut into sections.

An understanding of the relationship between contours and sea level can be achieved with the aid of a model and water. A clay or plasticine model representing an island is placed in a rectangular transparent fish tank, the side of which is marked with a scale graduated at regular intervals on a strip of masking tape. The base of the tank represents the dry shore at low tide and water poured into the tank represents the incoming tide. When the first mark on the scale is reached, an old ball point pen or pencil is used to draw a line round the model by making an impression in the clay or plasticine to record the water mark, which represents sea level (Figure 6.7). More water is then poured into the tank to represent storm conditions which flood the land around the coast. When the second mark on the scale is reached another line is drawn round the island to record the new water level. The process is repeated several times until the top mark on the scale is reached and the model is almost submerged. When the water is poured out of the tank and the model is removed from it, the pupils can see the lines drawn round the island at regular intervals to indicate increasing height.

If the model is replaced in the tank and a sheet of thick acetate is placed over the top of it, the lines as seen from directly above can be drawn on the acetate with a fibre tip pen. When the acetate is placed on the overhead projector the pupils can see on the screen the way in which a three-dimensional relief model is represented on a two-dimensional map.

Whenever possible pupils should themselves carry out a practical activity of this kind after they have observed a demonstration by the teacher. Pupils can be divided into groups and the members of each group can mould their own small plasticine model and place it in a sandwich box or other container with a flat base and open top. Having stuck a piece of masking tape on one side of the container and marked it at regular intervals, the pupils pour water into the container from a milk bottle and record the rising water level by lines at intervals on the model. After they have poured the

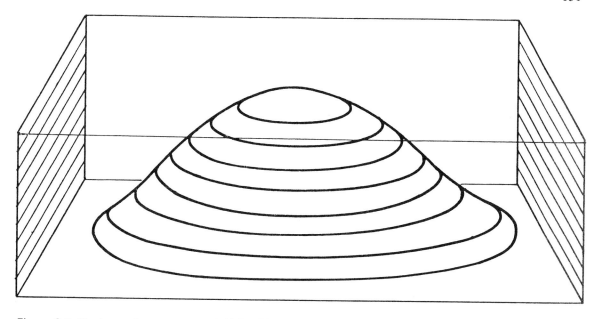

Figure 6.7. Contours drawn on a model island in a tank.

water from the container into a bucket, they place a small sheet of acetate over the top of the container and, looking down from above the model, draw the contours on to the acetate. The pupils write an account of what they have done and illustrate it with their own drawing of the model and the contour pattern.

When the pupils have completed this task, the teacher provides a definition of a contour using the correct terminology: 'a line drawn on a map through all points which are at the same height above, or depth below, sea level'. This definition reminds the pupils that contours are lines drawn on maps and do not appear on the ground. It is also a reminder that contours can be drawn below as well as above sea level, an important consideration in view of the inclusion of submarine contours on Ordnance Survey maps of lakes and coastal areas as well as on atlas maps.

Layer models

An alternative approach to teaching the concept of contours is to construct a layer model from contour patterns traced from a map. An area of gentle relief shown on the 1:10,000 map, the largest scale on which contours are printed, can be used and enlarged if necessary. The pattern formed by each contour is then traced on to a piece of thick cardboard or ordinary plywood. Each contour pattern is cut out from the cardboard or plywood and the complete set is used to build up the layer model (Figure 6.8). Marks have to be inserted on each shape to ensure that the next highest contour is correctly placed before it is glued in position.

The layer model should not be left in this form because its stepped or terraced appearance does not resemble the real landscape. The spaces between the various layers should be filled in using papier mâché or tightly screwed paper and covered with plaster or filler, which should be smoothed out so that the appearance of the model is similar to the relief of the landscape it represents. Finally the model is painted green and selected features are added to it in different colours: a stream in blue, a road in red, a railway in black, and so on (Figure 6.9).

It is important to avoid the excessive exaggeration of relief which occurs when the layers of a relief model of a small area are built up with very thick cardboard or plywood. This results in a

132 HANDBOOK FOR GEOGRAPHY TEACHERS

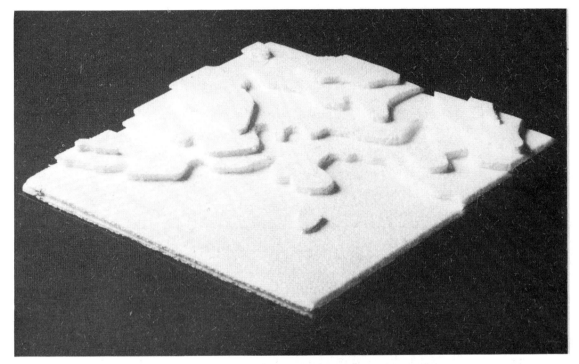

Figure 6.8. Relief model under construction with layers in position.

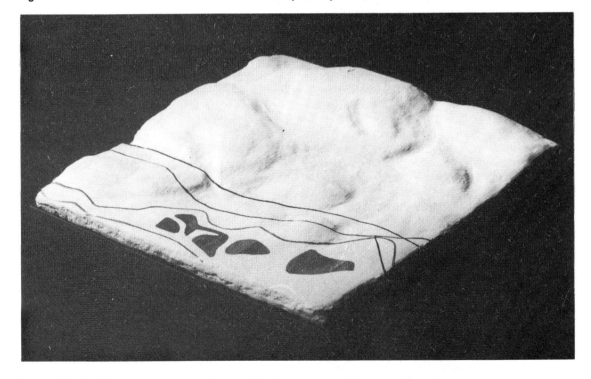

Figure 6.9. Relief model after completion with plaster and paint.

distorted impression of the landscape shown on the map and it does not help pupils to visualise it correctly. The vertical exaggeration is easily calculated. On a 1:10,000 scale map, for example, 1 cm represents 10,000 cm or 100 m on the ground. If the vertical scale is also 1 cm to 100 m, no exaggeration of relief occurs. Thus if card 1 mm thick is used to represent each 10 m in height, the relief of the model will exactly replicate the relief of the landscape. In practice the absence of any vertical exaggeration will usually fail to highlight variations in relief. If card 2 mm thick is used to represent each 10 m in height, the vertical exaggeration is x2, and if card 5 mm thick is used, the exaggeration is x5. The latter is probably the maximum which is permissible, otherwise the variations in relief are likely to be grossly exaggerated.

The side of an accurately constructed relief model represents a section through a landscape. A model is useful, therefore, in helping pupils to understand the concept of a cross section. Indeed they can benefit from drawing their first section across a map of an area for which they have already constructed a model. This enables them to see that a section simply shows in two dimensions what a model shows in three dimensions.

The chosen line of the section is drawn on the map and the straight edge of a piece of paper is placed alongside it. The edge of the paper is marked and numbered at each point where a contour meets it. When two adjacent contours have the same number a plus or minus sign can be inserted between them to indicate whether the land rises or falls. The piece of paper is removed from the map and placed along a base line of the same length drawn on graph paper. The heights recorded are plotted as a series of dots at the corresponding heights on the vertical scale and these are joined by means of a smooth, continuous curve. The positions of hills, rivers and other prominent features can be indicated by means of labelled vertical arrows. The horizontal and vertical scales should be shown and the grid references and direction added to each end of the section.

The extent of the vertical exaggeration should also be stated. This should not be so great that it unduly distorts relief, transforming a gently undulating hill and vale landscape into one which resembles high mountains and deep valleys. When calculating vertical exaggeration the working should be given in full. This is easily done by giving the horizontal and vertical scales as ratios and comparing them. Thus the ratio 1:25,000 simply means that 1 cm on the map represents 25,000 cm on the ground. If the vertical scale is 1 cm to 50 metres, then the ratio is 1:5,000. The vertical exaggeration is thus x5, which is generally regarded as the maximum if excessive distortion is to be avoided.

Aerial photographs

Aerial photographs may be used in conjunction with Ordnance Survey maps to assist with landscape visualisation. The processes involved in 'reading' and 'interpreting' photographs, however, are different from those required in map reading and interpretation. It is important to recognise this distinction when pupils are working with both maps and photographs in the classroom.

If a photograph of a landscape is taken from ground level, the effect of perspective means that features in the foreground are shown much more prominently than those in the middle ground and background. The scale along the lower edge of the photograph is different from that across the middle and along the upper edge. This variation in scale means that distances cannot be measured and indeed often can be only very roughly estimated.

In an oblique aerial photograph the effect of perspective is even more marked because the distances are generally greater and the variation in scale from foreground to background is usually considerable. Although a vertical aerial photograph closely resembles a map, it is different from a map in that it is taken from a specific viewpoint. Only the small part of the earth's surface which is directly below the camera is shown at a consistent scale, although in practice the size of the photograph means that the variation in scale towards the edge is minimal.

Photographs do not, therefore, possess the same geometrical properties as maps. They are useful aids in visualising the landscape shown on maps, but their characteristics are completely different. This helps to explain some of the difficulties which pupils often encounter when attempting to correlate aerial photographs with Ordnance Survey maps (Boardman and Towner, 1980).

The oblique aerial photograph forms an intermediate stage in the mental processes involved in transforming the three-dimensional landscape of the real world to the vertical representation on the map. When pupils are attempting to match the details on an oblique aerial photograph with the corresponding details on the map, they not only have to recognise features in an unfamiliar form but they also have to conserve mentally their correct spatial arrangement.

Pupils are often misled by the similar square or rectangular shape of map and photograph. The effect of perspective means that the area shown on an oblique aerial photograph is not square or rectangular when it is transferred to the map. The shape is not necessarily that of a triangle or trapezium, moreover, because the edges may be curved. For this reason it is advisable to delimit accurately on the map the edges of the area shown on the photograph.

The pupils' attention should be drawn to the differences between the scale of a map, which is consistent over the whole of the map, and the scale of a photograph, which varies with distance from the camera. When they have drawn on the map the boundaries of the area shown on the photograph, the pupils can measure the distance across the foreground, middle ground and background of the photograph. By comparing the photograph and map they will also learn to appreciate what distances measured on a map are actually like in reality. This should also help the pupils to overcome the tendency to overestimate the size of the area shown on the photograph when they are attempting to correlate it with the map.

The size of individual features is also affected by perspective. Buildings and fields in the foreground of a photograph receive prominence whilst those of similar size in the background appear less conspicuous. Pupils should compare the size of features at different places on a photograph with their size on the map. They should also do this if certain features are shown on the map by means of symbols, which usually take up a disproportionate amount of space on the map compared with the space the corresponding features occupy on the ground. The symbol for a church, for example, stands out as a prominent feature on a map, whereas the same church on a photograph may be surrounded by other buildings and hard to distinguish from them.

In their early attempts to correlate aerial photographs with Ordnance Survey maps, it is helpful if the pupils use photographs which were taken with the camera pointing towards the north. The similar alignment of map and photograph makes it easier to correlate features and read names on the map. It should be emphasised to the pupils, however, that the convention of having north at the top of a map does not apply to photographs. Pupils may need considerable assistance when attempting to correlate a map with aerial photographs which were taken with the camera pointing towards the east, south or west. To assist with the identification of features, it is helpful if the pupils mark an arrow on the map to indicate the orientation of the photograph before they delimit the area covered by the photograph.

Progression in map skills

The map skills which should normally be developed by secondary school pupils are summarised in Figure 6.10. The table suggests the skills which are appropriate for pupils in different age ranges. A number of cautionary remarks, however, need to be made about the table.

Some primary schools incorporate map reading into their programmes of environmental education; others do comparatively little. Children with experience of map reading, therefore, will arrive at secondary school already having learnt at least some of the skills listed in the column for the 11-14 age range, particularly those relating to direction, location and symbols. Children who attend middle schools to the age of 12 or 13 may have mastered many of the skills in this column. Useful outlines of map skills for children in primary and middle schools have been compiled by Catling (1980, 1981), and examples of the kind of mapwork of which children are capable in their last year of primary education will be found in Palmer and Wise (1982).

At the same time allowance should be made for the varying rates at which children develop. Slow learning children in the early years of secondary education may take considerable time to learn the skills listed as appropriate for the 11-14 age range. Conversely bright children may learn all of the skills quite quickly in their first year at secondary school. Figure 6.10 should be used as a flexible guide rather than as a rigid checklist, therefore, and modified as necessary to match the experience and ability of the pupils. The inclusion of a skill in the column for a particular age group does not preclude its use at an earlier or later age when this is considered appropriate.

Map skills develop continuously with constant practice. It should not be assumed that once a map skill has been learnt it will automatically be retained. Map skills should be practised whenever opportunities arise in each year of a geography course in the same way that numerical skills are regularly practised in mathematics. Mapwork should not be treated in isolation from other aspects of geography. Clearly the skills will need to be taught separately, but as soon as the pupils have acquired them they should be encouraged to apply them to the study of maps of landscapes.

The skills listed for the 14-16 age group in Figure 6.10 reflect the requirements of examination courses. Most syllabuses include the study of Ordnance Survey maps on the 1:50,000 and 1:25,000 scales. It should be recognised that these maps are probably the most complex documents that pupils are likely to encounter in geography courses up to the age of 16. Each map consists of a large number of point, line and area symbols, all of which are communicating specific pieces of information to the map reader. A discussion of the complex nature of cartographic communication through topographical maps is beyond the scope of this section but is available elsewhere (Boardman, 1985).

Questions on Ordnance Survey maps in public examinations fall into three broad categories: reading and calculation; transformation; and interpretation. *Reading and calculation* include such skills as identifying and naming the features shown by symbols at specified grid references, stating the direction from one place to another, measuring the distance between two places, and estimating the height of the land at a particular point. All of these basic skills should have been learnt lower down the school and subsequently practised.

Transformation requires information given on the map to be transferred into another form, such as a cross section or a sketch map of part of the map on the same scale or the whole of the map on a reduced scale. This category also includes the identification of selected features on an aerial photograph with the corresponding features on the map as evidence of the ability to orientate and correlate these two documents.

Interpretation covers the more complex skills such as description of the relief and drainage in a specified part of the map, the pattern of the main routes on the map, or the site and situation of a town or number of villages. Map interpretation frequently demands an ability to understand the relationship between the physical and human geography of the area shown on the map. Using map evidence alone, this is an advanced skill which requires a sound grasp of geographical ideas.

When pupils are learning to interpret a map they should be encouraged to search for specific kinds of evidence on the map by testing generalisations or simple hypotheses. Thus they may formulate a series of statements such as 'there is some relationship between rocks and relief'; or 'different drainage patterns develop on different kinds of rocks'; or 'settlements are mainly found at the junction between uplands and lowlands'; or 'relief exerts a control over transport routes'. The pupils then have to focus on specific aspects of the map and quote evidence from it either to support or to contradict each generalisation.

An intensive, detailed study of an area represented on an Ordnance Survey map involving the construction of a landscape model is recommended as part of a map interpretation course. A map extract on the 1:50,000 or 1:25,000 scale should be selected so as to provide contrasts in relief and drainage, show both upland and lowland areas, preferably developed on different rock types, and demonstrate contrasting aspects of settlement and communications. Following the procedure described earlier, the pupil traces the contours from the map, transfers them to sheets of cardboard and constructs a layer model. After infilling with plaster the model is painted, selected features such as roads and rivers are added, and rock strata are painted on the sides. The building of a landscape model as an integral part of map interpretation helps pupils, by means of practical work, to develop a deeper understanding of the ways in which various topographical features are represented by contour patterns on maps, and of the influence of topography on the human occupance of the area. It is suggested that a 16-year-old pupil should be able to submit a landscape model, accompanied by a written explanation of the geography of the area, for assessment as part of the coursework for the GCSE examination.

The skills listed for the 16-18 age group in Figure 6.10 are included in most A-level examination courses. Students are expected to become familiar with geology and soil maps as well as maps produced by the second Land Utilisation Survey. They should make more sophisticated use of vertical and oblique aerial photographs for the purposes of identifying and interpreting geographical distributions. Students are also usually expected to be able to perform certain statistical

SKILL	11 — 14 YEARS	14 — 16 YEARS	16 — 18 YEARS
DIRECTION	Plot the 16 points of the compass Give a compass direction from one point to another Align a 1:1,250 or 1:2,500 map with features using buildings as reference points Follow a route on the ground using a map Plan a route on a map	Align a 1:10,000, 1:25,000 or 1:50,000 map with features using landmarks as reference points Set a map using a compass Find the way between two points using map and compass State bearings in degrees Give and follow directions using a compass	Follow a route in an upland area using map and compass
LOCATION	Use grid squares to locate points on a map Relate position on the ground to location on a 1:1,250 or 1:2,500 map Give 4 figure and 6 figure grid references on maps Find places or features on maps from 4 figure and 6 figure grid references	Locate on an atlas map the area covered by a 1:25,000 or 1:50,000 map Relate position on the ground to location on a 1:10,000, 1:25,000 or 1:50,000 map	Locate on an atlas map the area covered by a geology or soil map
SYMBOLS	Identify and draw common conventional symbols used on maps Compile a key to illustrate groups of point, line and area features Compare symbols for the same feature on maps on progressively smaller scales Realise that the degree of generalisation on maps increases with decreasing scale Appreciate that some symbols on maps are disproportionate in size to the features they represent	Recognise and understand a wider range of symbols used on maps on different scales	Recognise and understand symbols used on a geology or soil map
SCALE	Draw a plan of a room to scale Make measurements from a prepared plan of a building Measure straight line distances on maps and convert measurements using scale Use scale lines, representative fractions and verbal statements of scale Mark a route on a map from a statement giving directions and distances Describe a route giving directions and distances from information shown on a map	Measure distances along winding routes on a map Calculate the gradient between two points Calculate the vertical exaggeration of a cross section Calculate the approximate areas of features using superimposed grid squares on a map	Delimit the area of a drainage basin on a map and calculate the drainage density Designate stream orders on a map and calculate bifurcation ratios

(continued)

PHYSICAL FEATURES	Mould a plasticine model and draw contours on it by immersing it in water at intervals Construct a relief model from the contours on a 1:10,000 map Read heights from contours on a map and estimate heights between contours Identify common relief features from contour patterns, e.g. valley, spur, hill, ridge Describe the shapes of slopes as steep or gentle, concave or convex Draw a section across the contours on a 1:10,000 map	Construct a landscape model of an area shown on a 1:25,000 or 1:50,000 map Identify the relief divisions of a landscape and describe landforms within them Subdivide the area shown on a map into drainage basins Describe the shapes and alignment of valleys within a drainage basin Describe the nature and pattern of rivers within a drainage basin Draw a section across the contours on a 1:25,000 or 1:50,000 map	Identify patterns of relief on topographical maps and suggest reasons for these patterns Identify drainage patterns on topographical maps and suggest reasons for these patterns Interpret geology maps and relate rock strata to relief and drainage on topographical maps Interpret soil maps and relate soil associations to relief and drainage on topographical maps
HUMAN FEATURES	Describe an area from symbols shown on a map Distinguish between point, line and area features in the same group, e.g. types of roads, water features Describe the features which would be seen when following a given route on a map Appreciate general relationships between physical and human features shown on a map	Describe the site, situation, form and function of settlements on a map Generalise about the location and distribution of settlements on a map Describe the type, density, direction and pattern of communications on a map Relate the pattern of communications to relief and drainage and to the location and distribution of settlements Suggest possible consequences of modifications to human features, e.g. road alignment, new housing	Relate land use patterns on land utilisation maps to relief and drainage on topographical maps Relate land use patterns on land utilisation maps to rock strata and soil associations on geology and soil maps Analyse urban and rural land use patterns using theoretical models Analyse distributions on maps using statistical tests Examine social and economic issues associated with the modification of features, e.g. motorway construction, industrial expansion
AERIAL PHOTOGRAPHS	Extract from an aerial photograph information which is not shown on a map of the same area, and vice versa Correlate features shown on an oblique or vertical aerial photograph with the corresponding features on a 1:1,250, 1:2,500 or 1:10,000 map Suggest relationships between the physical and human features shown on map and photograph	Correlate features shown on an oblique or vertical aerial photograph with the corresponding features on a 1:25,000 or 1:50,000 map Describe a landscape using the combined evidence of map and photograph	Correlate oblique and vertical aerial photographs with topographical, geology, soil and land utilisation maps, and use the evidence to interpret physical and human features
MAP DRAWING	Draw mental maps of the local area Draw freehand maps using conventional symbols Draw creative maps of imaginary areas, e.g. 'Treasure Island'	Draw and annotate sketch maps of areas or regions to illustrate selected distributions Draw maps from statistical data provided to illustrate selected distributions	Draw maps showing statistical data as point, linear and area distributions, e.g. dot, isopleth and choropleth maps

Figure 6.10. Outline of map skills for secondary school pupils.

tests on data obtained from maps and to present data on their own maps using standard cartographic techniques.

Conclusion

The discussion of mapwork in this section has necessarily been brief owing to restrictions on space; a more detailed treatment will be found in Boardman (1983). Although the suggestions for teaching have been concerned with the use of maps and related documents in the classroom, it should be emphasised that the main practical use of maps is out of doors. Younger pupils should always have an opportunity to practise map skills in the local neighbourhood whilst older ones should make full use of maps when carrying out fieldwork in more distant areas. In this respect the encouragement given to fieldwork in both GCSE and A-level examination courses provides an excellent opportunity for pupils to combine map reading and interpretation with the techniques required for studying geography in the field and for undertaking individual field study investigations.

References

Boardman, D. (1983) *Graphicacy and Geography Teaching*, Croom Helm.

Boardman, D. (1985) 'Cartographic communication with topographical maps', in Boardman, D. (ed) *New Directions in Geographical Education*, Falmer Press.

Boardman, D. and Towner, E. (1980) 'Problems of correlating air photographs with Ordnance Survey maps', *Teaching Geography*, vol. 6, no. 2, pp. 76-78.

Catling, S. (1980) 'Map use and objectives for map learning', *Teaching Geography*, vol. 6, no. 1, pp. 15-17.

Catling, S. (1981) 'Maps and mapping', in Mills, D. (ed) *Geographical Work in Primary and Middle Schools*, Geographical Association.

Palmer, J. A. and Wise, M. J. (1982) *The Good, the Bad and the Ugly*, Geographical Association.

6.3. Atlases and Atlas Mapwork

Herbert Sandford

Atlases of all kinds characterise the geographer at work. This section is concerned with the pupil's desk atlas, the relatively modest general world atlas that is bought in class sets and issued to individual pupils for class use and for homework. A great deal of what is written here, however, applies also to wall maps, map transparencies and globes.

The purpose of a school atlas

An atlas is often likened to a dictionary, to be constantly at hand and frequently referred to. This does, however, concentrate attention upon the *lexical* aspect of the atlas, that is, its function as a means of locating places. While the lexical function is essential for writing and talking about what is seen in maps, names cannot be more important than the places themselves in their locational context: the form of the land (its shape, height and slope) and its cover of natural and man-made entities.

The main purpose of general atlases is to portray real-world landscapes. Thus they contain a core of general-purpose (or of physico-political) maps, usually bearing names and thus becoming lexical maps, supplemented by thematic (or special purpose) maps, and supported by pictures, diagrams and text. The global landscape then becomes a framework within which all geographical data can find their place.

If the teacher issues pupils with an atlas without ensuring its use in class or at home, it becomes a costly but neglected volume. Perhaps it is assumed that pupils can and do use atlases competently without teaching or encouragement. However, while most children do indeed enjoy browsing through a colourful and pictorial atlas, they have no inborn intuition about using it more constructively than that. Indeed, the meaningful use of atlases appears to require more guidance and practice than is provided for Ordnance Survey maps. It is easy to be misled by the superficial skill of the pupil who locates Buenos Aires on the map. He may well have done no more than visually searched for a black word alongside a red dot on a green patch on a piece of paper. It would be instructive to ask him to describe the land around the city. The small scale, considerable abstraction, lack of standardised symbols, and original global shape of what is depicted are among the many things that make atlas maps particularly difficult to read and interpret.

A school atlas must serve the needs of school geography, and must adapt to a changing syllabus. In practice this means retaining most of the traditional content whilst adding the new. The 'traditional' content includes a world-wide map coverage, with special detail for the British Isles, and some standard thematic maps such as climate, population, and products. The 'new' content includes such socio-economic and environmental themes as conservation, pollution and natural hazards. New forms of ancillary and illustrative material are required: to the traditional picture and the conventional pie or bar graph are increasingly being added population pyramids, triangular diagrams, computer-generated grid maps and satellite images.

Atlas skills

An attempt to set out a possible programme of atlas mapwork throughout the secondary school, in which the various skills are developed over several successive years, is shown in Figure 6.11. It is not a formal statement of aims and objectives, skills and areas, tidily arranged in a matrix. Atlas mapwork is not an end in itself, but a set of skills to draw upon. It has therefore been thought more appropriate to employ a pragmatic and experimental approach to suggest the key atlas mapwork skills which need to be learnt in order to serve the needs of school

SKILL	11 — 14 YEARS	14 — 16 YEARS	16 — 18 YEARS
MAP CONFIDENCE	Select appropriate atlas map. Use the index to find a named place. Find the names of features mapped	Select appropriate atlas map using the contents page. Scan the map to answer questions and to solve problems	Select and use various reference, specialist and foreign atlases in the school library
SENSE OF SCALE	Make four lists of geographical phenomena: those on both map and ground (eg., in the case of the general atlas map, rivers), on the map only (eg. county boundaries), on the ground only (eg. crops and stock), neither on the map nor directly perceived (eg. geology and climate). Compare how towns and country are shown on maps at ever decreasing scale. Compare Apollo pictures with globe and world map, Landsat images with continental maps, conventional photos with regional maps	Locate on an atlas map the area covered by an Ordnance Survey map or by an overseas topographical map sheet. Calculate the plottable error	Use time-distance, cost-distance and other non-Euclidean scales
VERBAL MAPWORK	Recognise and understand map symbols (including typestyles) in the class atlas. Acquire a working vocabulary (highland, conurbation, etc.), distinguishing between terms (eg. seaport and riverport). Use realistically such basic concepts as near and far, large and small, high and low	Recognise and understand a wider range of map symbols in atlases. Extend glossary of terms for geographical features, eg. archipelago and fjord	Further extend glossary of geographical terms (eg. pays and puy). Describe and explain physical geography (eg. relief and drainage) and human geography (eg. settlement and routes)
NUMERICAL MAPWORK	Use verbal and linear scales and representative fractions on maps drawn to more or less equal area projections. Read off spot heights, depths and contours. Interpolate between contours, distinguishing between steep and gentle, concave and convex. Give directions by 16 points of the compass on maps on various projections (and make signposts to towns within Britain). Use geographical co-ordinates in degrees and fractions of a degree	Use scales on variable-scale maps (eg. world maps drawn to Mercator's projection). Calculate gradients. Calculate area by dot planimeter. Read off direction by degrees (and make signposts to overseas cities), distinguishing between geographical, magnetic and grid norths. Use geographical co-ordinates by degrees and minutes. Handle local time, Time Zones and the International Date Line. Know the main features of the common projections in the class atlas (rhumb lines, Great Circles, etc.)	Interpolate between and extrapolate from isolines (including contours) and re-contour on a different scale. Know the main characteristics of common projections in atlases and in the media (eg. Arno Peters'). Represent communication networks topologically and calculate indices of connectivity. Plot distribution of, eg. oil and coal fields in Europe, and compare their mean centres (centres of gravity). Compare maps using chi-square test eg. population distribution and precipitation in Africa

(continued)

(continuation)

COMPARISON	Compare two maps of the same area and on facing pages with the same scale, orientation and projection, eg. physical and political pairs	Compare two or more maps which differ by page, area, scale, orientation or theme (but not by projection)	Reconcile incompatible or contradictory data on maps
PRESENTATION OF DATA	Describe what the map shows by free search and by completing a textual passage with gaps, both verbally ('a' is near 'b') and numerically ('a' is 50km from 'b') Make 'Treasure Island' maps using conventional symbols correctly Compile a map of located composite bar graphs to show eg. the proportion of 'millionaire' cities out of all mapped settlements in each continent	Re-map at same, smaller or larger scale, as freehand sketch or by the grid method, a selection of data from the atlas map Describe verbally and numerically the positions of cities, routeways and regional divisions Draw sections along both straight and curving lines Trace river patterns and drainage basins eg. in relation to hydro-electric power and irrigation schemes	Draw sections showing subsoil and cover as well as surface shape and slope Make reviews and critiques of atlases and atlas maps Make regional divisions and sub-divisions according to physical and/or human criteria, and present these in map form Formulate hypotheses or models and test them through the mapped evidence and reasoning, and give an evaluated conclusion, eg. the high urbanisation in Australia
RELATED WORK	Understand the causes of day and night, the seasons, and the eclipses	Understand the phases of the moon and of the tides	Understand the varied length of daylight and of growing season

Figure 6.11. Outline of atlas map skills for secondary school pupils.

geography. Although it is convenient to allocate particular skills to particular age groups, the timing should relate more to the pupils' readiness and to the needs of geography lessons, and, indeed, to any parallel work being done with Ordnance Survey maps. Many young children are capable of quite sophisticated atlas mapwork, given adequate guidance, practice and motivation.

The pupils' atlas workbooks and teachers' atlas publications which are available in this country do not compare in quantity or variety with the books on Ordnance Survey mapwork. The few that exist tend to provide more practice with existing skills than guidance in attaining those skills, and are often merely a means of acquiring factual information rather than geographical understanding. However, some publishers are moving in an appropriate direction: examples are Nelson's *Atlas 80 Workbook* and *Atlas Scotland Workbook*. Cassell's *World Study Atlas Development Books* 1 and 2, and Schofield and Sims' *Notebook Atlas of the British Isles, Checkpoint* (for the *Our World* atlas), and *The Whole World Exercises* (with separately bound *Answers*). Arnold-Wheaton's *Young Geographer Study Atlas* 1 and 2 are excellent combinations of textbook, workbook and atlas. In most other cases teachers have to use what guidance there is in the atlas itself, usually very little, and supplement it as best they can.

Atlas mapwork should not be a clearly identifiable 'course' but rather a coherent body of concepts and skills acquired through constant use alongside text, picture and dialogue during all appropriate classwork and homework, so that the use of the atlas becomes part of the young geographer's second nature. During the years of secondary education the emphasis changes from simple map reading (a knowledge and understanding of the symbols and an ability to calculate distances, heights and so on) to map interpretation (the ability to deduce, induce and interpolate — to see beyond the symbols to the reality behind them). By the time they leave school pupils ought to have acquired at least the minimum skills needed by adult citizens: to be able to read and interpret for everyday purposes the small-scale maps in general atlases, family atlases and road atlases, and in the media, literary works, tourist brochures and so on. It could be noted here that excellent presents for parents to give their children might well include Hamish Hamilton's *AA Junior Atlas of Britain* with its fine mapwork guidance section (for use in the car) and Collins-Longman's *Junior School Atlas* and *Secondary School Atlas* (being general editions of their *Atlas One* and *Atlas Two*).

In the sixth form interpretative skills can be further developed so as to serve the specialist geographers who may indeed wish to continue their studies into higher education. Sixth form students ought to have at least access to library copies of some foreign atlases to help them in their special regional studies, such as Chiaomin Hsieh's *Atlas of China* (McGraw Hill) or Farley's *Atlas of British Columbia* (University of British Columbia). Even general world atlases published for foreign school pupils may contain invaluable material on their own country that is otherwise not easily obtained, good examples being the *Alexander Weltatlas* (Ernst Klett), *Atlas d'Aujourd'hui* (Larousse), *Diercke Weltatlas* (Westermann) and the *Nordisk Skolatlas* (Esselte of Stockholm).

Like all such statements, the programme outlined in Figure 6.11 is defective in obscuring the continuity of learning processes and in omitting those that are imperceptibly acquired, often as a result of natural maturation. As the pupils' information field grows from the local to the global, use of the atlas should expand their mental image of the earth so that they can mentally locate, orientate and size any map they might come across. They should slowly acquire a memory of shapes, features and names of general importance, all in their global context. They should learn to cope with the special problems of the small-scale map as a means of communication, with its rigorous generalisation resulting from the selection and omission of detail, its redundant data and its presentation containing a considerable amount of 'noise'.

Overcoming these last and similar problems might be considered as bridging the 'gap' between the atlas and the child. Another gap to be bridged is that between the atlas and the large-scale topographical map; in this respect some atlases help by including medium-scale maps. Finally, there is the gap between the atlas and the real world. Some atlases help by including at larger scales, aerial photographs of urban or rural areas; at medium scales, Landsat images; and at the smallest scales, Apollo space views of the earth and its hemispheres.

General maps

The programme, in the short form given here, pays more attention to general maps than to the thematic maps of climate, population, products and other features that often accompany them. The principles applied to learning to use the former, however, can be adapted to the latter. If a general map is sufficiently detailed to serve as a door to reality, then it is just as complex and difficult an image for the pupil as is an Ordnance Survey map. Many cartographers therefore 'simplify' atlas maps by leaving out many of the details. This simplification may be illusory and short-sighted, however, as the result is an imperfect reflection of the real world: what sense can we make of a jigsaw puzzle with many of the pieces missing, or a book peppered with misprints?

Special attention should be paid to the background colours of general maps as these may determine the choice of atlas. Some use landscape (environmental) colours showing farmland and forest, desert and moorland, and so on, with inprinted or overprinted hill shading (gradient shading). These maps provide an excellent *qualitative* impression of the relief, but, with few spot heights and no contours or altitude colours, they cannot provide a *quantitative* measure of height. Most general maps, however, use altitude (hypsometric) colours and so are quantified for relief, but even when hill shading is incorporated or added, they tend to give a false impression of the land rising in steps like a staircase.

Sometimes one general map may be replaced by a pair of maps, one called 'physical' and the other 'political'. With guidance and practice, children can learn to combine these mentally into a general image of the earth's surface even when they differ in orientation, scale and projection.

General maps almost always are the main source of names of places and so are therefore also lexical or name-bearing maps. These are not always as straightforward to use as they appear. Children need help in associating names with their features, and they are aided in this if they can unravel the typographical conventions, such as the use of italics for river names in many atlases. They have no intuition to tell them that the atlas cartographer almost invariably shows small settlements too large and large settlements too small, and that he omits, from densely settled areas, settlements of a size that he adds to sparsely settled areas!

Pupils need encouragement and practice to enable them to use the index accurately and fast enough for them to prefer it to scanning. They are helped if it has few (but geographically significant) names in a large typeface, if the parts of the index entry are arranged in columns (with the page number in its logical place before the co-ordinates), if it describes the kind of feature to be anticipated (as 'island', 'lake' or whatever), if it employs word-by-word indexing and avoids abbreviations, and, lastly, if the names are printed as they appear on the maps themselves.

Some atlases have unfortunately adopted letter-number co-ordinates instead of the geographical co-ordinates of latitude and longitude which are universal and the same in all atlases. Furthermore, latitude and longitude are so informative about time, distance, location, area, scale and shape.

Place names

Atlases vary considerably in the proportion of actual features they select to put on their maps as well as in the proportion of mapped features that they select to name. It is quite often the physical features that are under-represented and under-named, so that it is difficult to convince pupils that Finland is forty per cent lake and even more difficult for them to describe the distribution of these lakes as so few of them bear names. On the other hand, all settlements on the map are named, so that, if there is no room for the name, the settlement itself is omitted! This practice is one to which pupils must accustom themselves.

Any written or oral communication in geography depends upon the accurate and sure identification of places by name. It is as important for places in the pupils' atlas to be called by the same names that are used in their textbooks and by their teacher as it is for school geography generally to use the same names as the pupils encounter in literature, in newscasts, in the media generally, and in conversation with their parents.

Atlas and pupil are enriched when the latter, having read Alistair Maclean's *Bear Island* and savoured the wind and wave, cliff and reef, comes upon that same small Arctic island on the map. Equally the pupil is deprived if the scanning eye slips uncomprehensively past the little island's Norwegian name of Bjornoya! If we deliberately tried to confuse children, to stop them learning about the world in which they live, to prevent them sharing ideas with their parents, to cut them off from the great literature of their cultural heritage, we could hardly do so more effectively than by imposing foreign-language names of places on pupils by means of their atlas. Yet this is exactly what a few British publishers of school atlases are beginning to achieve by their increasing adoption of foreign-language names; for example, Foroyar for the Faroe Islands, Porto for Oporto, Yerushalayim for Jerusalem, and Al Qahirah for Cairo.

Perhaps no other aspect of school atlases is so misunderstood. Those who accept the standardised names of an international reference atlas, call these names 'correct' (however appropriately for their original context), and then assume them to be equally correct for all other purposes, are guilty of a logical fallacy. Perhaps, too, some publishers and their cartographers are taking shelter under the label 'correct'; it is easier to use someone else's list of names and cheaper to produce overseas editions of the same atlas.

Lively lessons

The use of an atlas is a natural part, and indeed a prerequisite, of so much geographical work that there is no need to create lessons for the express purpose of using one, yet it is helpful to offer a reminder that there are some less routine uses of school atlases. A class may be asked to use their atlases to plan (or describe) a journey abroad (by their family or by the school), or to follow the travels of some adventurer or globe-trotting personality on a wall map. When a distant part of the world is to be studied it is helpful to use the atlas to follow the voyages of its discoverers or else to imagine oneself making the journey today.

The Worldwise Quiz organised annually by the Geographical Association is helping to stimulate interest in knowledge of other countries in an entertaining manner. Indirectly the quiz is undoubtedly encouraging pupils to become more familiar with the maps in their atlases.

Worn-out atlases have their uses, too. Their maps may be used for jigsaw puzzles or for simple games of the Treasure Island kind, and some published board games can be adapted for use on actual atlas maps.

As a study of France is considered by linguists to be important for pupils learning to speak French, of Germany for pupils learning to speak German, and so on, geography teachers should not neglect opportunities to lend their spare set of atlases to their colleagues to enable them to carry on the good work!

Note

A guide to the selection of an atlas for secondary school pupils appears as an Appendix to this volume.

References and Further Reading

Sandford, H. A. (1978) 'Taking a fresh look at atlases', *Teaching Geography*, vol. 4, no. 2, pp. 62-65.

Sandford, H. A. (1980) 'Map design for children', *Bulletin of the Society of University Cartographers*, vol. 14, no. 1, pp. 39-48.

Sandford, H. A. (1980) 'Directed and free search of the school atlas map', *Cartographic Journal*, vol. 17, no. 2, pp. 83-92.

Sandford, H. A. (1981) 'Towns on maps', *Cartographic Journal*, vol. 18, no. 2, pp. 120-127.

Sandford, H. A., assisted by Young, G. (1981) 'Atlases and atlas mapwork' and 'Guide to the selection of an atlas for young children', in Mills, D. (ed) *Geographical Work in Primary and Middle Schools*, Sheffield: Geographical Association.

Sandford, H. A. (1983) 'Criteria for selecting a school atlas', *Teaching Geography*, vol. 8, no. 3, pp. 107-109.

Sandford, H. A. (1985) 'A new analysis and classification of school atlases', *International Yearbook of Cartography* pp. 173-196.

6.4. Satellite Imagery

David Carter

Introduction

Remote sensing refers to all aspects of the technology and methodology of acquisition, classification, intepretation and analysis of data obtained by instrumentation that is not in physical contact with objects or phenomena subject to measurement. For most earth and environmental scientists, this forbidding definition is made more manageable by restricting the field to data obtained by fixed or mobile sensing 'platforms' at various heights above the earth's surface. The format of the information may be either in the form of a data file or as imagery; most non-photographic imaging systems combine both formats.

Remote sensing especially from satellites has opened up possibilities of resource inventory and monitoring that are acknowledged by both developing and developed states as important to their programmes of social and economic planning. Improved appreciation of the detailed spatial and temporal scales of environmental changes such as desertification, deforestation and soil degradation, however, are outside the present capabilities of remote sensing from space. It is at this point that the complementary relationship between remote sensing and conventional methods of land and resource surveillance becomes evident. It is in no sense an 'armchair' methodology, but one which can stimulate and vitalise the traditional approaches of the field sciences and cartography.

Geographers have been more active than any other body of academics in appreciating and adapting remote sensing methods. It would be pedantic to argue that because remote sensing has become or is becoming established in most degree and other post-A-level courses in geography it should be adapted to secondary curricula. The more fundamental justification is that remote sensing methods are destined to grow to a position of pre-eminence in the various tasks of environmental data collection. Every school pupil should be aware of this and have the opportunity to handle a sample of the output available. Here is a chance for pupils to grasp a vital aspect of a contemporary development of space technology, and for geography teachers to integrate a topic that not only motivates most pupils but can also service many components of existing syllabuses.

Satellite images

Although nearly all research and professional analysis of remotely sensed data involves sophisticated computing facilities and photogrammetric equipment, it is the conventional image print that conveys most meaning for inexperienced pupils. This section is therefore written on the assumption that the teaching of remote sensing will be largely confined to the interpretation of hard-copy prints, supplemented perhaps by 35mm slides of selected images and library copies of published atlases of suitable scenes. (In some cases it may be possible to obtain microfiche records of image sequences, such as weather satellite pictures, as a basis for designing exercises.)

Original aerial photographs and prints of space images offer an infinite variety of interpretation tasks relating to both physical and cultural landscapes. Map extracts have long been used for this purpose, and it is not suggested here that basic map-reading skills should be devalued in favour of photo- and image-interpretation. For example, air photographs are free of symbolism and contain elements of 'reality' that may positively help students to construct logical arguments. In well-selected examples, detail need not be 'clutter' but a set of diagnostic clues (such as shadow lengths, traffic conditions, crop maturity, sediment storage marginal to beach foreshores or river channels) that help to sharpen deductive reasoning. Vertical air photographs are far more effective in this role than obliques, though it is only the latter that have been used by the examination boards. Satellite images available to date lack the spatial resolution necessary for detailed interpretation, but this is compensated by large area coverage on which intra- or inter-regional relationships are evident. Boundaries marking major changes of land use, the adaptation of drainage networks to structural controls and the spatial form of major conurbations are examples that can be readily used.

The accompanying plate is a Landsat image, in near infra-red, of south-east Wales and north Devon, at an original scale of 1:1 million (Figure 6.12). The sensing system is utilising only a small part of the electromagnetic spectrum, and thus the discrimination between different types of surface cover is different from that seen on a conventional photograph. Landsat imaging systems obtain a simultaneous record of any given field of view in

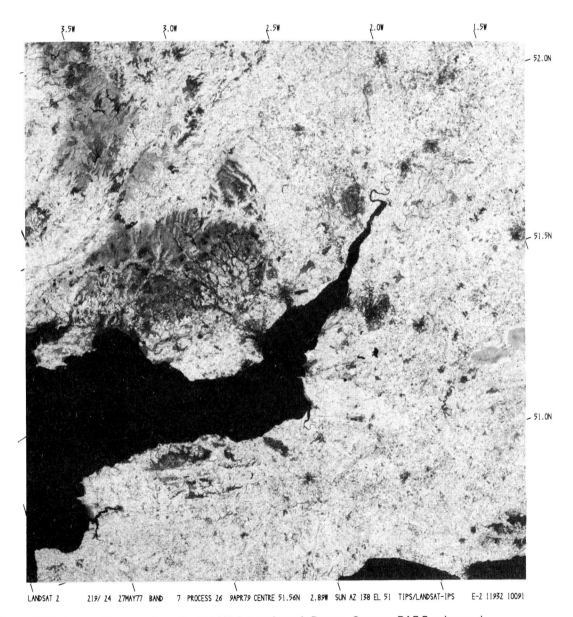

Figure 6.12. Landsat image of south-east Wales and north Devon. Source: RAE Farnborough.

several parts of the spectrum, so that this single scene is a part of a complementary set. The spatial resolution in this case is approximately 80 m, which effectively 'edits' the finer spatial detail. Thus, the image has a certain amount in common with an atlas plate, but it is not the product of cartographic generalisation and symbolic representation.

The land-use boundary between upland vegetation and lowland agriculture in south-east Wales and north Devon is clearly resolved, and the same is true of the spatial form of the major towns and cities. The motorway to the north of Cardiff can be seen, as well as the unmistakable form of the Llanwern steelworks to the immediate east of Newport. The control of major landforms on the gross pattern of land use indirectly enhances the relations between regional geological structures and their geomorphological expression. Major and minor scarps and other slopes show up as distinct linear or curvilinear features, particularly in south-central Wales and on Exmoor. The regular boundaries and sharp definition of areas of very dark tone in the upland areas are reservoirs and softwood forests. Their detailed shapes, as well as respective locations, will determine which is which. This is just one example of how satellite imagery can help to sharpen basic geographical skills. It also strongly encourages the use of other sources, such as maps and atlases, thereby introducing the concept of a geographical data base or reference system.

Newer satellites in the Landsat programme can achieve a much improved spatial resolution than seen here; they show, for example, the detailed mosaic of rural land uses clearly. Because the images are digitally recorded, it is possible to change the scale and extract sub-scenes of areas of particular interest. Microprocessor capacity for these purposes is the next obvious development.

Given just a limited awareness of aerial and space images much can be achieved with pupils using examples published in special atlas-style collections. The teacher also has the freedom to use slides or overhead transparencies of images that illustrate a particular topic. There are, for example, many magnificent spacecraft and satellite scenes of desert and semi-arid topography, some of which capture synoptic detail such as extensive patterns of sand transport and flash floods. Satellite images are particularly effective in recording the dynamic characteristics of physically hostile, remote areas that are often poorly mapped. Examples include standing water in zones of permafrost melt, fire scars in areas of forest and scrub, and patterns of snow and ice accumulation in mountainous environments. Furthermore, the relatively rapid repeat coverage of satellite imaging systems such as Landsat provides opportunities for temporal studies. Examples are the seasonal changes in land use cover and the longer term changes in urban-rural boundaries. Images from specific satellite systems are mutually consistent in terms of scale and format. If they are used to complement published maps they give a unified view of the earth's surface. The confusion of contrasting mapping styles adopted by different national and international cartographic agencies is removed.

'Satpack I and II' provide two ready-made interpretation exercises containing a background explanation of meteorological satellite systems and products: two visible and infra-red image prints, surface analysis weather charts, nephanalysis and upper atmosphere balloon ascent profiles relating to two synoptic situations. In each case, the data for the first day is fully analysed and the weather map, upper air ascents, etc, are completed using the satellite imagery. The data for the following day is left for student interpretation, but with guidance on approach. The two packs are complementary in terms of the weather conditions they illustrate, one in May and the other in October. Teachers' notes, giving answers to the student exercises, are included in each publication. There can be no better resource than this one to help both inexperienced students and teachers acquire an ordered and more confident approach to meteorological satellite data analysis, particularly that of cloud pattern interpretation.

Prints of Meteosat images, obtainable from the European Space Agency, provide an excellent basis for general teaching of global weather systems, based on the interpretation of distinctive cloud patterns. Meteosat is a geostationary satellite, positioned over the equator, that obtains daily imagery of Africa, most of Europe and the Atlantic Ocean in three distinct wavebands. Images for two-to five-day sequences, at different times of the year (preferably in both the visible and infra-red bands) give a versatile resource but need to be backed up with corresponding surface weather charts. Meteosat image sub-scenes (as well as other satellite pictures) are used routinely on BBC TV and ITV weather forecasts.

Some problems

There are, of course, some basic problems associated with the practical usage of remote sensing with pupils in schools and colleges. Many images can only be successfully interpreted if there is some general background knowledge of the environment concerned and specific familiarity with the location. With imagery of the local area or a field area, this can be overcome reasonably well. However, the geometrical inaccuracies of images are well-known, and it is therefore only possible to achieve qualitative interpretation of patterns and processes for which there is unambiguous evidence. All images are functions of the spectral and spatial resolutions of the systems by which they are obtained. Pupils need to have some background knowledge of the electromagnetic spectrum and physical laws relating to radiant energy. This requirement calls for some collaboration between geography teachers and their science colleagues. All types of imagery are potential sources of uncertainty, even confusion. Those obtained at microwave lengths provide difficulties of interpretation that exceed their intrinsic value, especially with inexperienced pupils. Near infra-red photographs and linescan images in both black and white and 'false' colours present psychological obstacles to easy interpretation, but experimentation does suggest that the 'conversion to type' is made quite quickly when there is an opportunity for direct comparative study.

The teacher needs to take pupils through this perception barrier if the versatility of remote sensing is to be really appreciated and exploited. The fact that images of a single scene taken in different portions of the electromagnetic spectrum provide complementary sets of geographical data is important as a teaching and learning objective. Once achieved, the large collection of published infra-red images of terrain and atmospheric conditions is accessible for a wide variety of illustrative purposes across the breadth of the geography syllabus. The numerous colour composite images from the Landsat multispectral scanner are the best examples of this resource. Many of them can be used in a variety of teaching strategies for both physical and human geography. They are particularly effective in demonstrating the relationships between natural environmental controls and cultural response.

Undoubtedly the crucial problem faced by the teacher in schools and colleges is the cost of acquiring even a small collection of original images, not to mention class sets of basic equipment such as hand stereo viewers. Air photographs and satellite image prints are expensive, though their unit cost would be reduced if there was vigorous demand. However, air survey companies and local government departments do dispose of unwanted collections from time to time, and it is sometimes possible for the assiduous teacher to acquire these at nominal cost. Fortunately, there are some useful annotated slide sets and overhead transparencies of remotely-sensed images which are a cost-effective substitute for hard copy originals.

Conclusion

Despite the variety of academic and practical problems involved, the gathering momentum of international remote sensing, particularly from space, must inevitably make an impact on secondary school geography courses. It has the strong advantage of providing great flexibility of use — for class exercises alongside or supplementary to map reading; for topical illustration of both objective data and value concepts such as environmental pollution and development strategies; for simple up-dating well ahead of textbook revision and literature sources; and for statistical and also for fieldwork projects. It is a visually stimulating medium that can make a positive contribution to the acquisition and refinement of basic skills. Above all, remote sensing can make a strong contribution to understanding the unity of physical and human geography. This is an opportunity for teaching and learning that is surely irresistible.

Note

A list of sources and publications on satellite imagery appears as an Appendix to this volume.

7. Structure and Progression in Geography

Trevor Bennetts

Whilst earlier chapters considered the planning of individual lessons and units of work, this chapter is concerned with the much longer time scale of the programme of geography throughout a secondary school. Such a programme will involve several courses, a considerable number of units of study and a very large number of lessons. In a typical six form entry, 11-18 comprehensive school, it may involve over 2000 hours of geography teaching each year, and must cater for a wide range of pupil ability and a curriculum extending over a seven year age span. The proper planning of such a programme requires attention to be given to various aspects of the selection and organisation of content which are absent or less significant in the shorter time scale: notably, a selection which is consistent with the overall aims and objectives of the programme, and provides satisfactory breadth and balance for each pupil, and a structure which is designed to facilitate pupils' learning. Learning geography in a secondary school is much more than a simple cumulative process of acquiring information, for it involves the progressive development of understanding, skills, attitudes and values, over a period during which pupils' capabilities change as a result of experience and maturation. Planning a geography programme throughout a secondary school is, therefore, a complex task.

The general pattern of course provision in geography varies from one school to another, especially in the first three years, when it may be taught as a separate subject or as part of a combined studies course, usually in association with history and sometimes also with other subjects such as English and religious education. Combined studies courses are fairly common in the first year of secondary schools, less frequent in the second year and comparatively rare in the third year. However, virtually all pupils take geography in some form or other during the first three years of secondary schooling. In years four and five, however, the subject is usually optional. Provision in these years is dominated by the requirements of public examinations in geography which, over the country as a whole, are taken by about half of the school population. But the actual proportion of pupils continuing with geography after the third year varies greatly between schools, probably reflecting the popularity of the subject in earlier years and the structure of the various option systems. In the sixth form, provision is usually dominated by A-level courses, although the subject also figures in one year courses, including some which have a prevocational emphasis, as in the TVEI (Technical and Vocational Education Initiative) and CPVE (Certificate of Pre-Vocational Education) courses. Figure 7.1 is intended merely to give a general impression of the present pattern of course provision involving geography. In the absence of firm statistics, the proportions allocated to combined studies courses in years one to three are very approximate. It should be clear, however, even from the evidence presented here, that curriculum planning in secondary schools should take account of the different 'tracks' which are available to pupils. Thus the course, or courses, planned for pupils in the lower school should provide an adequate and well balanced programme for those pupils who do not continue with the subject beyond the third year, as well as a sound foundation for those who do.

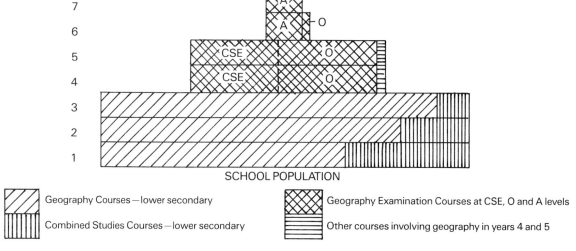

Figure 7.1. General pattern of course provision in geography in English secondary education.

Aims

The most fundamental criterion for the selection of content and activities for a course is that the learning which results is deemed to be educationally worthwhile. Such is the quantity and range of potential content available that selection is usually necessary. Curriculum planners therefore have the difficult task of determining priorities. Subject departments need to consider what sort of learning, associated with their field, is important for all pupils and what sort of learning may be more appropriate for some pupils than others. They also need to articulate their views in ways which give purpose and direction to their preparation and teaching. Planning logically presupposes some sense of direction and this usually leads to statements of intent. Educational goals can be stated at different levels of specificity. Aims are very general statements of purpose which tend to reflect broad educational values and assumptions. They have the important function of providing signposts for the way ahead. Objectives are more specific, especially when applied to particular units of study or individual lessons, although there are considerable differences of opinion about the level of specificity appropriate for different types of educational goal.

If the planning of a geography curriculum is to be effective, the links between aims and objectives, and between these statements of intent and other elements of the plan (for example, content, teaching strategies, resources, assessment and evaluation) should be made explicit. Furthermore, what is attempted in geography should, as far as possible, be related to what is attempted in the remainder of the curriculum. Following the publication of *The School Curriculum* (DES, 1981) and the related Circulars (6/81 and 8/83), most schools will have reviewed their curricula and set out in writing the aims which they pursue. There is perhaps now a stronger incentive for a subject department to consider carefully the nature of its contribution to the total curriculum of a school. Figure 7.2, which is a modified version of a model proposed by Graves (1979), indicates some of the elements which are involved in "curriculum planning in geography at a general level". It suggests that the aims of a geography programme are influenced by the overall curricular aims of a school and by the geography teachers' views about what their subject has to offer.

The planning of the programme is in the context of a school's organisational and curricular structure which allocates to the teaching of the subject a given amount of time, staff and accommodation. That structure determines whether geography is compulsory or optional for particular pupils in particular years, and whether it is taught as a separate subject or within a combined studies framework. When geography is associated with other subjects, geography specialists may have to negotiate with specialists in other fields over the content, teaching methods and structure of courses.

Such links may act as a stimulus or constraint on developments within the subject. Assessment of pupils' learning and evaluation of courses may in turn influence curricular decisions and bring about changes in the system. In particular, external examinations tend to exert a very strong influence on the aims and objectives and the choice of content and methods of courses designed for older pupils, and often have some effect on work undertaken in earlier years.

What aims are most appropriate for the contribution of geography to the secondary school curriculum? The Geography Committee of HM Inspectorate has recently suggested, as a basis for discussion, that the geography component of the secondary school curriculum should be designed to help pupils:

1. To develop their understanding of their surroundings and extend their interest in and knowledge and understanding of more distant places.

> The local area, being the pupils' immediate environment, is not only of personal concern to them, but also is an important potential source of direct experience from which to develop new understanding and through which to introduce specific techniques of enquiry. From this secure base of understanding pupils should extend their enquiries to a range of other places, including other parts of the UK and places much further afield.

2. To gain a perspective within which they can place local, national and international events.

> Such a perspective must be based on specific locational knowledge, especially where places are, and on the understanding of geographical patterns and processes.

3. To learn about the variety of conditions on the earth's surface; the different ways in which people have reacted to, modified and shaped their environment; and the influence of environmental conditions (physical and human) on social, political and economic activities.

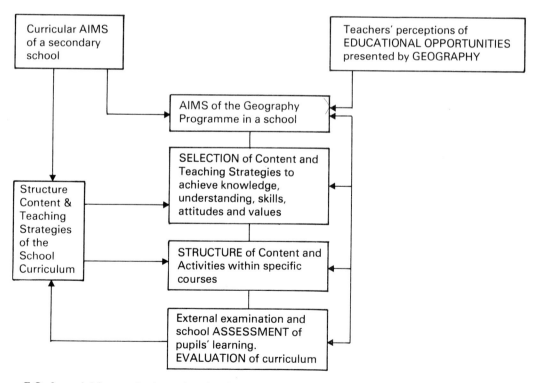

Figure 7.2. A model for curriculum planning in geography at a general level (based on Graves, 1979)

Attention should be given to a variety of 'natural' environments, different forms of land use and settlement, different levels of technology, and contrasting political and economic systems. As a result of their geographical studies pupils should gain an appreciation of the complex nature of the relationships and interactions between people and environments, and between people and places, and recognise the limitations of deterministic assumptions and explanations — both environmental and economic.

4. To appreciate more fully the significance in human affairs of the location of places and of the links between places, and develop understanding of the spatial organisation of human activities.

This should be a persistent theme in any geography course and should be applied to a wide range of human activities at a variety of spatial scales. Pupils should be able to apply their spatial understanding to such themes as population patterns, settlement patterns, the internal characteristics of towns, the exploitation of natural resources (eg. extractive industries), farming, manufacturing industry, communications (to include transport and trade) and other tertiary activities (eg. business activities, retailing, leisure activities).

5. To gain understanding of the processes which have produced pattern and variety on the earth's surface and which bring about change.

The study of physical and human processes is essential to the development of pupils' understanding of why places differ in the ways that they do, why particular human activities produce spatial patterns and how environmental and spatial relationships change over time. The processes of interest to geographical enquiry operate at different spatial scales, ranging from the immediate environment of an individual to regional, national, international and global scales. They also operate over different time scales. It is especially in the study of processes that geographers draw upon ideas from other disciplines — from the natural sciences, the social sciences and the humanities — to inform their own perspectives.

6. To develop a sensitive awareness of the contrasting opportunities and constraints facing different peoples living in different places under different economic, social, political and physical conditions.

Pupils need to be aware of different styles of life, and of the opportunities and difficulties presented by different environments and different locations. They should, for example, appreciate the contrasts in living conditions between rural and urban environments; accessible and inaccessible locations; rich and poor nations. Care must be taken, however, to avoid stereotyping and the acquisition of distorted images.

7. To develop an understanding of the nature of multi-cultural and multi-ethnic societies, a sensitivity to cultural prejudice, and values which reject racist views.

Geographical studies should be designed to enable pupils to develop a better understanding of the nature of, and reasons for, the cultural and ethnic diversity within their own society and other societies. They should be helped to appreciate the contributions which different communities can make to the social and economic life of a country, the links that exist between communities, and the severe problems which some minority groups face. Geographical education should aim to counter stereotyping, cultural prejudice and racism (Walford 1985).

8. To gain a fuller understanding of some social, economic, political and environmental problems which have a geographical dimension, reflect on their own and other people's attitudes to these problems and make their own informed judgements.

Reasonably detailed case studies can provide a suitable context for the proper consideration of attitudes and values in locational and environmental decision-taking.

9. To develop a wide range of skills and competencies that are required for geographical enquiry and are widely applicable in other contexts.

> Whilst pupils should learn techniques and acquire skills which are of specific value to geographical study and enquiry, for example, in the use of maps and in fieldwork, they should also develop more general competencies, often labelled as intellectual skills, study skills and social skills. They should be introduced to methods of enquiry which are not only of value in geography but can also be applied in other fields and in other situations.

10. To act more effectively in the environment as individuals and as members of society.

> Pupils should be educated to be spatially literate; to be able to 'orientate' themselves by 'reading' the landscape. Their education should help them to make spatial decisions, such as where to engage in particular activities and what routes to take for particular journeys. An informed concern about the quality of environments and the conditions which influence the quality of life in different places, combined with an appreciation of the processes which influence these, should help to prepare pupils to make their proper contributions as responsible members of the community and as adult members of society.

It may be helpful to make a few general points about the list which has been presented.

a. The aims are wide ranging and involve the acquisition of knowledge and understanding, skills and competencies, attitudes and values. They are intended to encourage a broad geographical education.

b. They are open-ended in the sense that each is capable of extension and development. There is no obvious limit to the scope for learning in respect of any of these goals; they are applicable to any stage in a pupil's secondary education. The levels of achievement which it is reasonable to expect will of course vary with the age, experience and ability of the pupils.

c. Many of the aims are interrelated and several may be relevant to an individual unit of study. Thus, whilst an investigation of the conflicting demands over the use of land in National Parks may focus on a particular set of environmental, social and political issues (aim 8), it would probably also serve other aims. Pupils would learn about the geographical conditions in the relevant areas of England and Wales; they would examine the relationships between people and the environment (including those who live and work in National Parks, those who are visitors, and those who may never visit them but nevertheless benefit from the exploitation of resources to be found there); they would need to take account of the location and accessibility of the different areas, which strongly influence some of the pressures on them; and the methods of enquiry and study which the pupils use could involve the use of many skills. Interests, indirectly aroused by such a study might at a later date influence the decisions some pupils make about where to spend a holiday or even where to live.

d. Whilst some of the aims have been couched in subject specific terms, drawing upon the various perspectives in geography (eg., the concern with places, with the significance of location, with the interrelations and interactions between people and their environment, and with human welfare in different parts of the world), other aims are closely related to broad goals which may be identified by a school as applicable to the total curriculum. Many of the general intellectual skills, study skills and social skills which are important in geography are also important in other subjects. Similarly, the treatment of social, political and environmental issues in geography should contribute to pupils' political education, enhancing their level of political literacy. In some schools, subject departments may be required to review their courses in order to take account of the general aims of the curriculum, and this may influence the goals which they emphasise. Consider a geography department in a school which has listed 'areas of learning and experience' (HMI 1977, 1985) which it considers to be essential to a broad and balanced curriculum, at least to the age of 16. Such areas may be conceived as extending 'across the curriculum' and, therefore, are the proper concern of all subject departments. The HMI paper *The Curriculum from 5 to 16* (HMI 1985) has proposed 9 areas: the aesthetic and creative; the human and social; the linguistic and literary; the mathematical; the moral; the physical; the scientific; the spiritual; and the technological. Although many geography teachers responding to the checklist might consider that their subject is most closely identified with the 'human and social' area, they would undoubtedly also wish to indicate their contribution to other areas. Geographical education is also much concerned with pupils' understanding of physical environments and processes, aspects which could be neglected if the emphasis is directed narrowly towards human and social conditions. Furthermore, the methods of enquiry and analysis which pupils can use to study geographical relationships and processes, both physical and human, can contribute to their scientific learning. A geography department may wish to consider its actual and potential contribution to several of the areas (Bennetts 1985). Curriculum decisions taken at a school level may thereby influence the way in which subject specialists frame their own statement of aims. It is reasonable that geographers should view their subject both as a source of educational aims and as a suitable medium for the implementation of other general goals.

e. The aims which have been stated can be translated into more specific objectives, which could be applied to a whole course or to parts of a course. For example, teachers will wish to identify the ideas (concepts, generalisations and models) which are associated with geographical understanding and the skills which are required for geographical investigations. Considerable progress has been made in the identification of such objectives. The Geography for the Young School Leaver Project (GYSL 1974-5) gave an influential lead by presenting clear statements of objectives (grouped as ideas, skills, and values and attitudes) for each of the curriculum units which it produced. HMI have attempted to identify some of the more important concepts and generalisations associated with the systematic themes that are most commonly studied in geography (HMI 1978). Such publications can be sources of ideas for geography teachers designing or revising their own courses. Not all educational intentions can be translated into specific objectives and certainly not into behavioural objectives. Where teachers and pupils are together exploring a field of knowledge or an issue which has no neat answer, it may be inappropriate to attempt to specify in advance precisely what pupils should know and be able to do at the end of a course. A syllabus which is too specific can be limiting. The Geography 14-18 Project chose to state the goals of their syllabus in fairly general terms, whilst giving considerable guidance on curriculum design and on criteria for assessment (Tolley and Reynolds 1977).

Selection of content

The frameworks of most school geography courses are described mainly in terms of content: a succession of topics, whose labels usually refer to places, types of environment, features of places and environments, types of human activity or human problems with a geographical dimension. Thus a topic on 'farming in upland Wales' would focus on a particular type of activity, in a particular type of environment, in a particular place; a topic on 'the development of squatter settlements in Third World cities' would be concerned with the difficult human problems associated with a prominent feature of many large cities in economically developing countries. Whilst it is possible to envisage a programme based on a framework of more abstract ideas, it is much easier, and probably more appropriate for geographical learning in school, to design courses composed of building blocks which appear to pupils to have substance, a content framework based on concrete experience rather than abstract ideas.

Nevertheless, the choice of specific content remains far from straightforward. Although the most important criterion is that the selection should effectively serve the educational purposes of a course, a set of aims such as that described earlier offers abundant opportunity for alternative selections. It is possible to identify some general areas of content which are necessary if these particular aims are to be fulfilled: for example, pupils would have to undertake some studies of their local area, of other parts of the UK and of areas beyond the UK; they would have to give attention to a variety of environmental conditions and human responses to these conditions; and they would have to include in their studies various types of human activity and some social, political and environmental problems. But this still gives an enormous scope for choice. It would be reasonable for much of the content to be in the form of case studies and illustrations which have been selected to enable pupils to develop geographical skills and understanding which can then be applied more widely. In other words, much of the specific content may be of secondary rather than of primary importance and its appropriateness should be judged as much on the basis of its suitability as part of the means by which broader goals are attained as for its unique interest.

In practice, the choice of specific content will also be influenced by such considerations as the availability of suitable learning resources; the interests and knowledge of teachers; and, for older pupils, the requirements of external examinations. The choice of illustrative content for any particular study may also be influenced by the content used in other parts of the course, or even by studies carried out in other subjects either because teachers are aiming at some sort of balance, for example, between case studies drawn from different economic and political systems, or because they wish to emphasise links between different topics and a careful selection of specific content may facilitate this.

Given this potential complexity, teachers may find it useful to check their selection of content against their aims, or criteria developed from these aims, with the help of a content grid. Various types of grid can be produced.

1. Each unit of work within a programme could be evaluated in terms of its contribution to the different geographical perspectives which are reflected in the earlier statement of aims (Figure 7.3). For example, an investigation of people's perception of different residential areas within a city would be concerned with a particular aspect of the relationships between people and environments (in this case probably the social as well as the built environment); a study of journey-to-work patterns would have a spatial emphasis; a more general enquiry into the distinctive character of a neighbourhood would focus on the idea of place; whilst an examination of the problems associated with the inner areas of cities could be approached from the perspective of human welfare. A unit of work may of course contribute to more than one perspective.

Some geography teachers may wish to identify other perspectives associated with the subject. For example, those who believe that geography courses should make a strong contribution to pupils' understanding of the physical environment, irrespective of relationships between people and their environment, could include an 'earth science' perspective, which would enable them to identify separately those parts of the course within which attention is given to landforms, hydrology, weather and climate, and soils and vegetation. A simple grading could be used to indicate whether the

Units of Work	Geographical Perspectives			
	People and Environment	Location, Links and Spatial Organisation	Character of a Place	Human Welfare
1				
2				
3				
4				
..				
..				
..				
..				

Figure 7.3. A geographical perspectives planning grid.

CONTENT	CONCEPTS					
	Man/Envt Relations	Relative Location	Distance	Networks & Interaction	Sphere of Influence	Dynamism
World Studies a b						
England, Wales and Scotland a b						
Northern England a b						
Fieldwork a b						

Figure 7.4. A key concept planning grid (based on the JMB, ALSEB et al joint 16+ geography examination syllabus).

contribution of a unit of work to a particular perspective is strong, moderate or weak. A geography department may attempt to achieve some balance between perspectives over the length of a course, or its philosophy may lead it to emphasise one perspective more strongly than others.

2. A similar format can be used where teachers wish to emphasise a limited number of 'key ideas', which they consider are widely applicable. For example, the joint 16+ geography examination syllabus of the northern group of examining boards (ALSEB et al 1983) recommends that the study of the subject content should be undertaken in relation to six geographical concepts: man-environment relationships, relative location, distance, networks and interaction, sphere of influence, and dynamism. As the subject content of the syllabus is divided into four sections — selected World Studies; England, Wales and Scotland; Studies in Northern England; and Fieldwork — a grid might be in the form of Figure 7.4, where general concepts are marked along one axis and the topics selected for the four areas of subject content would be listed along the other.

A study of the growth and distribution of population at the global scale would utilise the ideas of 'dynamism' and 'man/environment relationships'. A unit of work on the motorway system of the United Kingdom would probably focus on the concepts of 'distance, networks and interaction'. A unit on the National Parks of Northern England might give attention to the attraction to visitors of different types of landscape (man/environment relationships); to the significance of the location of the National Parks in relation to major centres of population and important routways (relative location, distance, networks and interaction); and to conflicting land use demands within National Parks (man/environment relationships). A fieldwork task might be concerned specifically with mapping the sphere of influence of a shopping centre. As each important concept will be relevant to a range of topics, it is useful for a teacher to consider the various ways in which the idea can be applied and the scope for extending the meaning which pupils associate with the concept.

3. Figure 7.5 brings together the two most common classifications of content used in geography syllabuses, one based on type of area and the other on systematic categories. Again, alternative or finer categories may be preferred. The physical environment section, for example, could be subdivided along traditional lines. Whilst the classification of places has an element of spatial scale at the two extremes — the local area and the world, studies concerned with the United Kingdom and other parts of the world should be at a variety of spatial scales, ranging from small sites, such as an individual farm or settlement, to regional, national and international scales. Thus spatial scale is to a large extent an additional dimension to take into account. It would be unrealistic to attempt to cater for every combination of area and theme shown on the grid, but the overall pattern derived from a geography programme could reveal emphases and gaps which teachers, on reflection, might wish to adjust. For example many geography schemes of work almost completely neglect the communist countries of the world. The pattern revealed might also suggest potential links which could be developed through the structure of a course. Content grids of the types described can, therefore, serve both as planning tools and as tools for evaluation.

Selection of learning activities

It has already been suggested that pupils do not develop understanding, skills, attitudes and values simply by being passive receptors of information. They need to engage in a wide range of learning activities which encourage and help them, for

PLACES

		Local Area	United Kingdom	Places beyond the UK			World Scale
				'Western' World	'Socialist' World	'Developing' World	
THEMES	Population						
	Settlement Patterns						
	Within Towns & Cities						
	Exploitation of Natural Resources (Extractive Industry)						
	Farming						
	Manufacturing Industry						
	Transport Systems and Movement Patterns						
	Other 'Tertiary' Activities						
	Physical Environment						

Figure 7.5. A place/theme content planning grid.

example, to be more observant of landscapes; to interpret, analyse and evaluate information; to undertake purposeful enquiries; to apply previous knowledge to new situations; to explore ideas, attitudes and values; to be reflective; and to make appropriate spatial and environmental decisions. All of the aims which were listed earlier have implications for teaching methods as well as for selection of content. The geography teacher, who is planning a course, will therefore need to consider not only what topics to include but also how they are to be tackled. Decisions about teaching methods and learning activities may appear to be most applicable at the more detailed levels of planning associated with teaching units or individual lessons. However, even when planning the overall strategy for a course, it is useful to consider what types of activity will be most appropriate and what implications this might have for the design and organisation of the course. A decision that fieldwork should form a persistent thread in a geography course, and that pupils should acquire fieldwork techniques and develop skills of enquiry over the length of the course, may influence the choice of learning activities and teaching methods to be used with particular topics. In some lessons the choice of learning activities will have clear priority over the choice of content.

Organisation of content and learning activities

Having determined what sort of content and learning activities are most suitable for a given set of goals, the teacher designing a curriculum must attempt to incorporate an appropriate selection of these within an overall plan, which indicates when particular topics are to be studied and how pupils' learning is to be advanced. It is a process which involves both further selection — a more specific identification of objectives, content and methods is required at this stage — and the placing of the

---

AUSTRALIA

1. Distribution of population

2. The 'Empty Heart' — a land of desert and semi-desert
   case study: Alice Springs

3. Mining in arid and semi-arid areas
   case study: the Pilbara District of Western Australia

4. The Grasslands of Northern Australia
   case study: a cattle station
   transport problems

5. The Queensland Coast
   the sugar cane industry
   the Barrier Reef

6. South East Australia
   a transect from the coast to the Murray-Darling Basin
   case study: a wheat and sheep farm

7. The great cities of Australia
   case study: Sydney

8. The Australian Iron and Steel Industry

9. The Snowy Mountains Scheme — hydroelectric power and irrigation

10. South West Australia — the remote corner

11. An overview — understanding the distribution of population

Figure 7.6. An area-based framework for part of a second year geography course.

MANUFACTURING INDUSTRY

1. What is manufacturing industry?
   primary, secondary and tertiary activities

2. What are the requirements of manufacturing industry?
   a. the need for markets, capital, materials, power and labour; and the importance of transport
   b. case study of a factory — analysed as an open system

3. Where do manufacturing industries locate?
   an analysis of types of location (eg. market orientated, raw materials orientated and footloose industries) and the influence on location of the basic requirements of industry and other factors

4. Why do manufacturing industries often cluster together?
   case studies of (a) an industrial estate, (b) an industrial region

5. What brings about changes in industrial location?
   a. changing patterns in a particular industry (eg. iron and steel)
   b. changes within an industrial region (eg. SE Lancashire)
   c. changing patterns in a town

6. Where would you locate a new factory?
   a role playing game requiring application of ideas

7. What effects can manufacturing industry have on an area?
   a. beneficial effects — the 'multiplier' effect
   b. harmful effects — problems of noise and pollution
   Who should control? Who should pay?

Figure 7.7. A systematic framework for part of a fourth year geography course.

components within a programme. Three important aspects of the design of such a programme are *grouping, sequence* and *progression*.

The grouping of related topics, within either area-based or thematic-based frameworks, is common in geography courses. The precise sequence of topics within such frameworks may or may not be significant. Figure 7.6 illustrates an area-based approach applied to part of a course for second year pupils. Australia has been selected as the focus of study for half a year. Whilst no attempt has been made to achieve a 'regional coverage' of the country, the selection of content would enable pupils to gradually build up a picture of the varied conditions within Australia and in a form which is probably appropriate for pupils of this age. Topics are at a variety of spatial scales and include several case studies which would enable pupils to gain an appreciation of what it would be like to live in specific localities. But each case study is set in a more general geographical context. Although the course is described here in terms of specific content, the objectives for each topic would reveal ideas and skills which it is intended to develop. The value of grouping topics which are concerned with the same part of the world is that the geographical context is strong and the potential links between topics can be exploited to facilitate learning.

In this example the sequence of topics is by no means haphazard. The initial topic, the distribution of population, introduces a theme to which most of the other topics contribute. Furthermore, the sequence might be envisaged as a route taken by a traveller, a very useful device when the teacher wishes to draw attention to gradual changes in environmental conditions over geographical distances. But, obviously, it is not the only sound sequence which could be devised. The study might just as well begin at Sydney as arrive there at a late stage. The route might follow different directions. One important consideration is that the sequence should hold the interest of pupils by providing suitable variety of content and activities. In this case, it is less important to take a particular route than to ensure that some of the potential links between topics are exploited and that, at the end of the study, the important threads are drawn together.

Figure 7.7 illustrates a grouping of topics within a thematic framework, in this case a substantial unit of work on manufacturing industry for fourth year pupils. The organisation of content in this form enables pupils to explore a theme in some depth, and to develop and apply a set of general ideas. The sequence of topics in this particular study is more significant. Pupils, at an early stage in the unit, are introduced to important ideas which are reinforced and extended in later topics. They begin with a comparatively simple study of a single factory and later consider more complex distribution patterns. They look at the locational influence of the factors of production, in an apparently static situation, and then examine the reasons why changes occur. Near the end of the unit, when they have a reasonable grasp of some of the basic locational concepts, they investigate the impact manufacturing industry can have on an area, including some of the social and environmental problems which can arise. Although this would appear to be a rational sequence, it is certainly not the only sequence which could be defended. Pupils could even begin with a specific social or environmental issue and from there proceed to examine more general questions about the location of manufacturing industry, the causes and consequences of changes and the role of government and other organisations. In contrast to the unit on Australia, the illustrations and case studies for such a thematic scheme may be drawn from widely scattered places. The gain in thematic coherence, based on strong conceptual links, may be at the cost of a less secure geographical context for each topic. In a long course, a combination of area based and thematic based units may have more to offer than a structure which uses only one of these ways of organising content.

In geography courses, the grouping of related activities is much less common than the grouping of related content, but it does occur. Perhaps the most obvious example is the concentration of various fieldwork activities into a residential course, a device which has considerable practical attractions. Examples from the classroom are mainly concerned with teaching particular techniques. It is not unusual for geography courses designed for fourth and fifth year pupils to contain a unit which focuses on the techniques required to read and interpret 1:50,000 and 1:25,000 Ordnance Survey maps, especially when the external examination includes a compulsory OS map question. Similarly, many A-level examination syllabuses have stimulated the development of teaching units on the use of statistical techniques. Unfortunately, a concentration on techniques in one section of a course is sometimes associated with neglect of their application in other sections. Such narrow treatment of techniques is usually at odds with the reason for introducing them into a syllabus in the first place. But this need not happen. Some techniques can be taught when they are required; others are best taught in a separate section of the course and then applied selectively when appropriate.

Although the organisation of content has been illustrated here by outlines of courses concerned with studies of Australia in the second year and manufacturing industry in the fourth, either of these broad areas of content could be tackled in any year of a secondary school geography programme. There is no strong educational reason why Australia should be considered especially appropriate for the second year and no reason why pupils' study of an important theme, such as manufacturing industry, should be limited to a particular stage of their education. Indeed manufacturing industry is likely to figure, in some way or other, in most years of a geography programme. How then should such learning be structured? In what ways should the study of an area or theme in later years be significantly different from that undertaken at an earlier stage? In what ways can pupils build upon their previous experiences and learning? These questions lead us to the idea of progression.

Progression

Progression is a more subtle idea than sequence, for whereas the latter is easily related to such discernible components of the curriculum as content and activities, the former has rather more to do with what pupils learn and the quality of that learning. The idea of progression is most useful when applied to pupils' gradual extension of skills, understanding and appreciations. Ideally, a programme should be designed to facilitate progression in learning by building systematically on pupils' previous experience and achievements. Such a programme would, of course, take account of the interests and capabilities of pupils and the ways in which these develop as a consequence of maturation. The matching of pupils' capabilities with appropriate learning experiences and demands is a considerable professional challenge. For one thing, it is necessary to identify the strands within a subject which deserve special attention and consider how they can best be developed within a teaching programme. In geography, that entails identifying the more important ideas and skills, the different levels of understanding and attainment which can be achieved, what is involved in the acquisition of such objectives and what learning difficulties may be anticipated. At the same time, however, it is important that teachers have an appreciation of what it is reasonable to expect pupils of a given age and ability to achieve, so that they try to avoid frustrating pupils, either by grossly underestimating their capabilities or by presenting them with tasks which they really cannot manage.

Most pupils experience considerable intellectual development during the secondary phase of education. At the age of 11, many are still in most senses children, intellectually as well as physically, but during succeeding years they pass through the stage of adolescence to become young adults. This development is marked by significant changes in their styles of reasoning and in the quality of their thinking, which their schools can do much to support and foster. For most pupils this is a period during which they greatly extend their capacities to arrange and classify objects and events, to form concepts and to explore relationships. But in the earlier years of secondary schools the thinking of most pupils remains, for the most part, tied to concrete experiences; they still need to relate their ideas to particular objects, events and situations, which are either present or can be recalled by memory. As pupils mature they begin to loosen these bonds and show themselves capable of making sense of more abstract ideas. Their thinking begins to extend beyond an understanding of relationships between objects or events to an understanding of relations between ideas which themselves represent relations. This is the sort of understanding necessary to cope with such concepts as the water balance, relative humidity, lapse rates, location quotient, conservation of resources and economic development, or to have a proper understanding of such theoretical models as those of Von Thunen, Christaller and Weber.

Examples of progression which involve the development of skills in interpreting and analysing maps were provided in the previous chapter. These will help to indicate some other considerations which should be taken into account when planning geography curricula. The following example outlines a planned progression in a programme on routes and route networks for more able pupils.

a. At 14+ pupils should be able to:
  i. distinguish between different types of routeway on a map from the symbols and colours shown in a key;
  ii. measure distances along a straight line and along winding routes;
  iii. follow a route along the ground using a map;
  iv. mark a route on a map from a statement giving directions and estimated distances;
  v. describe a route (giving directions and distances and naming relevant features) from information shown on a map;
  vi. identify shortest and most convenient routes through simple networks.

b. At 16+ pupils should be able to:
  i. describe the pattern of communications shown on a map; including the general geometry and density of a route network;
  ii. relate the pattern of communications to relief and drainage and to the distribution of settlements;
  iii. use simple techniques to measure some of the characteristics of networks (eg. the density and connectivity of a network and the accessibility of individual nodes);
  iv. examine the likely consequences of proposed alterations to route systems.

c. At 18+ students should be able to:
  i. use more demanding techniques to measure and describe the characteristics of communication networks and communication surfaces (eg. the use of isoline maps to describe time distance and cost distance);

ii. analyse the factors which influence the characteristics of networks;

iii. describe and analyse the patterns of movement through networks (eg. use of flow maps and interaction models);

iv. analyse the development of communication networks over time, including the influence of technological changes and economic developments, and evaluate network development models (eg. the Taaffe, Morrill and Gould model);

v. analyse the social and economic issues associated with the development and modification of communication systems.

The example is based on the idea that learning is likely to be more effective when pupils are helped to gradually extend their understanding and skills, with each new phase of learning building on the achievements of earlier phases. It illustrates what Bruner has described as a spiral curriculum (Bruner 1960). However, this example also assumes that older pupils are less likely to be hindered by certain qualities of the types of learning favoured by secondary schools which can present difficulties for younger pupils. Prominent among these qualities are *complexity, abstraction, precision* and appreciation of *attitudes and values* (Bennetts 1981).

Complexity may be an attribute of a situation which pupils are required to investigate, or of the information and ideas which are presented to them, or of the tasks which they are set. There is an obvious increase in complexity from identifying a type of relief feature, such as a valley or a scarp slope, to describing a landscape which consists of an assemblage of relief features. Some maps are more complex than others in the detail which they display and the extent to which this is packed into a given space. For example, the Ordnance Survey 1:50,000 maps are in some respects more complex than either the larger scale 1:25,000 maps or the smaller scale 1:250,000 maps. An analysis of a route network is potentially more complex than the study of a single route, and networks themselves vary greatly in the number of links and interconnections which they have. Road networks in densely populated areas are usually more difficult to analyse and describe than those in sparsely populated areas. An analysis of changing spatial patterns over a period of time, as in the development of a transport network, may call for understanding of a wide variety of factors. The understanding of a controversial issue, such as the routeing of a new motorway, may be complicated by the involvement of technological, environmental, economic, social and political considerations which are often interrelated.

In order to cope with an increase in complexity, we usually introduce a degree of *abstraction* into our thinking. We select from the detail available, we generalise from particulars and we try to reason without the direct support of concrete evidence. All of these processes involve some distancing from the world which we can observe. We use abstract representations on maps to deal with the complex details which we wish to record. One of the prerequisites for interpreting a 1:50,000 map is an understanding of the symbols and conventions, the ways of depicting relief and the decisions what to include and what not to include on such maps. One of the advantages of introducing young pupils, in lower secondary as well as in primary schools, to maps of areas with which they are familiar, is that they can draw on their experience to make sense of the scale of the map, to appreciate the differences between the classes of roads, to recall what a particular relief feature looks like and to recall what it is like to walk up a steep slope which is revealed on the map by an arrangement of contours. Isolines are a form of abstraction and their interpretation — whether they are contours, isotherms or isochrones — can present problems for some pupils. As suggested earlier, concepts vary considerably in their level of abstraction. Accessibility, connectivity and density are all abstract ideas which pupils may find difficult to understand. There is little point in using indices to measure such attributes of nodes and networks if pupils cannot understand the basic underlying ideas. On the other hand, in some circumstances, the use of an index, followed by a discussion of what it reveals, may help pupils to develop a better understanding of an idea. Conceptual models, like individual concepts, vary in their level of abstraction. It is easier for pupils to appreciate the significance of a descriptive model, such as the hydrological cycle, than to understand an equilibrium relationship expressed by an equation, as in the case of the water balance. The explicit use of abstract models is usually more appropriate with older pupils and even then they must be handled carefully. There is evidence that many able students, who have obtained good grades in their A-level geography examinations, have a familiarity with such models as Central Place Theory and Gravity Models without understanding the underlying principles of these models.

Learning difficulties may also be created by demands for greater *precision* from pupils. Such demands may be for precision in the use of language, for example, as a result of introducing specialised vocabulary or formal definitions. It is useful to remember that language can be a barrier as well as an aid to learning. Pupils may be able to reproduce a definition without understanding its meaning. Many techniques involving the use of maps and diagrams require precision. Drawing a cross section from a contour map is an example. The need for precision is perhaps even more obvious in the use of statistical techniques. Part of the challenge for the teacher is to decide when greater precision is required and what level of precision is useful and feasible. Much may be gained from detailed discussion between teachers of geography and mathematics.

An involvement of *attitudes and values* in the content of learning introduces a further dimension to be taken into account. The study of controversial issues, such as the routeing of a motorway, the siting of a new airport, or the contraction of public transport services in rural areas, is especially appropriate for older pupils, not only because such issues are often complex and older pupils tend to be more skilled at evaluating evidence and the consequences of alternative courses of action, but also because an appreciation of social and political issues requires some sensitivity towards, and understanding of, other people's beliefs, attitudes and values and a willingness to reflect on one's own. At its most developed, such an ability is an indication of a mature mind.

An understanding of the ways in which pupils develop intellectually and an appreciation of some of the basic causes of learning difficulties which lie within the content and activities of geography courses, can help teachers to improve the match between pupils' capabilities and educational demands.

Structure in the geography curriculum has been presented here mainly in terms of the selection and organisation of content and activities, and the progressive development of skills, understanding, attitudes and values. The two parts of this structure are interrelated and both should be designed to serve the aims and objectives of the curriculum. The first part is the more easily envisaged and the more easily described. It is the part over which teachers have the closer control. But the second part is perhaps the more fundamental, because it is more directly concerned with what pupils learn and how their learning advances. The fact that learning cannot be guaranteed is no reason why progression should not be planned. It just requires a more pertinent and a more flexible plan than a content based syllabus can reveal.

Note: The views expressed in this chapter are those of the author and do not necessarily reflect the views of the Department of Education and Science.

References

Bennetts, T. H. (1981) 'Progression in the geography curriculum', in Walford, R. (ed) *Signposts for Geography Teaching,* Longman.

Bennetts, T. H. (1985) 'Geography from 5 to 16. A view from the Inspectorate', *Geography,* 70, 4, pp. 299-314.

Boardman, D. (1983) *Graphicacy and Geography Teaching,* Croom Helm.

Bruner, J. S. (1960) *The Process of Education,* Harvard University Press.

DES (1981) *The School Curriculum,* HMSO

Graves, N. J. (1979) *Curriculum Planning in Geography,* Heinemann.

GYSL (1974-75) *Geography for the Young School Leaver: Teacher's Guides,* Nelson.

HMI (1977) *Curriculum 11-16. Working Papers by HM Inspectorate: A Contribution to Current Debate* (The Red Book), DES.

HMI (1978) *The Teaching of Ideas in Geography: Some Suggestions for the Middle and Secondary Years,* HMI Series: Matters for Discussion, 5. HMSO.

HMI (1985) *The Curriculum from 5 to 16.* Curriculum Matters, 2, HMSO.

Tolley, H. and Reynolds, J. B. (1977) *Geography 14-18: A Handbook for School-Based Curriculum Development,* Macmillan Education.

Walford, R. (ed) (1985) *Geographical Education for a Multi-cultural Society.* Geographical Association.

8. Diverse Abilities

8.1. Mixed Ability Groups

Richard Kemp

To most teachers the term 'mixed ability' means any teaching group that contains pupils drawn from the full range of ability. Thus mixed ability teaching is in complete contrast to streaming. Between the two extremes, however, are various types of banding (Figure 8.1). In practice many of the aims and principles of mixed ability teaching apply just as much to a situation of broad or even narrow banding. In this section 'mixed ability' is defined as any situation in which there are pupils with considerable differences of ability being taught together in one group.

While people may argue about the merits of mixed ability as a method of organising teaching groups what few people dispute is that as a method of teaching it requires very considerable skill, effort and commitment on the part of the teacher. There are no easy answers, as is clear from some of the comments that teachers make (Figure 8.2). There is certainly no magic formula, and all this section can do is to point out some of the possibilities and pitfalls.

Mixed ability teaching is also an issue which provokes considerable feeling. Teachers tend to be either very much for it, or very much against it. Academic and social arguments can be produced on both sides of the issue. This is not the place to discuss the advantages or disadvantages of mixed ability teaching. This section takes as its starting point the fact that mixed ability grouping exists. The question, therefore, is how can teaching children in mixed ability groups be made most effective?

A way of thinking

To teach mixed ability groups successfully really involves a complete way of thinking about how teaching is prepared, carried out and evaluated. Some of the elements involved are shown in Figure 8.3. All of these should mesh together if a teaching style is to be successfully geared to mixed ability groups. For example, if a teacher prepares well-designed resources for mixed ability groups, and provides class activities suited to all abilities, but then follows this up with assessment which merely tests factual recall, much of the mixed ability benefit will be lost. The pupils who are weak in factual recall, no matter how much they enjoy the classwork, will feel a sense of failure as a result of an inappropriate method of assessment.

Similarly the teacher cannot expect to involve and stimulate all levels of ability unless proper thought has gone into the preparation and use of teaching resources. It is essential to recognise that mixed ability teaching is not easy. The teacher

Slow learners + bright pupils

Assignment ③.

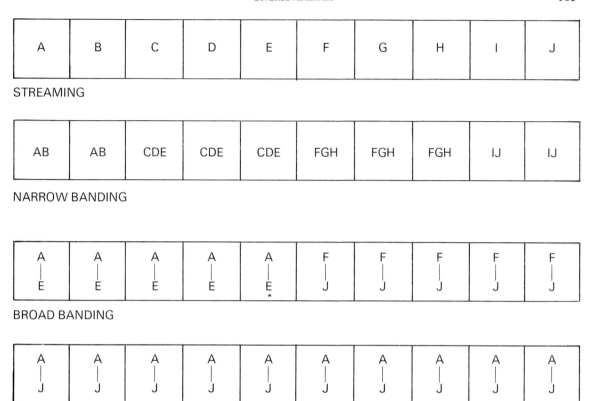

Figure 8.1. Ways of organising pupils within a year group. (Letters represent pupils of different abilities.)

must be prepared to adjust and modify his or her teaching to suit the particular demands of a group or situation. Flexibility and self-awareness are vital.

An important question to consider with any mixed ability group is the level at which to pitch the teaching. Teachers often say 'at the middle'. This is a fundamental misconception, because to pitch at the middle is to risk excluding up to about a third of the group from the start. An aim of successful mixed ability teaching must be to involve everyone. It is vital, therefore, to start by pitching the lesson at a level to include everyone, and then to build up from that initial starting point. This means considering the characteristics of those pupils who are slow or reluctant learners.

To be effective mixed ability teaching must fulfil certain conditions:

The teaching *must*
— stimulate all the pupils in the group.
— provide good learning opportunities for pupils of all abilities.
— involve all pupils in the class activities.

The teaching *must not*
— allow any pronounced polarisation of ability groups within the class as a whole.
— allow any pupil or group of pupils to feel left out of class activities.

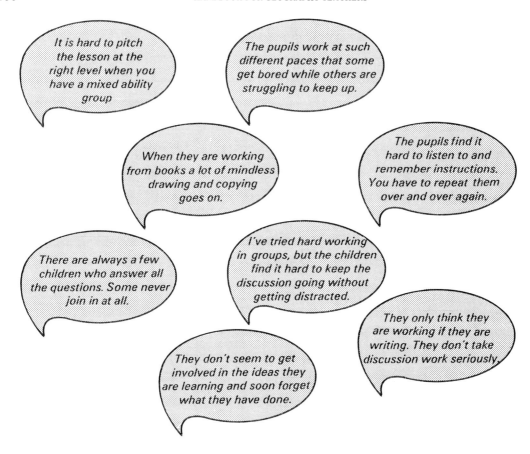

Figure 8.2. Some teachers' comments about mixed ability teaching.

Some key points

Interest gaining. I once spent two days following one pupil around all her lessons. It was a salutary experience. Most of the lessons were frankly boring, not because the content of the lessons was necessarily inherently boring, but because little thought had gone into making the lessons interesting and alive. I was particularly struck by the uniformity of the lesson openings. By the end of the second day I longed for something different, anything to break the monotony of the 'now get out your books' approach. With any class one should aim for as much variety as possible and include interest gaining activities; with a mixed ability class, including its quota of slow and reluctant learners, it is absolutely vital.

The beginning of any lesson is particularly important. One of the best lessons I have observed started with the teacher holding up a piece of white rock. 'What is it?' she asked the class. 'Chalk?', 'Limestone?' came the tentative replies. 'No', she said, 'it's coal'. There followed a silence, punctuated with mutters of disbelief. Then she explained that she had painted a piece of coal white. The pupils thought that was rather ridiculous, but it made for a marvellous lesson on the properties of coal and how coal was formed. She had everyone's attention right from the start.

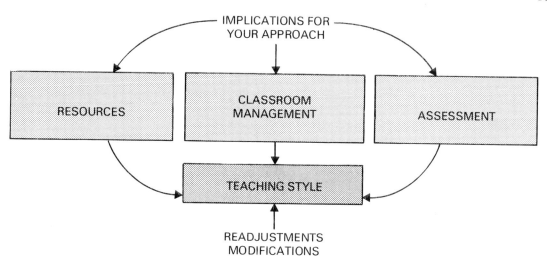

Figure 8.3. Some elements of mixed ability teaching.

Use slides, photographs or other resources at the beginning of a lesson to focus the pupils' attention on what they are going to study. Do not say 'Today we are going to study rivers'. Instead show a slide of the Grand Canyon and ask the pupils how wide and deep it is, and how they think it was formed. Then lead on to river processes.

Lesson structure. Some pupils in a mixed ability group will have a short motivation span, so structure the lesson to cope with this. Divide the work into a series of limited objectives. For some children an hour seems like a lifetime, so set targets that are short enough to maintain motivation. Structure the lesson so that there is variety. Do not stick with the same sort of activity all the time. Beware of 'death by a thousand worksheets'. If necessary have a short break half way through the lesson, or a complete change of activity, such as from individual work to group or class work.

Language and presentation. In any mixed ability class there will be pupils whose grasp of English, both oral and written, is not as fully developed as that of the other pupils. It is important to appreciate that some pupils are easily confused or excluded by unnecessarily complicated language. In a mixed ability class it is vital to keep the language simple but also imaginative.

Written instructions are a case in point. It is all too easy to slip into teacher language. Do not write 'compare and contrast' but 'what are the things that are different between'. Do not say 'write an account of' but 'in your own words write down what you think about'. In short, use language that does not deter any pupil from taking part in class activities. Language should be a means of access, not exclusion.

Consider, too, the way that technical terms are used in geography lessons. To a geographer the word 'ice' means something precise and geomorphological. To many pupils it does not have the same geomorphological meaning at all! It is important to ensure that every pupil knows exactly what is being talked about when everyday words are used as geographical terms.

For some pupils in a mixed ability group the written word alone, no matter how interesting, is not a source of stimulation. As a subject geography is well placed to use visual aids and other resources. With a mixed ability class it is especially important to use a variety of visual, oral and written stimuli in lessons.

Classroom interaction. One of the aims of mixed ability teaching is to create a co-operative atmosphere within the class. If pupils of differing abilities and skills can work together they can all benefit from the experience. This can only happen if there is a degree of interaction within the classroom. Pupils working individually or in silence cannot be interacting. There will be times when a mixed ability class will be working as individuals and in silence, but there will be other times when pupils will need to work in large or small groups. Group work inevitably means a certain loss of control for the teacher, and yet it is something that has to be accepted on occasions. Children involved in enjoyable and stimulating activity will rarely be silent, so within limits the teacher should not be afraid of productive noise!

Skills development. Much work in school geography draws heavily on skills that fall into what may be called the cognitive-intellectual domain. These skills involve, for example, intellectual analysis, factual recall or interpretation. Because cognitive-intellectual skills tend to dominate it is easy to forget that there are other important domains, such as the affective and the practical-technical. It is important not to make judgements about pupils' abilities on cognitive-intellectual grounds alone. Many so-called 'less able' pupils possess considerable skills that all too often go unrecognised.

When teaching a mixed ability class a key point to remember is that everyone in the class can offer something. It is thus equally important to bear in mind that the ability to write or speak with fluency is not the only way of judging a person's ability.

The teacher should look for a wider range of skills to encourage and develop. Earlier in this section it was suggested that mixed ability teaching is as much a way of thinking as a collection of teaching techniques. Unless one is thinking about the different attributes, needs and demands of all the different pupils, one cannot really be thinking 'mixed ability'.

Some strategies

On strategies for teaching mixed ability groups there are no simple prescriptions. The right strategy for any situation depends on the teaching requirements, the classroom, and even the 'chemistry' of the group members in their relationships with one another and the teacher. It is only possible, therefore, to make some suggestions and offer some guidelines.

It has been emphasised that all pupils in a mixed ability group must be given proper attention. An important element in any strategy is to pitch the lesson so that it involves everyone, including any slow or reluctant learners. It is equally important to ensure that during the course of the work all pupils are stretched towards the highest level of their particular abilities. The strategies outlined below, and summarised in Figure 8.4, suggest some ways in which the work in a lesson can be developed upwards.

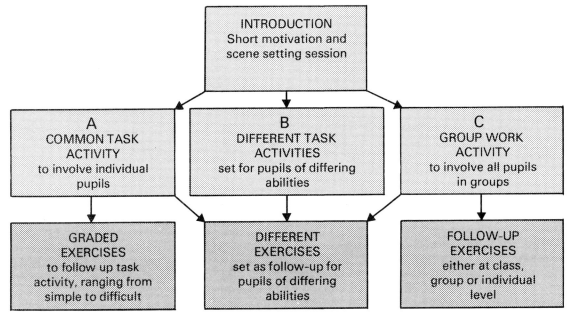

Figure 8.4. Some strategies for teaching mixed ability groups.

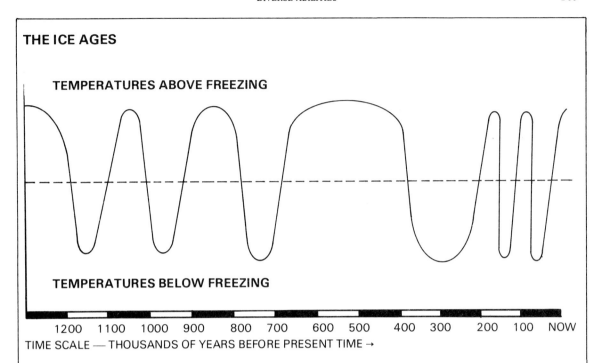

THE ICE AGES

The climate in Britain has not always been like it is today. The line on the diagram shows what has been happening to temperatures in Britain over the last one million years or so.
At times temperatures have been below freezing for almost all the year — these times are called the ICE AGES, or GLACIAL periods.
At other times, as at present, temperatures are above freezing for most of the year, and the climate is much warmer — these periods are called INTERGLACIALS.

1 (a) Draw your own copy of the diagram.
   (b) What does the dotted line stand for? Write the answer by the side of your diagram.
   (c) How many ice ages have there been in all? Number them by the side of your diagram, starting with the one that happened earliest.

2 (a) Why are those periods of cold climate called ice ages?
   (b) When was the longest ice age, and how long did it last?

3 (a) Write a sentence to explain the meaning of the words *ice age* and *interglacial*.
   (b) What is another word meaning the same as an ice age?

4 Imagine that another ice age came to Britain in the near future. Describe what effects this coming ice age might have on (a) vegetation and wildlife, and (b) on our way of life.

5 15,000 years ago, at the end of the last ice age, people were still living in caves. Today men and women go up into space and we have the use of a whole range of complicated machines and equipment. Do you think we would be able to stop another ice age if it threatened us? Have you any ideas how it might be done?

6 What happens to sea level during an ice age? Explain why.
   What effects could changes in sea level have on the shape of the land?

Figure 8.5. Example of a graded worksheet.

The introduction. The start of a lesson should normally be a short, sharp session to set the scene and motivate the pupils. It might include, for example, some slides, photographs, a videotape or an exhibit. The introduction can then be followed by strategy A, B or C.

A. Common task activity. The pupils carry out a common programme of work. The programme could start with a task activity that all pupils could begin with the minimum of explanation. While this task is being carried out the teacher is freed to go round the class sorting out any problems. The follow-up to the common task activity needs to include a series of graded exercises which range from the straightforward to the more advanced. Near the end one or two 'stretcher' questions can be included to give the more able pupils a chance to really stretch themselves. An example of a worksheet using this strategy is shown in Figure 8.5.

A common myth that needs to be punctured is that the less able pupils always work more slowly than the more able. Not infrequently the reverse is true, as the pace of work depends on the depth to which the exercises are tackled. Two points follow from this. First, all the exercises set must be capable of some sort of answer from all pupils, though the depth of the answer will, of course, depend on the ability of each pupil. Second, the work should be structured so that it makes sense even if the pupil does not complete all the exercises. Pupils work at different paces; to expect them all to work at the same speed is unreasonable. In a mixed ability situation teachers sometimes have to curb their fetish for making pupils finish! Work from one lesson need not necessarily be carried over to the next lesson. Spending the first twenty minutes of a lesson finishing work from some days ago is hardly motivating.

B. Different task activities. A number of tasks at different levels of difficulty are set to suit pupils with differing abilities. The most usual method is to have the different levels of tasks explained on workcards or worksheets, which the teacher allocates as appropriate. Different exercises at various levels of difficulty may be set to follow up either strategy A or strategy B. In both cases these involve careful planning and organisation.

The teacher should take care to avoid allowing his or her judgement of the pupils' abilities always to determine the level of the task or exercise that is allocated. Feedback from pupils is important. Have they found the task easy or hard? Would they like to try something more or less challenging next time?

C. Group work activity. The strategies outlined above involve pupils mainly in individually orientated activity, although informal collaboration may well take place. Strategy C involves starting off in a group work situation. Group work activity can then be followed up either as a whole class, or within the groups, or individually, or in a combination of these ways.

A point to consider with group work is the composition of the groups. It may sometimes be a good idea to deliberately mix groups across friendship bands, so that the pupils are obliged to work with others they do not know so well. In a mixed ability class pupils often form friendships with others of similar backgrounds and abilities. Similarly, left to their own choice, boys will often form groups with other boys, and girls with other girls. 'Mixed' groups formed by the teacher can help to foster co-operation between pupils of differing attitudes and abilities. At other times, however, allowing pupils to form their own voluntary groupings can lead to a more lively response to the task set.

Conclusion

Mixed ability teaching, although rewarding, is hard work, and there are no easy answers. For the teacher of mixed ability classes the watchwords must be flexibility in approach and critical self-awareness in evaluating one's own performance. In this respect the support of one's colleagues is valuable. If departmental staff can meet regularly to discuss and exchange ideas, at least two benefits follow. First, the individual teacher becomes aware of, and is reassured by, the fact that his or her colleagues are experiencing the same kinds of difficulties. Second, ideas and suggestions are shared, and resources are pooled, so that the individual teacher's repertoire is enlarged. No one can be expected to have brilliant ideas all the time. Mixed ability teaching is likely to be most effective where there is a strong system of departmental support to back up individual teachers. Working as a team is not only more efficient and usually more satisfying, but, being practical, it also helps to reduce the heavy workload demands of mixed ability teaching.

8.2. Bright Pupils

Neville Greyner

'Bright' or 'gifted' pupils show a high level of general intellectual ability or specific academic aptitude which may be accompanied by creative thinking. Yet in their survey, *Gifted Children in Comprehensive and Middle Schools,* HMI (1977) found that teachers were often indifferent to the needs of their gifted pupils. This was partly because they tended to associate giftedness with elitist ideas, partly because they considered that gifted pupils were able to look after themselves, and partly because, in the welter of problems facing them, those of the more able pupils were a lower priority than those of the less able pupils in their classes. When stating a preference for a particular secondary school for their child, parents often express the wish to secure that he or she is sufficiently stretched intellectually. If under-achievement and under-performance are to be minimised it is essential that able children are identified and given adequate opportunities to develop and use their abilities to the full.

Recognising bright pupils

A survey of 200 teachers carried out by the Teacher Education Project (Kerry, 1981) showed that the characteristics of bright pupils which are most frequently recognised by teachers include the ability to grasp concepts, think out problems, ask intelligent questions and use their own initiative. They have lively, enquiring minds, assimilate facts quickly, relate new work to previous knowledge, and are able to draw conclusions (Kerry, 1981, p. 9). Yet gifted pupils do not always conform to this stereotype and are frequently difficult to recognise. Awareness of their intelligence often makes bright pupils feel awkward with the result they do not easily fit in with their peers. The latter, in turn, often notice something different about gifted children and may actively resent or reject them. Evidence for the extent of the problem can be found in the survey by HMI (1977) and in the work of Painter (1977).

Failure to attain rapport with their peers, especially in large classes, may lead bright pupils to become frustrated or bored. They may well become lazy or even pointlessly 'naughty'. The latter course of action is often a desperate attempt at obtaining the approval of their peers. Such pupils rarely display much evidence of their true abilities and frequently under-perform. Lack of recognition of a bright pupil's abilities may result in low motivation and acute laziness.

Further problems may arise with pupils whose ability enables them to coast their way through the work in their early secondary school years. Later in their school careers, when the work becomes more demanding, these pupils have to 'learn how to learn'. Teachers of mathematics and languages are familiar with this problem, which is the cause of pupils developing what is often described as a 'mental block' in one subject or another. It can occur in geography, for example, in the transition between a traditional fifth form course, where a good memory can produce high grades, and a modern sixth form course, where more practical, critical, decision-making skills may need to be learnt.

The learning characteristics of gifted children have been analysed by the Saskatchewan Department of Education (1978). The attributes identified include: keen powers of observation; powers of abstraction, conceptualisation and synthesis; interest in inductive learning and problem solving; interest in cause-effect relations; ability to see relationships; interest in applying concepts; liking for structure and order; verbal proficiency; questioning attitude; intellectual curiosity; power of critical thinking; creativeness and inventiveness; high energy, alertness and eagerness; independence in work and study. Not all gifted children display all of these characteristics, but the possession of some of them suggests that further investigation may be worthwhile.

Because of the nature of the subject and its overlap with other disciplines in both methodology and content, it is rare to find a pupil who is gifted only at geography without also displaying considerable ability in other related subjects. It can happen, however: an example from the writer's experience is that of a 12-year-old pupil, otherwise undistinguished at the time, who spent many hours forging Ordnance Survey maps of imaginary landscapes. Not only were his maps virtually indistinguishable from the real thing, but the physical and human landscapes he depicted demonstrated a remarkable awareness of the relationships between physical and human geography.

Catering for bright pupils

The identification of more able pupils does not necessarily imply that they should be taught separately. The advantages of separate groupings for more able pupils will be apparent to both teacher and pupils: the standard of discussion is raised with the level of expectation and there is scope for work of greater breadth and depth. Such groups will not necessarily work through the syllabus more rapidly, however, because more able pupils often show a greater awareness of the ramifications and implications of what they are studying. There is no need, therefore, to separate the more able pupils from the others in a subject such as geography purely on the grounds of rate of progress through a syllabus.

Indeed such separation may have adverse effects on both the more able and the less able alike. It is useful for the more able to explain their ideas clearly and simply to their peers. As all teachers know, there is nothing that makes for good and efficient learning better than being called upon to teach. Such a strategy should nevertheless be used sparingly and sensitively in order to prevent the further isolation of the more able as 'different'. The main disadvantage of separating the more able pupils, however, lies in the adverse effect it can have on the pupils in the lower range of ability.

They may be left with a limited view of the levels of attainment that are possible, and they will not experience the stimulus of real enthusiasm that the more able pupils can bring to a class.

It is essential that the general approach to teaching bright pupils does not enccurage them to be passive receivers of information. Neither should worksheets be seen merely as a set of exercises to be worked through. The main teaching strategy to adopt with more able pupils, as with all other pupils, is to identify their abilities. Foremost in the mind must be kept the awareness of the pupil as an individual with needs, aspirations and qualities of his or her own, to identify enthusiasm and to foster them.

The Education Standing Committee of the Geographical Association affirmed in 1978: 'In common with other subject teachers, geographers are evolving programmes of work for children of varying ability following a common course of study ... Within the context of modern geography syllabuses, individual children can readily be offered more challenging material than many of their fellow pupils, while retaining a common framework for all to study'. Yet it should be remembered that there are many activities in the classroom in which all pupils can participate together and respond at levels which vary according to their own levels of attainment. Figure 8.6 attempts to show how a modern geography course can provide enrichment not only for the brighter and more able pupils but also for pupils with individual gifts or skills. This is especially the case where the course incorporates elements of a humanistic approach as advocated by Fien (1983).

Role playing, decision making exercises and simulations can provide enrichment for pupils of all levels of ability together. The advantage of games and simulations as a learning activity in mixed ability classes lies in their capacity to involve the whole class while providing useful enrichment for the brighter pupils. Most games can be followed by pupils of differing ability since pupils learn from participation according to their own aptitudes and perception. A good game provides opportunities for decision making at different levels of complexity. The brighter the pupil, the more complex will be his or her grasp of the implications of each decision. Group activity in games provides experience of co-operative learning as well as friendly competition, an important element in the education of bright pupils.

Pupils' abilities	Examples of skills and activities
Logical and analytical	Enquiry learning and values enquiry Decision making exercises Writing computer programs
Mathematical or numerical	Statistical methods Simulations involving calculations
Descriptive or poetic writing	Humanistic fieldwork and surveys
Oral skills	Role play exercises Debates and discussion
Graphic skills	Mapwork Sketching
Craft skills	Hardware models

Figure 8.6. Matching abilities with geographical skills.

Resources for enrichment

The resources available for teaching the more able pupils include a wide range of published material. For pupils in the 11-14 age group the *Oxford Geography Project* (Rolfe et al, 1979-80) consists of three textbooks, a teacher's guide, slide sets and worksheets with exercises that are graded in difficulty. The three books in the *Patterns in Geography* series (Rice, 1973-78) contain well designed material with supplementary filmstrips and there are structured exercises in the accompanying workbooks. *A Sense of Place* (Beddis, 1981) consists of three colourful and attractively presented books which, although designed for pupils of a wide range of ability, contain many exercises suitable for more able pupils, the more simply graded tasks being found in the separate workbooks.

For pupils in the 14-16 age group, the books in the *Geography and Change* series published by the GYSL/ODA project (1981-82) contain many ideas appropriate for more able pupils and are accompanied by several computer programs. The five packs of resources produced by the Geography 14-18 Project (1978-80) involve an enquiry-learning approach and are well suited to more able pupils. The series of booklets produced by the Geography 16-19 Project (1984-85) should capture the interest of sixth-form students through the use of the issues-based approach and related study of attitudes and values.

Slides and filmstrips are valuable visual aids for all pupils and they have considerable potential for the more able, who are often very perceptive and possess keen powers of observation. The interpretation of a photograph is an important geographical skill and the level of sophistication depends less on the photograph itself than on the learning or interpretative activity involved. Everything depends, therefore, on the ways in which photographs are used with bright pupils.

Questions help to structure the interpretation of photographs and wherever possible 'open' rather than 'closed' questions should be asked. Questions in physical geography might include the following: What clues can you find as to the type of rock which underlies the landscape? How would you describe the slopes? How might they have been formed? Can you tell from the weather and vegetation what the climate is like? Questions on the human response might be framed along the following lines: What clues suggest the part of the world which is shown in the photograph? How is the land used? Why do you think the features shown are located where they are? How might the people earn their living? What problems might they face? Questions on the interaction between people and their environment might include: Which features shown in the photograph would you describe as natural? Which have been made by people? Are there any features which you find it difficult to put in either class? What signs are there of the influence of people on their environment? What signs are there of a strong influence of the environment on people? Are there any features or activities shown which suggest that people are facing problems in this area? Do you think that the alterations made to the scene are beneficial or harmful and in what ways?

Attitudes and values can also be considered during the questioning. Examples of questions might be: What do you imagine the people think of their surroundings? What do you suppose are their greatest hopes? Are they aware that they are being photographed? Do you think that they would approve? Would you regard this as a pleasant place in which to live, or work, or have a holiday?

Other teaching methods can be used in addition to questioning. A useful technique is to show a slide to the pupils for one minute whilst they look at it without making notes. At the end of the timed minute, they are permitted a further two minutes in which to write a list of as many features as they can recall from the photograph. The teacher then builds up a blackboard summary of the features listed and notes the number of pupils who recorded each feature. The pupils have another three minutes in which to discuss whether the features listed most frequently were those that the photographer intended them to notice.

Duplicated copies of a relationships diagram such as that shown in Figure 8.7 can be distributed to the pupils. A slide is projected and the pupils are asked to decide how many of the relationships in the diagram are illustrated in the photograph. Pupils can discuss the relationships in groups, each producing its own summary diagram.

Caption writing is an activity which helps to highlight the differences between the geographer's perception of a scene and that of people with other interests. Pupils examine photographs in magazines such as *The Geographical Magazine*, *The New Internationalist* and the Sunday colour supplements. They then write captions for each photograph in the style of each magazine.

Groups of pupils can also devise their own tape-slide sequences. They are given a cassette tape recorder and a set of between six and ten slides on a particular topic. They construct the tape-slide sequence on the topic using reference material provided by the teacher or obtained from a library. The groups subsequently present their sequences to one another.

Whilst slides and filmstrips are essential resources for use with more able pupils, they should not be used excessively. Otherwise their impact will be lost and with this, their value as a teaching aid. This is a particular danger with

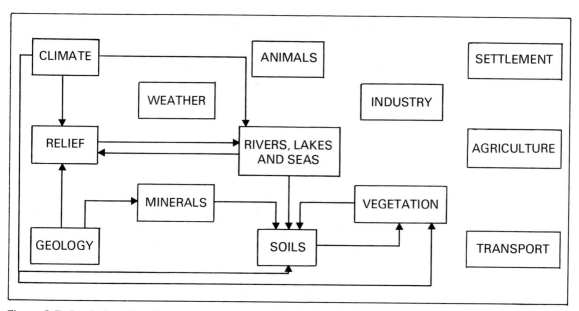

Figure 8.7. A relationships diagram.

filmstrips, where it is often tempting to show a large number of frames in rapid succession. The teacher should always aim to finish with the pupils wishing to see more.

Films and videotapes also constitute valuable resources but unfortunately their commentaries are all too frequently patronising in tone. Sometimes they obscure or ignore the geographical content of the image shown on the screen. It is essential to preview any film or videotape in order to ascertain its strengths and weaknesses as a teaching aid before showing it to a class. If the soundtrack is too poor to use, a substitute soundtrack can be recorded and played back on a tape recorder. Alternatively more able pupils can listen to the original sound track and then write or record their own commentary, or they can view the film or videotape with the soundtrack turned down and subsequently produce their own commentary.

Another way of encouraging more able pupils to be critical is by asking them to write a review of a film or videotape. The pupils can read a critical review in *Teaching Geography* or *The Times Educational Supplement* as an example and can be given a questionnaire to complete on the good and poor features of the film or videotape. This is a valuable exercise in its own right and it also provides insights into the pupils' attitudes and reactions.

Conclusion

Catering for the more able pupil in a modern geography course should not require a major change in teaching strategy provided that teachers remember they are dealing with individuals who have their own characteristics and needs. A useful discussion of the general characteristics and needs of bright pupils is contained in *Teaching Bright Pupils in Mixed Ability Classes* (Kerry, 1981), which also outlines teaching strategies and provides other suggestions for teachers. What is essential in geography lessons is that the pupils' abilities are recognised and that appropriate enrichment material and activities are provided to develop those abilities and thus enable pupils to realise their potential. Sources of enrichment material in geography and further suggestions for activities are to be found in *Geography for Gifted Pupils* (Grenyer, 1983).

References

Fien, J. (1983) 'Humanistic geography' in Huckle, J. (ed) *Geographical Education: Reflection and Action*, Oxford University Press.

Grenyer, N. (1983) *Geography for Gifted Pupils*, Longman for Schools Council.

Her Majesty's Inspectorate (1977) *Gifted Children in Comprehensive and Middle Schools*, Matters for Discussion: 4. HMSO.

Kerry, T. (1981) *Teaching Bright Pupils in Mixed Ability Classes*, Macmillan Education.

Painter, F. (1977) *Gifted Children: their Relative Levels of Scholastic Achievement and Interests*, Knebworth Pullen.

Saskatchewan Department of Education (1978) *Background Information: Gifted Students*, Saskatchewan Department of Education.

Pupils' Books

Beddis, R. (1981) *A Sense of Place*, Oxford University Press.

Geography 14-18 Project (1978-80) *Geography 14-18 Units*, Macmillan Education.

Geography 16-19 Project (1984-85) *Geography 16-19 Units*, Longman.

GYSL/ODA (1981-82) *Geography and Change*, Nelson.

Rice, W. F. (1973-78) *Patterns in Geography*, Longman.

Rolfe, J. et al, (1979-80) *Oxford Geography Project* (2nd edition), Oxford University Press.

8.3. Slow Learners
David Boardman

The terms 'slow learner', 'less able', 'remedial' or 'pupils with learning difficulties' are variously used to describe the less successful and under-achieving pupils in secondary schools. The Warnock Report, *Special Educational Needs,* widened the concept of special education and made recommendations intended to influence the education of around 20 per cent of children. The Report observed that 'a teacher of a mixed ability class of 30 children even in an ordinary school should be aware that possibly as many as six of them may require some form of special educational provision at some time during their school life and about four or five of them may require special educational provision at any given time' (DES, 1978). It is important, therefore, that geography teachers are able to identify pupils who are experiencing learning difficulties and make appropriate provision for them.

Identifying slow learners

Pupils with special educational needs include those whose general ability and rate of progress are considerably below average as well as those with specific learning problems. It is often difficult to distinguish between these two groups because their difficulties form part of a continuum. The term is also used to describe pupils of average ability who are under-achieving because they are experiencing some kind of emotional problem, which may be associated with unsociable behaviour, inability to relate to other pupils and their teachers, or disinclination to work in a group.

Slow learning pupils generally take longer than their peers to remember factual information, to understand unfamiliar ideas and to master new skills. They are likely to need considerable time to use their earlier experiences and previously acquired concepts to deal with new learning tasks. In particular they usually experience difficulty in comprehending abstract ideas unless these are related to their own personal experiences. Their span of concentration tends to be limited so that they are easily distracted and do not listen for long in class.

In a survey of 200 teachers carried out by the Teacher Education Project (Bell and Kerry, 1982) it was found that the majority of teachers recognise as slow learners those who need constant extra help, extra explanations of subject matter and specially clear, step by step, repeated instructions. They lack concentration, do not absorb information and have difficulty in comprehension (Bell and Kerry, 1982, p. 10).

If slow learning pupils repeatedly experience failure in their school work, they may develop poor attitudes towards it and show little motivation. They may arrive late for lessons, forget to bring equipment and fail to attempt homework. They may be absent from school more frequently than their peers, with the result that they tend to fall still further behind in their work. These generalisations conceal a very wide range of individual differences, however, and slow learners may show considerable potential for learning in favourable circumstances. They are not necessarily slow at all kinds of learning, and it should be recognised that slowness is not necessarily a disadvantage in some learning activities.

Reading and writing

Slow learners need carefully planned learning experiences if they are to achieve success and make progress. The general principles of teaching slow learning pupils apply just as much to geography lessons as those in English and mathematics. Each new piece of work should be introduced by means of concrete experiences that are within the pupils' range of understanding. The work should be divided into small, carefully graded steps which pupils can master one at a time. The pace of work should be adapted to match the level of attainment of the pupils. They should have plenty of opportunity for repetition and revision so that the new ideas and skills they learn are adequately reinforced. Clearly all teaching has to be systematically planned and carefully structured in such a way that each stage of learning depends on mastery of the previous stage.

Slow learning pupils often have reading ages which are several years behind those of their peers and their vocabulary is correspondingly more restricted. Teaching materials which demand high reading levels will prevent the full participation of slow learners. Geography teachers will find it helpful to consult their specialist remedial colleagues about the reading levels attained by slow learning pupils. The benefits of relating remedial help in reading to the pupils' learning in other subjects of the curriculum are emphasised by the Bullock Report, *A Language for Life* (DES, 1975).

Teachers often have to use their own subjective judgement and general impression when deciding whether or not to use a particular book with a class, but it can sometimes be helpful to apply one of the measures of readability that are available. The readability of a passage is the level of difficulty of the reading material and it is largely determined by the length of words, the number of syllables they contain and the length and structure of sentences. The procedures for using different kinds of readability formulae are described in *Readability in the Classroom* (Harrison, 1980). It should be noted, however, that all formulae measure only the readability of the text and do not take into account the links between text and illustrations which form an essential feature of modern geography textbooks. In a well designed book the illustrations should help the reader to understand the text but a poor layout can actually increase the difficulties of the slow learner.

Several series of school textbooks have been written with the reading levels and vocabulary of average to below average 11-14-year-old pupils in mind. Examples are *Steps in Geography* (Bateman and Martin, 1980), *Harrap's Basic Geography* (Greasley et al, 1979), *People and Places* (Crisp, 1975) and *Geography 10-14* (Kemp, 1984). *Harrap's Geography Project* (Higginbottom et al, 1984) is a series designed specifically for less able pupils. *A Sense of Place* (Beddis, 1981) consists of textbooks accompanied by workbooks containing exercises which are graded in difficulty for use with mixed ability classes.

Resource sheets and worksheets produced within the school should also be readable by the pupils for whom they are intended. Those for slow readers should preferably be typed using a plain, bold typeface and adequately spaced. The very large typeface of a jumbo typewriter may be helpful for younger pupils. If a word processor is used the clarity of the printing should be carefully checked. If it is necessary to write resource sheets by hand, clear, legible, lower case letters should be used, capitals being reserved for beginning sentences and proper nouns. Unclear handwriting on a resource sheet increases the pupils' difficulties in understanding a prose passage.

To enable slow learners to progress at their own pace, a worksheet should begin with fairly easy tasks using words which are familiar to the pupils and then proceed to increasingly difficult exercises. If the questions are structured so that the work is in small, graded steps within the capabilities of slow learners, they will experience initial success, which produces satisfaction, develops confidence and provides an incentive for further effort. A remedial specialist may often be able to suggest places where a change of wording or simplification of a sentence in a worksheet will help the slow reader. It is not necessary to say 'the climatic conditions deteriorate with increasing altitude' when the same meaning is conveyed by 'the weather is worse higher up'.

Some suggestions for preparing resource sheets designed for slow learners are provided in Figure 8.8. If these are followed the standard techniques used in geography worksheets are just as applicable for slow learners as for other pupils. Key words can be left blank in a sentence or paragraph, the pupils being asked to select the correct word or phrase from two options provided or from an accompanying list. Subsequently they can attempt to insert their own words or phrases without a list from which to choose. Assistance with sentence construction can be given by asking pupils to match the first half of a sentence with the correct second half from several provided. When pupils construct their own sentences and build them into

---

1. Layout
   (a) Break up the text into sections with sub-headings, which may be written in the form of questions.
   (b) Use different type faces or different colours to distinguish sub-headings, questions, instructions, etc.
   (c) Place maps, diagrams, drawings, etc immediately after the section of text to which they relate.

2. Text
   (a) Group sentences on the same topic into a paragraph, making the links between ideas clear.
   (b) Ensure that sentences follow each other in a logical order, usually the sequence in which events occur.
   (c) Avoid a high density of text, often the result of including too much factual information.

3. Vocabulary
   (a) Keep the vocabulary simple, using words with which the pupils are familiar.
   (b) Underline a new word to draw attention to its spelling and if necessary its pronunciation.
   (c) When it is essential to introduce a technical term, define it clearly and concisely.

4. Grammar
   (a) Keep sentences short but vary their length slightly to avoid jerkiness.
   (b) Place the main clause first in a sentence and use subordinate clauses sparingly.
   (c) Avoid difficult negative and passive constructions, very long subjects and unusual word orders.

Figure 8.8. Writing resource sheets for slow learners.

paragraphs, drafting and re-drafting is often helpful. Lists of words and rough notes can be written on the left-hand page of the exercise book, and the sentences and paragraphs on the right-hand page.

Talking and listening

Pupils who do not read or write very well may nevertheless be quite good at communicating orally with others. Slow learners should be encouraged to talk so that they experience the stimulation of spoken language. Pupils talk to one another naturally when they are engaged in practical work in the classroom, such as drawing and model making. Slow learners often talk more freely in pairs and in groups, both of which encourage listening as well as talking. In the more formal classroom setting slow learners may be inhibited from volunteering answers to the teacher's questions or contributing to discussion. Slides and photographs are often a stimulus for questions and discussion as well as valuable visual aids for pupils with reading difficulties.

Portable cassette tape recorders are very useful for developing facility with spoken language and slow learners should have opportunities for using them. Errors in sentence construction and grammar are easily and quickly erased and corrected on tape. Pupils can record real or simulated interviews and reports of work they have undertaken, such as a first-hand investigation of the local environment. Oral communication can help to develop the confidence of slow learners and improve their mastery of language. The geography teacher should accept that information and ideas communicated on tape are just as valuable as those presented on paper.

Discussion in pairs and small groups encourages pupils to listen as well as talk, and thus helps them to learn to appreciate the different viewpoints held by others. News items, current events and environmental issues often provide opportunities for discussion in geography lessons. The geographical content of the topic may be less important for slow learning pupils than the potential of the topic for contributing to their general education and developing their personal, social and communication skills. Fieldwork and site investigations form an integral part of local issues and it may sometimes be possible to involve pupils in discussions or interviews with members of the local community. Permission to tape record these may be granted, but if not pairs of pupils can carry out their own simulated interviews and record these. Sometimes it may be possible for pupils to be directly involved with the community in environmental conservation or improvement schemes, such as canal towpath restoration, footpath clearance or simply litter removal.

Role-playing simulations can help to ensure that discussion in the classroom is positive and constructive. Simulations designed for slow learners need not be complicated but all pupils need to be adequately prepared for playing roles. Even in fairly simple simulations the pupils have to be given sufficient background information. This can often be provided orally by the teacher, perhaps supplemented in note form on the blackboard or on role cards. A simplified map or a set of photographs will often be needed to set the scene. The simulation should be carefully structured and the various stages planned in advance, so that the pupils are aware of the specific purpose of each activity.

Working with numbers

Geography lessons with slow learners may involve the use of simple numerical techniques. The growth of mathematical concepts is a gradual process in slow learning pupils, however, and they may experience difficulty in handling numbers. Many basic concepts in mathematics depend upon an understanding of language, and so it is again advisable for geography teachers to check with remedial specialists that the words used are part of the pupils' vocabulary. Slow learning pupils progress more quickly if mathematical concepts and operations are presented, in the first instance, through the use of concrete materials that they can manipulate. If pupils are encouraged to describe orally what they are doing, the language they use will help them to remember the concepts and operations so that concrete materials can subsequently be discarded.

Methods of teaching numerical techniques in geography lessons should be the same as those used in mathematics and remedial lessons. If different methods are used in lessons in different subjects, slow learners will soon become confused. The procedure for teaching and presenting data on bar graphs and pie charts, for example, should be the same in all subjects. The steps should be carefully graded; thus pupils should learn to draw block graphs, in which one unit is represented by one square on a sheet of squared paper, before they draw bar graphs, which require an understanding of the concept of vertical scale. Consultation with mathematics teachers is clearly important to ensure that pupils have been taught the appropriate numerical techniques before they use them in geography lessons or those in other subjects.

The data used in geography lessons should be kept as simple and straightforward as possible for slow learners. The numbers used should be easy to handle in order to simplify calculations. Whole numbers should be used in, for example, scale measurement, until the pupils are able to handle decimals and fractions successfully. Positive numbers should be used in preference to negative numbers in temperature charts until the pupils have grasped the concept of negative quantities. Pupils should be able to see the geographical ideas which emerge from data without taking up a disproportionate amount of time making arithmetical calculations. Sometimes data may have to be selected by the teacher for the purpose of illustrating particular ideas; for example, in order to show the relationship between population density and distance from the centre of a town, the values plotted on a graph should permit the identification of a clear pattern.

Like other worksheets, those which present data and require the use of numerical techniques should be structured in graded steps to enable slow learning pupils to progress at their own pace and achieve some measure of success. Short, simple tasks at the start of a worksheet can be followed by longer, more difficult exercises. The instructions should be clear and the data should be presented as attractively as possible so that it stands out from the text. It can be helpful to use colour and pictorial symbols as well as charts and diagrams. Data collected by the pupils themselves in fieldwork, such as a count of traffic or types of shops, will be meaningful to them when converted into graphical form.

The application of mathematical concepts and skills to work in other subjects of the curriculum is emphasised in the Cockcroft Report, *Mathematics Counts* (DES, 1982). The report recommends that mathematics courses and those in other subjects should be designed so that pupils master specific mathematical concepts and skills before they are needed in other subjects. Slow learning pupils in particular derive considerable benefit from applying the mathematical ideas they have learnt to everyday situations. If full use is made of the environment for teaching purposes the pupils will see the advantages of mastering certain skills so that they can apply them to real life problems.

Conclusion

It has been possible here to outline only some of the approaches to teaching slow learning pupils. A more detailed consideration of geographical work with these pupils is to be found in *Teaching Slow Learners through Geography* (Corney and Rawling, 1985). This publication gives considerable attention to skill development in slow learners and there are contributions from teachers who use a wide range of successful teaching strategies and effective learning activities. The general characteristics and needs of slow learning pupils are discussed in *Teaching Slow Learners in Mixed Ability Classes* (Bell and Kerry, 1982). This contains practical exercises which can be carried out by student teachers and includes suggestions for further reading.

References

Bell, P. and Kerry, T. (1982) *Teaching Slow Learners in Mixed Ability Classes*, Macmillan Education.

Brennan, W. K. (1979) *Curricular Needs of Slow Learners*, Evans/Methuen Educational.

Corney, G. and Rawling, E. (eds) (1985) *Teaching Slow Learners through Geography*, Geographical Assocation.

Department of Education and Science (1975) *A Language for Life* (The Bullock Report), HMSO.

Department of Education and Science (1978) *Special Educational Needs* (The Warnock Report), HMSO.

Department of Education and Science (1982) *Mathematics Counts* (The Cockcroft Report), HMSO.

Harrison, C. (1980) *Readability in the Classroom*, Cambridge University Press.

Her Majesty's Inspectorate (1984) *Slow Learning and Less Successful Pupils in Secondary Schools*, DES.

Hinson, M. and Hughes, M. (eds) (1982) *Planning Effective Progress*, Hulton/National Association for Remedial Education.

Lunzer, E. and Gardner, K. (eds) (1979) *The Effective Use of Reading*, Heinemann Educational.

Pupils' Books

Bateman, R. and Martin, F. (1980) *Steps in Geography*, Hutchinson.

Beddis, R. (1981) *A Sense of Place*, Oxford University Press.

Crisp, T. (1975) *People and Places*, Nelson.

Greasley, B. et al (1979) *Harrap's Basic Geography*, Harrap.

Higginbottom, T. et al (1984) *Harrap's Geography Project*, Harrap.

Kemp, R. (ed) (1984) *Geography 10-14*, Macdonald Educational.

9. Special Considerations

9.1. Multi-cultural Society

Rex Walford

The need for geography teachers to take account of Britain's multi-cultural society is inescapable. Whether a teacher is in Launceston or Lambeth, Bexhill or Bradford, the same responsibility applies. The presence (or absence) of black, brown, yellow or other ethnic minority faces is not the point at issue: to say 'I don't have any Irish/Muslim/African children in my class and therefore multi-cultural education doesn't concern me' is a perverse response.

Geographers are likely to be significantly involved in the implementation of two of the aims of the curriculum which were laid out in the DES policy document *The School Curriculum* (1981):

> to instil respect for religious and moral values, and tolerance of other races, religions and ways of life.
>
> to help pupils to understand the world in which they live, and the interdependence of individuals, groups and nations.

Such respect and understanding should apply as much to groups within the local community and within Britain as to those in other countries and continents.

Unfortunately in some respects, concern for multi-cultural education has become both confused and confusing. A multiplicity of viewpoints are expressed, even though they stem from the same basic ideals. Polemic and naïve views exist side by side in the courses, the conferences and the literature. Well-publicised disputes within the area have brought some issues into sharp focus, but have at the same time alienated other teachers.

The emergence of controversy in this area should not be seen, however, as an excuse to ignore it. It is important to recognise that some apparently innocent and well-meaning attitudes and practices give offence and dismay to minority groups and that self-examination is a necessary preliminary to dialogue and rational consideration of issues. In schools and colleges where the topic has so far lain dormant, it should be the responsibility of geographers to actively bring about its consideration — through the very subject matter of what is taught in much of human geography.

Anti-racism

Most observers and participants in the field under discussion distinguish between the wider concern of 'multi-cultural education' (though no one phrase commands universal agreement) and the more specific task of opposing racism in schools.

The term 'racist' is usually taken to mean

> attitudes, procedures, social patterns and institutional organisations that intentionally or unintentionally create and maintain the superiority of one racial group over another

and is often the result of the belief that

> people of a particular race, colour or national origin are inherently inferior,

so that their identity, culture, self-esteem, views and feelings are of less value and can be disregarded or treated as less important.

(from the Report of the Rampton Committee, 1981).

It would be hard to find a school or college which would not assent to a desire to oppose racism in general terms, but there is much less agreement about how opposition should be expressed or organised.

The amount of unconscious and unintentional racism expressed in their own institutions has been acknowledged by some teachers with surprise. A vital first step in developing opposition to racism is therefore the encouragement of 'awareness' and the willingness to openly discuss attitudes and feelings.

In recent years a number of local authorities (of varying political hue) have published guidelines about the establishment of anti-racist policies in schools. Documents produced by the Inner London Education Authority and by Berkshire LEA have been among the most widely-circulated and influential. These documents have sought to help schools develop their own stance on the matter. Examples can be cited of both success and failure in this enterprise.

It is likely that geographers may already be active in the support of such policies, given their involvement with the topic through their own lesson material. The policies have involved matters ranging from the practical consideration of how to deal with disparaging racial taunts in classroom and playground, to the special care needed in the assessment of pupils for whom English is not a first language. The need for such policies has been clearly felt in schools where ethnic groups are a prominent part of the school population, but it may be salutary for other schools to consider where prejudice and ignorance of ethnic groups exist.

The difficulty in the genesis and implementation of such policies lies in the problem of securing consensus and understanding. Some teachers — not racists — may believe that a quasi-legal policy is counter-productive, and that any kind of positive discrimination in favour of minorities is unwise. If an anti-racist policy in a school is to be effective it probably needs authorship and action through discussion rather than through imposition.

The Council of the Geographical Association adopted an anti-racist policy statement on behalf of the members of the Association in 1984, following widespread consultation. It was published in the January 1985 issue of *Geography* and is reproduced in Figure 9.1.

The Council of the Geographical Association affirms its intention:

1. to encourage all geography teachers
   a. to consider what forms of behaviour might justifiably be regarded as racist;
   b. to develop their own and their pupils' awareness of racism — both in the textbooks and materials they use, and in their own and their pupils' contribution to lessons;
   c. to condemn such racism;
   d. to consider how geographical education may best seek to counter racism.

2. to support geography teachers in efforts to create greater awareness and concern amongst their colleagues about racism, and to develop overt anti-racist policies in their schools;

3. to examine the Geographical Association, its committees, its Branches, its publications and activities, for racism and racist practices, and to act against these;

4. to encourage all GA Branch Committees to develop their own and their members' awareness of racism and to support anti-racist courses for local teachers.

Figure 9.1. GA anti-racist policy statement.

Content, resources, strategies

The wider issues of multi-cultural education include a concern about the content of lessons, the accuracy and viewpoint of resources which are used, and the development of particular lesson strategies.

In these areas geography, as a school subject, is likely to be in the forefront of scrutiny, since much human geography taught in schools deals directly with the consideration of other groups and peoples, and with such questions as migration, development, living standards and cultural patterns.

Most geography teachers have always been aware of the sensitivity and discretion needed in order to avoid making 'people of other lands' feel alienated or patronised; the task is even more delicate when it includes the consideration of ethnic minorities in our own society. Their presence and response to mishandled information may be immediate and it would be a mistake to believe that they automatically identify with their forbears (or indeed, even their contemporaries) in the lands from which they originally came.

The consideration of other cultures in the classroom has long been an integral part of many geography lessons but it is hardly likely to be objective in all respects. For instance, views about the place of women in Muslim society are likely to be tinged with a Eurocentric perspective; some customs of other societies may seem to be not only unfamiliar but undesirable. Such a view should not preclude a consideration of other societies on their own terms, however; the explanation of practices by those who do support them (or by literature which does) ought to be an important component in many discussions. A consideration of *more* than one point of view in a topic is an integral part of any democratic education worth the name.

Difficulties arise when this produces uncomfortable dissonance from a teacher's own views. Although a teacher may well choose to express a viewpoint, the need for a dispassionate professionalism makes high demands. I recently saw a class of 15-year-olds view a BBC video on Brazil; the film finished with the commentary words 'And so the rich get richer as the poor get poorer'. The teacher, switching off the set, asked for comment and was told forcibly by one pupil 'If the poor worked harder they could be rich too'. The teacher — I thought with admirable restraint — sought other opinions but found the class in total agreement with the first speaker. She then offered an alternative viewpoint, courteously expressed, but found the class quite resolute in their view of the fecklessness of the Brazilian poor...

One way to explore issues like this is through case studies about individuals and families. These can give the pupils an opportunity to empathise with other people at a concrete and personal level. In this particular instance a study of a Brazilian 'poor' family who *do* work hard could be appropriate. Teachers may find biographies and biographical fiction a useful resource for such lessons.

In the discussion of other cultures and peoples it should, at least, for geographers, be axiomatic not to work from hackneyed or seductively picturesque stereotypes. Such material can be found not only in the yellowing pages of old textbooks but also in the alluring photography and lush prose of travel brochures.

The introduction of an informed and personable visitor from another culture can be a useful corrective to a mish-mash of misconceptions. It is only necessary for such a visitor to answer the simple questions of the curious for erroneous ideas to be put at rest. A friendly, personal appearance (especially in areas with few minority students) may be worth many hours of textbook study.

Recent articles and analysis from a number of sources have drawn attention to the predilections and lack of balance in the way in which textbooks and resources present material about other cultures. Such work has certainly sharpened the scrutiny of geography texts, though one suspects that it is always going to be easier to be critical of a textbook than to produce a satisfactory alternative text which would satisfy a larger audience. There is no doubt, however, that some books distort by major sins of omission and by the inaccurate representation of factual material (on maps, for example).

The objective textbook is never likely to be a reality; most materials present a particular image of a country mediated through the eyes of its author. But if the author is alert to the need for 'openness' then at least contentious issues may be covered with impartiality. In the current climate, those who provide classroom materials should scrutinise their balance and viewpoint with more care and understanding than formerly. One ultimate safeguard for the teacher is to use more than one source wherever possible in order to ensure that a single viewpoint or image (however cogent) does not become the key mediator of a culture.

Concern about resources leads towards the scrutiny of teaching and learning strategies as a whole. The ultimate effect of a textbook or a film or a slide is related to the way in which it is used. Even so-called 'biased' material (for example, a blatant publicity film distributed by an Embassy) can form the basis of useful study if the teacher places it in context.

Activities such as the use of particular cultural artefacts, the consideration of prose and poetry from minority writers, the development of role play and simulation exercises to increase empathy, can each have a useful part to play in making another group or society more familiar. I know of one teacher who considers that the organising of a social occasion in which the class went out for a Chinese meal together was a major turning point in establishing a more friendly approach towards that minority group from the rest of the class.

Geography, in teaching about other regions and nations, is always making an oblique comment on the need for groups of differing cultural and racial origin to understand each other. In Britain the realities of a multi-cultural society may have very much more direct meaning in some classrooms than others but the responsibilities for creating understanding lie equally in *all* classrooms. Geography teachers have a key role in developing such an understanding.

The brevity of this section should not be taken as an indication of the marginality of the topic; the issues are complex and cannot be fully taken up within the limitations of space inevitable in a publication such as this. For a fuller discussion readers are referred to the Report of a Geographical Association Working Party, *Geographical Education for a Multi-Cultural Society* (Walford, 1985). The Working Party drew on a variety of viewpoints all over Britain and consulted extensively in the community of geographical education. The Report includes a record of some of the official statements on the topic, a checklist for teachers who wish to examine their own practices, and some examples of relevant classroom scenarios in which multi-cultural issues are raised. A lengthy bibliography indicates many other resources and articles which discuss the topic from particular viewpoints.

Some starting points for further reading

Berkshire LEA (no date) *Education for Equality*, Berkshire LEA, Shire Hall, Reading.

Hicks, D. (1981) *Minorities — a Teachers Resource Book for the Multi-ethnic Curriculum*, Heinemann.

Hicks, D. (1981) *Bias in Geography Textbooks*, University of London Institute of Education.

Minority Rights Group (1983) *Teaching about Prejudice*, Minority Rights Group, 36 Craven Street, London WC2.

Stenhouse, L. et al, (1982) *Teaching about Race Relations*, Routledge and Kegan Paul.

Twitchen, J. and Demuth, C. (1981) *Multi-cultural Education*, BBC Publications.

Walford, R. (ed) (1985) *Geographical Education for a Multi-cultural Society*, (Report of a Working Party) Geographical Association.

Journals

Contemporary Issues in Geography and Education, Issues 1 and 2, from Mailing Secretary, University of London Institute of Education, 20 Bedford Way, London WC1H 0AL.

Multicultural Teaching, from Trentham Books, 30 Wenger Crescent, Trentham, Stoke-on-Trent, Staffordshire.

Multicultural Education, from National Association for Multicultural Education, PO Box 9, Walsall, West Midlands.

9.2. Third World Studies

Michael Morrish

Third World studies is one of the most controversial areas of the geography curriculum. This arises from the virtual impossibility of separating economic and social conditions from the cultural, historical and political factors that shape them. In addition the marked inequalities in wealth distribution throughout the world introduce a moral dimension to the teaching of development issues. Questions of bias will be raised too, and the dangers of a 'Eurocentric' approach represent a further hazard for the unwary teacher. Nevertheless, this lack of certainty is exactly what makes Third World development such a valuable and vital subject for school geography. Only by tackling contentious material are children's opinions likely to be mobilised, and having raised such matters the teacher is bound to accept and respond to all contributions from the class. Such openness encourages a positive atmosphere in which pupils can begin to examine their own attitudes and important lessons may be learnt.

It is probably most common for the geography of developing countries to be taught towards the end of a lower school course. By this stage pupils should be familiar with basic geographical techniques and concepts. Development studies are usually organised systematically, with sections on population, agriculture, health and disease, industrial growth, transport and so forth. General principles will be illustrated by a variety of case studies from one or more major regions of the world. For this age group (13-14-year-olds) it is most appropriate to provide a solid basis of information on which legitimate viewpoints may be built. Care must be taken not to overload the class with complex ideas or to present everything as a 'problem'. This can lead to a demotivating sense of helplessness which causes many children to switch off. Of course, for some pupils this will be their only opportunity to consider the poorer countries of the world: a major aim of the work, therefore, should be to acquaint them with the nature and scale of the poverty gap, and to offer some reasons for it.

Pupils taking public examinations at 16+ will almost certainly study the Third World in greater detail. While it is possible to introduce more complicated analyses and explanations at this level, the framework of the course should be similar to that lower down the school, with an emphasis on the real world. In the sixth form it is reasonable to expect a more sophisticated response, and here difficult topics can be discussed with greater confidence and effectiveness. Once again, however, concepts need to be firmly anchored in a sound regional context.

A question of aid

For the purposes of illustration a number of teaching strategies that can be applied to the subject of international aid are outlined below. The assumption is that the section on aid will come towards the end of a Third World course, since it represents an appropriate synthesis of many ideas that will have been introduced earlier. It is also a topic which can be taught by drawing substantially on the personal resources of the pupils: their values, feelings, perceptions and prejudices.

There are a variety of techniques which can be used to lead a class into the initial consideration of an issue such as aid. One of the most effective is

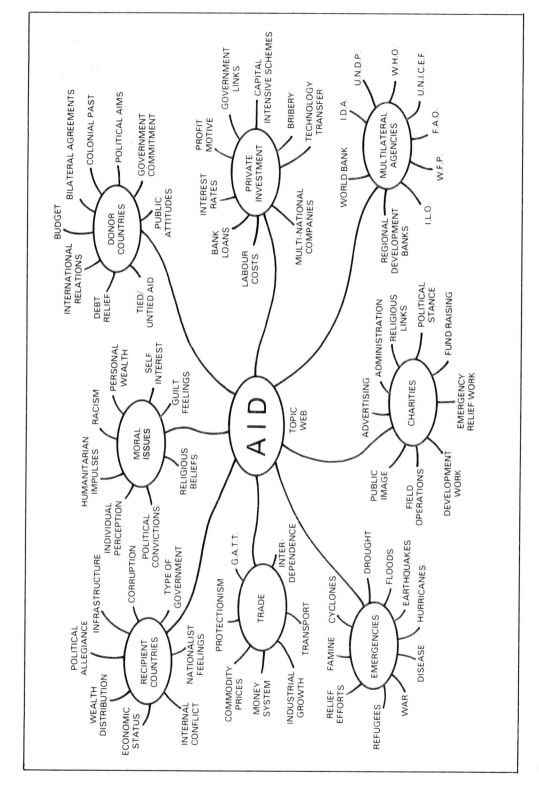

Figure 9.2. A topic web on aid.

brainstorming. Members of the class are asked to write down as many things as they can think of that are associated with aid. The results can then be pooled and collated by the teacher, who uses them to build up a topic web such as that in Figure 9.2. Teachers are strongly recommended to go through this procedure themselves and construct their own topic web before the lesson. Figure 9.2 shows one version of a topic web on aid which one might expect to produce with the help of a sixth form set. The results from a middle or lower school form will necessarily (and quite rightly) be simpler.

Another approach is to give the class a series of open-ended sentences that they are asked to complete in the way they consider most suitable. Once again the emphasis is on input from the pupils and an absence of ideas imposed from above. Individual suggestions are then offered to the class for discussion and comparison with other possible endings.

A set of questions related to aid might be as follows:

1. Many people in the Third World are poor because......
2. Developing countries need.....
3. International aid helps to.....
4. Charities like Oxfam are.....
5. Third World governments want.....
6. Rich countries rely on the Third World for.....
7. European powers used their colonies as.....
8. Unemployment in Britain is a result of.....
9. Politicians are most interested in.....
10. In future developing countries will not.....

A further possible approach is to use the technique known as 'labelling'. Each pupil is given a blank outline map of the world accompanied by a series of words. They are asked to write each of these on the map in the place that they think it applies to most strongly. The choice of words can be adapted to a particular topic, and the following might be a suitable selection for aid: wealthy, agricultural, isolated, communist, generous, needy, overfed, war-torn, industrial, comfortable, poor, democratic, aggressive, hungry, secure, helpless, powerful, underdeveloped, declining, dangerous, affluent, repressed. It may prove revealing to ask the class to identify certain important countries on the map, or even to name the continents, as a way of assessing their knowledge of world geography. If this is obviously lacking then it would be sensible for them to refer to a simple political map of the world when carrying out the exercise.

Before embarking on any detailed work associated with foreign aid, it is advisable to make a study of aid within Britain. The pupils' perception of aid may be redefined if they are reminded of the importance of social security, unemployment pay, urban renewal grants, government loans to industry and so forth. In the middle or upper school a map showing the distribution of Development Areas and Enterprise Zones in Britain would be a good starting point. The discussion could be extended to include industrial grants in the European Economic Community and subsidies to farmers under the Common Agricultural Policy.

For lower school pupils the most easily recognisable form of aid within Britain is the work of charitable organisations. It should be pointed out that not all of these are concerned with aiding the helpless: the National Trust is a useful example to consider in this respect. As a homework task the class could be asked to list ten charities in Britain, briefly explain what each one does, and then decide how they would split a notional sum of £100 among them, giving their reasons. Alternatively, each pupil could collect ten charity advertisements from newspapers or magazines, trying to get as wide a range as possible. The advertisements should then be pooled and displayed, prior to discussing their style and effectiveness. The class might also be asked to divide them into groups, based on their aims and relative importance.

When it comes to teaching about international aid it is best to adopt a two-pronged approach. Firstly, it is necessary to define the different forms of aid and consider the arguments for and against its existence. Secondly, it is essential that an actual aid scheme is studied, so that the theoretical assumptions can be applied to a real-life situation.

Two check lists summarising the pros and cons of giving aid are shown in Figure 9.3. These should provide a sufficiently detailed basis for evaluating the issues involved with a 16+ or A-level set. To cover the second part of the work, a magazine article about a British-funded development scheme in Sri Lanka is critically assessed from a number of different viewpoints.

For

1. The rich, industrialised countries have a moral responsibility to help the poorer nations.
2. The colonial powers exploited the developing world in the past and should now make amends.
3. Aid alleviates suffering and improves standards of living.
4. Countries which are undergoing a crisis need aid to survive and recover.
5. Aid promotes international understanding and helps to maintain peaceful co-existence.
6. Developing countries need aid to support development schemes that they could not finance totally themselves.
7. Aid provides technical assistance that is not available from within developing countries.
8. Economic development in the Third World will create new markets for industries in the North.
9. Aid provides infrastructure that makes economic activity more efficient and productive.
10. The developing countries themselves want aid and are asking for more of it.

Against

1. Aid undermines initiative and determination at all levels, from the individual to the government.
2. Aid is used to maintain the status quo and discourages new policies or institutional change.
3. The main people who benefit from aid are the rich, ruling minority: it never reaches the really poor.
4. The distribution of aid is mismanaged due to inefficiency and corruption: this causes political unrest.
5. Most schemes funded by aid are inappropriate to the needs of developing countries because they reflect Western values of capital investment and high technology.
6. Aid disrupts traditional patterns of life and promotes unsuitable aspirations, tastes and values.
7. The developed countries give aid to divert attention from the need to restructure the world trade system.
8. Aid ties the recipient to the donor and makes developing countries economically dependent on the North.
9. Aid is a weapon of political domination and represents a form of neo-colonialism imposed by the West.
10. Aid does not guarantee development: after 35 years of assistance many Third World countries are much worse off.

Figure 9.3. Arguments for and against international aid.

An example of an aid scheme

Since textbook material can be rather dated, it is particularly helpful to draw upon contemporary news items as a basis for discussion. This emphasises the topicality and relevance of the subject, and encourages students to take an interest in current affairs. However, any media report should be approached with caution, and it is important to point out the political bias that exists in much editorial policy. Students should be aware that no news story can hope to give a full picture of events, and that this is especially true of short television bulletins.

When using written extracts the teacher must be careful to choose articles that have an appropriate language level for the intended class, although it is no bad thing if they include one or two technical terms that require explanation. The article from *The Geographical Magazine* (Figure 9.5) and map (Figure 9.4) make a profitable topic for discussion with a fifth or sixth form set. The article is brief enough to be read quickly in class, but is rich in ideas and issues associated with major aid schemes. A simple comprehension exercise could be carried out prior to the discussion, which would serve to clear up any basic points of understanding that may initially arise. Nevertheless, a structured question and answer session led by the teacher will be necessary to get full mileage out of such a resource, and this should lead naturally into a more wide-ranging discussion. A reasonably comprehensive list of questions that could be raised by direct reference to the article and map is shown in Figure 9.6.

In the final stage of the teaching unit it is desirable to give the pupils an opportunity to consider and express their personal viewpoints on the issues they have been studying. This has the double benefit of encouraging them to clarify their own values and reminding them that a range of opinions will exist on any topic. Some of them may change or modify their attitude in the light of comments made by others.

Although this can be achieved through a teacher-guided discussion it is perhaps preferable to give the event a more formal structure, such as a debate. A suitable motion might be 'This house believes that charity begins at home'. This is a sufficiently well known saying that everyone should understand what is being debated. It also provokes

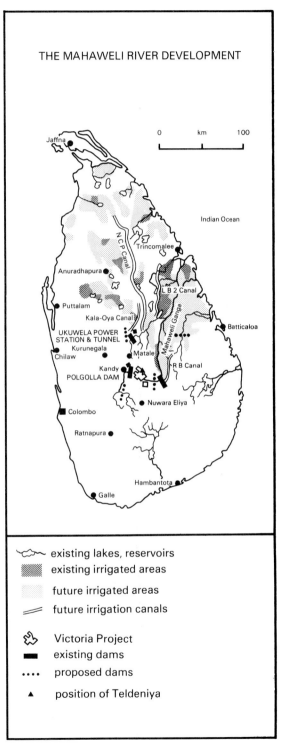

Figure 9.4. The development scheme in Sri Lanka.

Progress has its price

by John Madeley

Sri Lankans have been building dams for centuries and the new Victoria Dam, partly funded by Britain, is the biggest yet. However, the project is controversial.

IN SRI LANKA later this month, following the planned completion of the Victoria Dam, centre piece of the country's biggest ever river development project, about 3230 hectares of the picturesque Dumbara Valley should be under water. The valley lies on the edge of the ancient city of Kandy and through it runs the 335-kilometre long Mahaweli River on its way to the Indian Ocean. People have inhabited the area for many centuries, living off fertile rice paddies and the river itself.

The Victoria Dam has been partly-funded by the British government. In fact, it received, at the time, Britain's biggest aid grant ever. The whole Mahaweli River Development Programme was designed to supply hydroelectric power and much-needed irrigation in the battle against food and energy shortages, and also to create employment at the same time. Three huge dams have been built across the Mahaweli to triple electricity-producing capacity; the Victoria Power Station itself is expected to generate 780 million KW hours annually. Part of the development plan is also concerned with the resettlement of one and a half million people, almost a tenth of the country's population, to areas which are to benefit from improved irrigation. About 129,500 hectares of rainforest in the east of the country are being cleared and irrigated.

Although the whole scheme would appear to be of benefit to the country as a whole, and to the people who live off the land, it has aroused much anger within Sri Lanka and criticism from outside. Six kilometres from the dam wall, the town of Teldeniya, a town with 2000 years of history, will be drowned. It was, until recently, home to 10,000 people. Along with it will go another five towns, 123 villages and 400 hectares of fertile rice-growing land. Several schools, a prison and 14 Buddhist temples will also 'go under'.

Before waters from the Mahaweli can be impounded to fill the Victoria reservoir, every house, tree and bush is being flattened and removed from the reservoir bed. Most of the people have already left; in all about 6000 families, comprising 45,000 people, have been removed from homes where they and their forefathers lived for generations. Known in Sri Lanka as 'Victoria's victims', they were not consulted before the Mahaweli programme began and have seen no public enquiry into the flooding. Many received less than the equivalent of £100 in compensation.

'In the entire history of Sri Lanka's dam construction work' wrote Dr. L. D. Mediwake, a resident of Dumbara valley, four years ago, 'never has a project been undertaken where so much highly developed productive wealth would be destroyed and dislocate so many thousands of families at a time when rebuilding and reconstruction work is at its most expensive...'

Sri Lankans have been inveterate dam builders for centuries and many see the Mahaweli programme as an attempt to recapture the spirit of former glories. Yet the wisdom of the project is increasingly being called into question; so expensive has it become since being launched in 1977, that the project threatens to bankrupt the Sri Lankan economy.

Before the project started, representatives of a Dutch company, Netherlands Engineering Consultants, warned the Sri Lanka government that it was 'impossible to find examples from other countries where such a high rate of land development and resettlement had been achieved'. The World Bank warned that the country could not afford the scheme, which in 1977 was expected to cost around £700 million.

But the Sri Lankans were successful in persuading aid donors to give £400 million to the project. Britain's Overseas Development Administration gave £100 million aid for the Victoria Dam and Sweden, West Germany, Canada, Kuwait and the European Economic Community all offered generous sums.

The money raised made the Mahaweli programme the world's largest aid scheme, and would have meant that Sri Lanka only had to find £300 million from its own purse. Between 1977 and 1983, however, the cost of the project trebled to around £2000 million. Instead of having to find £300 million, Sri Lanka has had to provide finance of £1600 million, over five times as much as expected. It has done this by raising taxes, borrowing more money, cutting subsidies and services — which hits hardest at the poor.

The 122-metre high Victoria Dam itself was expected to cost £136 million in 1977; it will now cost around £250 million. While the dam constructors have regularly increased their prices to reflect rising costs, the aid amounts remain much the same.

Some of 'Victoria's victims' have opted to stay in the Kandy district. But most have gone to the new land, and are trying to re-build their lives. They are allocated one hectare of land for cultivation — a bigger area than many had in the Dumbara valley — plus 2023 m^2 for a house. Problems, however, are acute.

The area where most Teldeniya people are resettling is mainly rain forest and scrub which the settlers themselves have to clear and turn into food-producing plots. For this they need basic supplies. But the area is relatively isolated with generally poor roads; there have been problems getting tools, house-building equipment and seeds through to people. New towns and villages are also behind schedule.

For the settlers there are problems of adjustment. Some families with teenage children were split up in the resettlement — the teenagers were reluctant to move to an isolated area. Few of the parents find it easy to come to terms with this separation.

Although the resettlement area where most of the Teldeniya people are going is only 96 km from their home, it is in a different climatic zone, and many people find difficulty in adjusting to the weather. 'It is much hotter and more humid here' said one settler, adding that the soil was less fertile.

These problems were compounded in 1983 when, because of the escalating cost of the Mahaweli programme, the government cut spending in resettlement zones by a third. With less money available for new roads, supplies will continue to be a headache — as is the programme itself for the government.

John Madeley is editor of International Agricultural Development

Figure 9.5. A magazine article as a resource. Source: *The Geographical Magazine,* May 1984.

1. What type of aid is a grant?
2. Why would Britain give such a large grant to Sri Lanka?
3. Who is likely to have planned and designed the Victoria Dam?
4. Who will receive most of the construction costs of the dam?
5. What are likely to be the gains and losses of the dam scheme?
6. How might the electric power from the new dam be used?
7. How will new jobs be created by the scheme?
8. Which people are being moved as a result of the scheme?
9. In what ways will they suffer economically and socially from their displacement?
10. What problems might they face in their resettlement areas?
11. What effect will the irrigated agriculture have on the land?
12. What types of crops will be grown on the newly irrigated land and will they be for home consumption or for export?
13. Why are the resettlement areas being neglected in terms of investment and development priorities?
14. Why have the costs of the scheme to the Sri Lankan government increased so much?
15. Could the money devoted to this programme have been used more effectively in some other way?

Figure 9.6. Some questions about aid.

argument by promoting a deliberately self-interested stance in relation to the Third World. Four of the more articulate or outgoing members of the class can be selected as main speakers, but everyone will find it helpful if some time is devoted to discussion in small groups before the debate commences. Adequate time must also be allowed for comments and questions from the floor after the main speeches have been delivered.

In summing up the debate it would be appropriate to remind pupils that they are all in a position to do something themselves about aid, if they choose to. Reference could be made to schemes such as sponsored runs, flag days or community service as a way of making a personal contribution to the welfare of others.

Resources for teaching about the Third World

The most useful single source of materials for teaching about the Third World is the Centre for World Development Education (CWDE), 128, Buckingham Palace Road, London SW1W 9SH. This is an independent agency funded partly by the Overseas Development Administration. Each year it produces a comprehensive catalogue of educational resources relating to world development issues: books, leaflets, posters, slide sets, information packs, games and pictorial charts.

A network of Development Education Centres throughout the country serve a similar purpose to CWDE. Details are available from the National

Association of Development Education Centres, also at 128, Buckingham Palace Road, London SW1W 9SH.

Free materials about the British aid programme are available from the Information Department of the Overseas Development Administration, Room E920, Eland House, Stag Place, London SW1. It also produces a guide to sources of information and material which lists more than a hundred organisations associated with world development.

The following books all provide stimulating ideas and new approaches to the teaching of development issues:

The Development Puzzle (7th edn, 1984), co-published by CWDE and Hodder and Stoughton Educational.
Priorities for Development (1981) and *People before Places: Development Education as an Approach to Geography* (1985), both published by the Development Education Centre, Gillett Centre, Selly Oak Colleges, Bristol Road, Birmingham B29 6LE.

Learning for Change in World Society (revised edn, 1979) and *Ideas into Action* (1980), both published by the World Studies Project.

For background material to supplement textbooks, three collections by Paul Harrison are recommended: *Inside the Third World* (1979), *The Third World Tomorrow* (1980) and *Inside the Inner City* (1983), all published by Pelican.

An up-to-date reference source is the *World Development Report,* published annually by the World Bank.

CWDE produces two short summaries of the Brandt Report, *North-South: A Programme for Survival,* and its sequel, *Common Crisis,* which provide concise introductions to the contents of these important documents. An alternative viewpoint is presented in the booklet, *Beyond Brandt* published by Third World First.

The New Internationalist is a good source of challenging ideas and is available monthly on subscription from 374, Wandsworth Road, London SW8 4TE.

9.3. Sex Bias and Differentiation

Patrick Wiegand

One of the main aims of education is to provide equality of opportunity for girls and boys. Indeed in the United Kingdom this is enshrined in legislation. However, there is ample evidence to suggest that, far from reducing inequality between the sexes, schools actually reinforce, in a number of subtle ways, the stereotyped views that society holds about what constitutes 'girlishness' or 'boyishness' (Marland, 1983). This section discusses ways in which geography teachers might contribute unwittingly to discriminatory treatment of boys and girls and attempts to suggest ways of countering sex bias (which largely operates against girls).

A distinction will be made below between sex bias and sex differentiation. Sex *differentiation* refers to differences in behaviour between the sexes, for example, apparent differences in aptitudes, attitudes and attainment. Sex *bias* exists where children are put at a disadvantage on account of their sex.

It must be said at the outset that although sex differentiation and sex bias have been recognised for many years in science and mathematics education it is only in the 1980s that geographers have turned their attention to it (Bale, 1982; Larsen, 1983; Slater, 1983). What little writing there is about sexism in the geography classroom, including this present contribution, is tentative, speculative and exploratory. There is very little *evidence* about sex bias or sex differences between boys and girls specifically related to the *geography* curriculum. However, it would be very unlikely if geography teaching did not exhibit at least some of the processes shown to be present in much of the general literature on sexism and schooling, such as, for example, *Learning to Lose* (Spender and Sarah, 1980) and *Sexism in the Secondary Curriculum* (Whyld, 1983).

Sex differentiation

What evidence is there about differences between boys and girls that might affect their progress in geography? Test results do appear to reveal consistent differences in the aptitudes of boys and girls. It is generally agreed, for example, that girls appear to perform better on test items requiring verbal ability whilst boys do better on spatial visualisation items (involving, for example, the ability to mentally rotate two or three dimensional objects) (Maccoby and Jacklin, 1974; Wittig and Peterson, 1979). However, the differences between the sexes are not large and they do vary according to the age of the pupils being tested. Indeed the differences that do appear may be further masked by the effect of other factors, such as social class.

There is also a great deal of overlap in both verbal and spatial ability. That is to say there may be more variation between some girls than there is between most girls and most boys. It is probably also true to say that more attention has been given to studies that do reveal differences between the sexes than to those that do not.

It is not clear whether the observed differences between the aptitude of boys and girls are innate or whether they are the result of social learning. By the time boys and girls enter secondary school at the age of 11, parents, teachers, books, toys and the media will all probably have played a part in establishing sex stereotyped behaviour.

Boys and girls do appear to be different in two areas which are of concern to geography teachers: map reading and environmental experience.

Boys appear to do better at mapwork tests than girls, but the evidence is not conclusive and the studies that have been done are mostly small scale and unpublished; they are reported in Boardman and Towner (1979). There has been no research specifically concerned with establishing precisely where sex differences exist in map reading. Boardman and Towner (1979) administered a conventional map reading test involving a 1:50,000 Ordnance Survey map extract and an aerial photograph to a sample of 578 pupils from 12 schools. Boys obtained higher scores than girls for all sections of the test. Follow-up interviews after the test indicated that the girls enjoyed map reading less than the boys and participated less in out of school activities related to map reading.

Studies of children's maps of their home area have tended not to report sex differences. An exception is a study in Mexico City by Lynch (1977). Most girls' maps focused on their homes whereas the boys' were more likely to focus on the local playing field. Girls made their own homes and those of their friends distinguishable on their maps by the use of careful (and often romantic) embellishment. Boys, on the other hand, represented their homes by simple crosses or squares. Girls revealed a more detailed knowledge of local shops whereas boys showed greater knowledge of street names and were able to locate places on their map more accurately.

It would seem, therefore, that boys may come to the geography classroom with some advantages. They appear to be better at map reading and may have a more substantial knowledge of their local area. It must be stressed, however, that the evidence is slender and that greater differences may exist among boys and girls separately than between them. If boys *are* advantaged at the beginning of the course, though, what evidence is there about the attainment of boys and girls at the age of 16 and 18? The evidence here is more straightforward and is based on entry and achievement in public examinations, as reported in the *Statistics of Education*, published annually by the DES.

Figure 9.7 shows that during the decade 1971-80 more boys than girls *entered* geography examinations at all levels. *Success*, however, was more varied. At CSE boys were consistently more successful than girls. Boys were also more successful at GCE O-level. Over the decade the difference was slight (1%). However boys have only been more successful since 1975. Before then, girls had greater success. It is only possible to speculate why this should be so. The significance of the mid-1970s will not escape geographers. The subject in school was feeling the effect of the 'quantitative revolution' at this time and geography was becoming more 'scientific'. Several boards were also beginning to use objective tests — a type of assessment thought by some to favour boys (Murphy, 1980). However, not all the evidence on objective items with respect to geography points to an advantage for the boys (Wiegand, 1982). The question of the girls' greater success after 1975 becomes more interesting, however, in the light of an exact parallel with history examinations, where boys have also done better since 1975. One trend which both history and geography shared over the decade was the increased use of resource material (such as maps, photographs, documents, etc) in the examination room. Perhaps the clue to the change lies here.

At A-level it is the *girls* who are consistently more successful. Their pass rate is on average 5% higher than for the boys. This differential is sizeable — greater in fact than the difference between boys and girls in the 'modern languages' and 'scientific and technical' groups of subjects. Fewer girls take geography at A-level but those that do are more likely to pass.

Sex bias

What are the sources of bias in the school geography classroom? Much sex bias takes place in the area of the 'hidden curriculum'. Geography teaching materials, for example, may be sexist through their use of language. The word 'man' has for long been used in geography to denote humanity at large — as for example in 'man-

	Percentage entering for geography		Percentage entering for all subjects	
	Boys	Girls	Boys	Girls
CSE	58.0	42.0	53.0	47.0
GCE O	55.7	44.3	51.3	48.7
GCE A	60.9	39.1	58.3	41.7

Figure 9.7. Examination entries by boys and girls, 1971-80. Source: Statistics of Education (DES).

environment relationships'. Many publishing houses now have explicit guidelines for their editorial staff concerning the equal treatment of the sexes in their publications. Members of both sexes should, for example, be represented as whole human beings with human strengths and weaknesses, not 'masculine' or 'feminine' ones. Women and girls should be shown as having the same abilities, interests and ambitions as men and boys.

Pictures in geography texts may also reveal sex bias. Many 'geographical' pictures of course do not show people at all as they illustrate landforms or are aerial photographs. Those that do, however, often show men more than women. A survey of photographs from 15 modern geography textbooks for the 11-16 age group revealed that there were on average three times more photographs of males alone than of females alone (Wright, 1985).

Monk (1978) examined a selection of role play activities and simulation games in geography and found that more often than not women had minor roles. Decision making was largely the prerogative of men. Monk recommended that authors of new games should include a greater proportion of roles for women.

Changes in the school geography curriculum may affect boys and girls in different ways. Sutherland (1981) points to evidence concerning the greater sensitivity of girls to feelings and emotions. We might, as a result, expect girls to be more responsive toward humanistic geography, but this affective approach may not receive much attention in school, not least because of the difficulties of examining such work. One way in which an affective dimension could be included in the geography curriculum would be through fieldwork projects and school based assessment but (ironically) the predominant attitude to projects is within the objective, *positivist* tradition (Tolley, 1984).

Some teaching strategies in geography may disadvantage girls. Larsen (1983) has pointed out that fieldwork activities are not always designed to allow girls to produce their best work. They may not be well equipped for what is sometimes seen by the boys as an endurance test. As more computers are used in the geography classroom it is becoming clear that boys are usually first in the pecking order for using the hardware. Many girls exhibit a reluctance to run programs whereas the boys appear confident and authoritative.

Action

What action can geography teachers take to allow for *differences* between boys and girls? There is a dilemma here. If you develop materials on the basis of, say, boys being better at map reading and having a greater knowledge of the local environment, you run the risk of reinforcing the very stereotypes you are trying to avoid. It would be as undesirable to establish an overt 'remedial' programme for either the boys or the girls as it would be to construct a course around 'boys' interests' or 'girls' interests'. In any case, knowledge of what appear to be *group* differences must be tempered by knowledge of *individuals'* strengths and weaknesses, irrespective of sex. One approach for the teacher is to acknowledge that sex differences exist but to focus attention on the needs of individuals rather than groups.

How can sex *bias* in geography teaching be avoided? Three possible strategies are suggested based on an examination of three aspects of the curriculum: course content, teaching materials and classroom practice.

Bias is present in the content of many geography courses. The study of work, for example, usually places more emphasis on heavy, male-dominated industries than on the female-dominated service sector. It does seem appropriate to pay rather more attention than we do at present to the growth of women's employment, the distribution of women within the labour market, part-time work, job sharing and the effect of new technology on women's jobs. Many women, for example, work as secretaries or on assembly lines doing repetitive work. Word processors and robots could either free them to do more interesting work or severely reduce employment opportunities. Similarly, studies of change and development in local industry, shopping and transport could be studied from the point of view of how women might be affected differently to men. Many writers are against a separate 'geography of women' arguing that women should not be seen as a separate phenomenon but 'adding women in' does at least provide a starting point.

Several checklists are available for testing for bias in published materials (for example, Whyld, 1983) and these are also useful for checking home-produced worksheets. Checklist items applicable to geography texts might include:

1. Are men and women only depicted in traditional male and female occupations?
2. When men and women have the same job are they shown to be performing it equally? For example, in a book on farming are the tractors always driven by male farmers while the chickens are always fed by female farmers?
3. Is the language sexist? For example, 'man and the environment' could be replaced by 'people and the environment'; 'man-made' by 'artificial' or 'synthetic'; and 'businessman' by 'industrialist' or 'business executive'.

Detecting bias can be a useful classroom activity for pupils. Textbooks may need to be supplemented with more specific non-sexist materials. *The New Internationalist* sometimes contains material very suitable for posters or classroom discussion.

Many teachers use extracts from the diaries of travellers and explorers to add interest to regional studies. It may be worth deliberately using selected passages from women's autobiographies for this purpose. Virago Press have a 'Travellers' series including, for example, Mary Kingsley's *Travels in West Africa* and Isabella Bird's *A Lady's Life in the Rocky Mountains* which may prove useful. Source material might also be found in the travel writing of (for example) Freya Stark and Dervla Murphy. Neither should the contribution of women regional novelists be forgotten (for example, The Brontë sisters or Daphne Du Maurier).

It is in the everyday conduct of class teaching, however, that most sexist messages seem to be unconsciously transmitted (Delamont, 1980). And men seem to be most guilty! Is it necessary, for example, to organise classes on the basis of sex divisions? — 'Two lines please — boys and girls' or 'Boys on the left of the class, girls on the right'. It is also worth recognising, however, that stereotyped views of the opposite sex are often held firmly by adolescents themselves. Many boys and girls are uncomfortable working together in pairs or small groups. To deliberately mix pupils may cause more problems than it solves (Barnes and Todd, 1977).

The aim would appear to be to attempt to involve the girls more in classwork not by coercion but by fostering the kind of environment where girls feel confident to contribute equally. It may be helpful to deliberately raise awareness of sexism amongst pupils. Cartoons, jokes and newspaper cuttings are good starters. 'When I grow up I will be a housewife and could not care less about the length of the Rio Grande' writes a little girl in an answer to her geography test in a 'Peanuts' cartoon. Differences between boys' and girls' attitudes towards what constitutes a safe environment is an issue requiring sensitivity, however, although open discussion of feminism in the classroom is to be encouraged.

The treatment of sex bias and sex differentiation in the geography curriculum is problematic but it is worth remembering that sex discrimination is unlawful as well as undesirable. It would be illegal, for example, for a department to exclude boys or girls from say, a field course solely on the grounds of sex. It is also worth remembering that the geography department cannot go it alone. Reducing sex bias requires a 'whole school' approach and policy. But a start has to be made somewhere. Perhaps the most important first step is the recognition that a problem exists.

References

Bale, J. (1982) 'Sexism in geographical education', in Kent, W.A. (ed) *Bias in Geographical Education*, University of London Institute of Education.
Barnes, D. and Todd, F. (1977) *Communication and Learning in Small Groups*, Routledge and Kegan Paul.
Boardman, D. and Towner, E. (1979) *Reading Ordnance Survey Maps: Some Problems of Graphicacy*, Teaching Research Unit, University of Birmingham.
Delamont, S. (1980) *Sex Roles and the School*, Methuen.
Larsen, B. (1983) 'Geography', in Whyld, J. (ed) *Sexism in the Secondary Curriculum*, Harper and Row.
Lynch, K. (1977) *Growing Up in Cities*, MIT Press.
Maccoby, E. E. and Jacklin, C. N. (1974) *The Psychology of Sex Differences*, Stanford University Press.
Marland, M. (ed) (1983) *Sex Differentiation and Schooling*, Heinemann.

Monk, J. J. (1978) 'Women in geographic games', *Journal of Geography*, vol 77, pp 190-2.

Murphy, R. J. L. (1982) 'Sex differences in objective test performance', *British Journal of Educational Psychology*, vol 52, pp. 213-19.

Slater, F. (1983) 'Sexism and racism. Parallel experiences: an exploration', *Contemporary Issues in Geography and Education*, vol 1, pp 26-31.

Spender, D. and Sarah, E. (eds) (1980) *Learning to Lose: Sexism and Education*. The Women's Press.

Sutherland, M. (1981) *Sex Bias in Education*. Blackwell.

Tolley, H. (1984) 'New directions for project work in A-level geography', in Orrell, K. and Wiegand, P. (eds) *Evaluation and Assessment in Geography*. Geographical Association.

Whyld, J. (ed) (1983) *Sexism in the Secondary Curriculum*. Harper and Row.

Wiegand, P. (1982) 'Objective testing in geography at 16+,' *Geography*, vol 67, pp 332-6.

Wittig, M. A. and Petersen, A. C. (1979) *Sex Related Differences in Cognitive Functioning*, Academic Press.

Wright, D. R. (1985) 'Are geography textbooks sexist?' *Teaching Geography*, vol 10, pp. 81-4.

9.4. Pre-vocational Education

Clive Hart

Introduction

It is a reflection of the recent arrival of pre-vocational education that it has yet to attract a precise and positive definition. Instead, there are assorted descriptions of the target groups for which it is intended, various statements of aims by which it is characterised, and several proposals for actually implementing programmes of pre-vocational preparation in schools and colleges. But while the initial lack of pre-conceived constraints and tightly specified objectives is perhaps no bad thing, there is uniform agreement among its proponents that pre-vocational education is decidedly a cross-curricular undertaking with no room for the vested-interest teaching of single subjects. Thus the introduction of pre-vocational provision presents a challenge to established patterns of teaching. It also offers some new opportunities for teachers and learners alike.

In this section the role of geographical education within programmes of pre-vocational preparation is examined and certain suggestions are made. These suggestions are based on the general nature of pre-vocational education as it has been advanced in key documents such as *A Basis for Choice* (FEU, 1979). In this connection, the reader is asked to bear in mind that although formal pre-vocational preparation is likely to take place under the auspices of the nationally-validated Certificate of Pre-Vocational Education (CPVE) for students in the 17+ age category, and initially within the TVEI (Technical and Vocational Education Initiative) pilot scheme for younger pupils, it is not confined exclusively to these two channels. The ideas presented here are intended for implementation by teachers of geography within any pre-vocational context. Accordingly, they may need tailoring to suit the requirements of specific courses leading to the award of recognised certificates or profiles of student performance. Hence, the focus of this section is not on particular courses, but on a strategy for geographical education as a medium for pre-vocational preparation.

Most work on the purpose and nature of pre-vocational education has been published in connection with provision for young people in the 17+ age group, and a list of helpful references will be found at the end of this section. The increase in the number of students of this age embarking on full-time post-compulsory education has risen dramatically since the early 1970s, and has finally focused effective attention on the need for appropriate courses for them.

The description of the target group prepared by the Joint Board for Pre-Vocational Education (established in 1983) provides a working indication of who and what is involved in pre-vocational preparation at 17+. The target group consists of young people:

— who after completing compulsory schooling, have chosen, and may benefit from, full-time education in a school or college as preparation for adult life, including the world of work;

— who are not wishing, at this stage, to proceed to A-level study;

— who are interested in progressing to specific vocational education, training and/or work, but may or may not be yet committed to a particular career or occupation, and may or may not have the qualifications to further their progress towards it.

This description clearly emphasises the work-related orientation of pre-vocational education, but it also draws attention to preparation for adult life. To a large extent, the target group identified by the Joint Board concurs with that recognised four years earlier in *A Basis for Choice,* but with the important omission of a direct reference to ability. Thus, pre-vocational preparation through the CPVE is not intended purely for those students who have been traditionally classified as intellectually less able. Instead a substantial range of ability, experience and aspirations is likely to be encountered. However, for as long as qualifications in academic subjects continue to retain their superiority as passports to an extremely large number of occupations at all levels, it is likely that pre-vocational courses will tend, in the main, to attract students who neither wish to proceed to a higher level of academic study, nor who would be likely to succeed if they were to do so.

The contribution of geography

The impact of *A Basis for Choice* on the philosophy of pre-vocational education has been so great that it has, in effect, provided the underpinnings for it. Essentially, the document proposes that pre-vocational courses should comprise three identifiable entities, namely, core, vocational and job-specific studies. The core (or more usually 'common core') is defined as 'those studies to which all students of this age should have right of access, and that learning which is common to all vocational preparation'. Further, it is recommended that core studies should occupy a major proportion of the time available, preferably in the order of 60 per cent. Vocational and job-specific studies are each envisaged as absorbing 20 per cent of course time and are defined respectively as that learning which is particular to a given vocational sector and that which is particular to a given job. All current

---

1. To bring about an informed perspective as to the role and status of a young person in an adult society and the world of work.

2. To provide a basis from which a young person can make an informed and realistic decision with respect to his or her immediate future.

3. To bring about continuing development of physical and manipulative skills in both vocational and leisure contexts, and an appreciation of those skills in others.

4. To bring about an ability to develop satisfactory personal relationships with others.

5. To provide a basis on which the young person acquires a set of moral values applicable to issues in contemporary society.

6. To bring about a level of achievement in literacy, numeracy and graphicacy appropriate to ability, and adequate to meet the basic demands of contemporary society.

7. To bring about competence in a variety of study skills likely to be demanded of the young persons.

8. To encourage the capacity to approach various kinds of problems methodically and effectively, and to plan and evaluate courses of action.

9. To bring about sufficient political and economic literacy to understand the social environment and participate in it.

10. An appreciation of the physical and technological environments and the relationship between these and the needs of man in general, and working life in particular.

11. To bring about a development of the coping skills necessary to promote self sufficiency in the young people.

12. To bring about a flexibility of attitude and a willingness to learn, sufficient to manage future changes in technology and career.

Figure 9.8. A basis for choice: core aims. Source: Further Education Unit (1979).

initiatives concerned with constructing pre-vocational courses tend to reflect this division of purpose in one way or another, with the core being central and dominant. As far as the contribution of geography to such courses is concerned, it is in the areas of core and vocational studies that most opportunities present themselves; a contribution to job-specific studies perhaps not being a particularly realistic proposition.

The core

A proper understanding of the nature of core studies recommended for pre-vocational courses is best acquired from consulting *A Basis for Choice* itself. Here, it is sufficient to note that the ABC core comprises a series of twelve aims and attendant objectives which are expressed in terms of both the observable performances to be expected of students, and the learning experiences which they should be offered. The key feature of the core, then, is that the aims are not the preserve of any single discipline, or discrete groups of disciplines. On the surface, the aims bear little relationship to traditional curricular divisions, and may even be seen as the potential agents of erosion of single-subject teaching. They are listed in Figure 9.8.

Where does this leave geography and its teachers? The answer is twofold.

Firstly, it lies in the capacity of modern geographical education to satisfy those educational aims of a generally applicable and transferable nature. Secondly, it depends on the willingness of its teachers to accept the loss of single-subject autonomy and to co-operate in the teaching of a core curriculum which has been specifically designed to fulfil the sorts of needs of the target group as outlined previously. In reality, this dual requirement is not as problematic or radical as it may seem on first acquaintance.

The features of geographical education which lend themselves to core teaching and learning derive largely from the strategies which are increasingly employed in the teaching of the subject as an independent discipline. Many of these strategies are of an experiential nature and their use is facilitated by work that is orientated towards guided student enquiry. Data collection and processing, role play, simulation, problem-solving and decision-making exercises are relevant examples of activities which provide a most appropriate context for the achievement of many of the objectives specified in *A Basis for Choice*. The Geography 16-19 Project, in conjunction with a group of teachers from Sheffield, identified approximately forty objectives which could be satisfied through the teaching and learning techniques that have become an everyday part of modern geographical education. In addition, the same study recognised that certain topics which occur in many geography syllabuses could also act as suitable vehicles in the achievement of some of the ABC objectives, especially where prescribed circumstances such as 'the local area' were involved. A list of the teaching strategies and topics which figure prominently in the Sheffield study is shown in Figure 9.9.

A document from the FEU entitled *Skills for Living* (1982) shows that enhancement of many basic skills and personal competencies can be assisted if teaching and learning is orientated towards encouraging active student enquiry. Such a strategy demands considerable student involvement in the process of learning, and provides an authentic framework for skill acquisition and improvement across a range of circumstances. From a geographical education viewpoint, this claim may be illustrated by reference to students engaged upon a field enquiry. Through this work they may gain experience of how to recognise a problem for investigation, of how to organise the collection and analysis of data, and of how to evaluate and communicate their findings. Group tasks developing social skills are likely to feature in such work, and associated problem-solving and decision-making may assist in improving their capacity to tackle problems methodically and effectively. Quantification, map and diagram interpretation, and the representation of statistical data may also be involved — all within the context of one coherent investigation. In many occupations, all these skills would be valuable.

Vocational studies

Vocational studies concern that learning which is particular to a given vocational sector. *A Basis for Choice* emphasises that pre-vocational courses should help students 'to develop an informed orientation while keeping their employment and further education options open'. Informed orientation would appear to be the central element of this statement, and it is an aspect of vocational studies to which geographical education can make a most valid contribution. It is interpreted here as

STRATEGIES

— Role play

— Decision-making

— Data collection and recording (primary and secondary)

— Measurement techniques — surveying; map-making

— Values enquiry

— Construction of measuring apparatus for fieldwork

— Aesthetic assessment — building and landscape evaluation

— Oral questionnaires

— Simulation

— Studies of perception

— Quantitative techniques

— Analysis of statistical diagrams

— Analysis of Press articles

— Enquiry into local issues

— Problem-solving

— Analysis of conflict situations

— Conducting surveys

— Photograph analysis

TOPICS

— Employment structure

— Industrial classification

— Spatial analysis (of job opportunity)

— Regional disparity

— Spatial justice

— Various environmental conflicts

— Energy supplies

Figure 9.9. Selected teaching strategies and topics
Source: Geography 16-19 Project (1981).

referring to, or encompassing, knowledge about employment opportunities and about the wider environmental characteristics and demands of employment and an adult working life.

How informed orientation may be attained with the help of a geographical input is almost totally dependent on the adoption of an approach to the subject that is relevant to this facet of pre-vocational preparation. Clearly, any old geography will not do, and it is also worth noting at this stage that attempts should not be made to isolate vocational studies from core studies. It is the development of the skills and abilities advocated in the core that provide the necessary tools with which to become orientated. In this manner vocational studies offer a valuable context for the achievement of core aims, and this connection between them should not be overlooked.

What is required is an approach to informed orientation which allows geographical knowledge and understanding to play a part which is both meaningful to, and seen to be meaningful by, the target group. Such an approach needs to be topical, not distanced from the observed real world of everyday life, and not developed from a hypothetical or highly theoretical point of origin. Knowledge about employment opportunities, and about the wider environmental characteristics and demands of employment and an adult working life, has an immediacy about it which, in the interests of motivation, should be maintained at all costs.

An effective point of entry into vocational studies for geographical education is provided through the approach to geography advocated by the Geography 16-19 Project. This approach focuses on the inter-relationships between people and their various environments and takes as its starting point enquiry into questions, issues and problems of relevance in the world today. In the case of contributions to vocational studies, the opening up of questions and issues related to the 'environment' of the world of work and the characteristics of a working life would seem to present a most apposite platform from which to work. Questions and issues may be selected according to several criteria (eg. the local situation), but they should all be relevant to the aim of developing informed orientation. For example, in the context of a particular industry, questions might be concerned with its changing character and status; issues with its location and the effects of possible relocation; and problems with any adverse impacts it may have on the environment and quality of life in adjoining districts. In depressed areas, the potential effect of positive initiatives may be examined. The creation of enterprise zones and the introduction of other measures designed to attract new employment opportunities would provide relevant substance for study for many young people.

How work of this nature may also satisfy core aims can be demonstrated by reference to ABC Aim 9. A study of appropriate aspects of regional policy would involve students in a consideration of economic and political issues, and in a discussion of the values and beliefs behind those issues. Many decisions about the use of space are strongly influenced by certain value systems so that simply as part of explaining locations, distributions and patterns, it is necessary to understand the significance of those systems. From the point of view of understanding the social environment this question and issue approach to geography can thereby make a telling contribution.

At the local scale, too, study centred on questions and issues has some important advantages. The immediate locality can be used to stimulate interest, and it can also act as a source of real information, appropriate contacts and helpful resources. Enquiry into local matters regarding employment can thus provide a direct means of helping students to broaden their basis for occupational choice, even if it does lead to a decision to look for work elsewhere. Finally, it should be remembered that contributions to vocational studies from geography will not take place in isolation from activities such as careers guidance, visits to employing organisations, job-specific studies, and so on. Awareness among all contributing agencies of what each has to offer will be important in preparing a balanced and coherent programme of vocational orientation.

Resources for the geographical contribution

Space does not permit a detailed review here of all the resources appropriate to the role of geography in pre-vocational education, but it should be noted that few (if any) resources will have specific relevance for only one contributing agent. Figure 9.10 is an example of the type of resource which is easily acquired and which may well catch the geographer's eye. In connection with the ABC common core, it could be used in relation to the

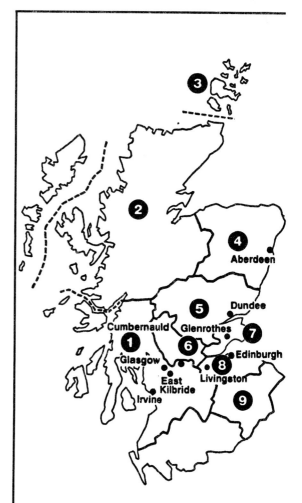

Numbers of Electronics Companies in Scotland —

by city or town:		by region:*	
Cumbernauld	6	(1) Strathclyde	62
Glasgow	24	(2) Highland	9
E. Kilbride	12	(3) Islands	1
Irvine	2	(4) Grampian	5
Livingston	14	(5) Tayside	7
Edinburgh	10	(6) Central	1
Glenrothes	21	(7) Fife	16
Dundee	7	(8) Lothian	13
Aberdeen	12	(9) Borders	13
		* in addition to those in the named towns	

Source: Scottish Development Agency, 1982.

Britain's Silicon Glen still growing

Bill Johnstone talks to Mr. George Younger, Secretary of State for Scotland, on Scotland's electronics success

Question: more people are employed in the high technology industries in Scotland than in the traditional industries like shipbuilding and steel making. Most people would find that surprising. What effect is it having on the level of employment in Scotland?

Answer: We have roughly 40,000 people employed in the electronics industry in Scotland. That's direct employment, not ancillary employment. That is more than shipbuilding and more than steel. We have been building up a very prestigious electronics industry in Scotland for quite a number of years now and we have got a collection where pretty well all the big names in the electronic world are represented in Scotland. Taken as a whole, I believe the Scottish electronics industry is second only to California as a concentration of high technology industry.

Presumably the person who is employed in the electronics sector is not the same type of person who would have been employed in the other types of traditional industries?

Well, there has been a lot of re-training and we have found that a lot of Scottish labour retrains extremely well. For instance in some of the mining areas miners have re-trained for electronics industries.

But in the electronics industries a substantial proportion of the manual labour force tends to be made up of women.

It is true that there is a predominance of women who go in for assembly or sub-assembly work. They are naturally suited for that kind of work. But there are quite a lot of men too. It is part of the reason I suppose why the unemployment rates haven't been quite as bad for women as they have for men.

What sort of reputation has Scotland got among the companies which you would wish to attract?

Outstandingly good. I've been over to California a few times and Scotland is the place that they are all talking about for a European base now.

Our Locate in Scotland organization has an office in San Francisco. We have a prestigious committee of Scots in northern California who can open doors for us.

I think you'd find, it is fair to say, that in Silicon Valley the place they are talking about for a European base is Scotland.

Presumably they are also attracted by the grants and the loans?

I think it helps. Most of them say its a very peripheral part of their decision making. The things that they really like are the environment, training and university links.

What are you doing to ensure that the academic world provides industry with the type of high technology or electronics graduates which it will need?

We have eight universities in Scotland but we've backed that up through our own central institutions which are the equivalent of polytechnics in England, funded by us through the Scottish Education Department. We have put a lot of money into microelectronics training. The Manpower Services Commission together with my department did a study of the supply of manpower to the electronics industry and they've taken steps to remedy some of the shortfalls which they found.

For instance the number of Tops awards for electronic technicians have been substantially increased. Then we have also set up information technology centres to provide the young unemployed with an opportunity to learn basic computing skills. Two of these centres are already open and a further 14 are planned.

The Engineering Industry Training Board has diverted its resources towards electrical engineering training, particularly in Scotland. We have a number of government schemes for giving assistance to innovation including funding to assist companies in re-training work forces in new technology.

Do you think you could ever get to the stage where the rate of job creation in the high technology sector is higher than the rate of loss from the conventional industries?

I hope we shall get to that stage but it won't be quickly. The number of jobs that we have been losing has been outstripping the number of jobs we've been gaining by quite a lot. That's why the unemployment rate has been going up, although its an interesting fact that this is the first recession that any of us can remember when unemployment rates in Scotland have not risen as fast as they have in the UK as a whole.

Part of the reason for Scotland feeling the bite of any recession was that a number of Scottish factories were annexes of American companies? That could still be said to be the case? They are still annexes, this time of American and Japanese companies. Scotland could still catch a chill.

It is certainly true that in recession you do get the big multi-nationals that cut back. But the great majority of our recessionary job losses have been big, rather old-fashioned Scottish concerns. Nevertheless, there is a truth in saying that the multinationals cut back and we've had our share of them.

But we have through this recession had a huge balance in our favour from the multinationals, because right through these grim times our multinationals field have been expanding. Although we've had a few that have gone we have had many more that have expanded.

There are plenty of examples. National Semiconductors has just completed a huge expansion, IBM has just announced another expansion, Motorola is in the middle of its third expansion, General Instruments announced its expansion a few months ago, Hewlett Packard has just completed an expansion and the list goes on.

Does the fact that General Instruments cut back its staff by 25 per cent concern you. Do you think that this could be a trend?

No, that was not unexpected. It was a small part of their concern which everyone knew was getting out of date and it wasn't part of the new project and the people concerned are going to be offered new jobs back.

Has the announcement of those redundancies and those at Times and the dispute there tarnished the reputation of Scotland as a home for high technology?

I hope not. I was particularly worried about it for that reason because it hit the headlines and it gives the impression that this is a general picture in Scotland. The general picture is that everyone is entirely satisfied.

Figure 9.10. A newspaper article as a resource. Source: *The Times*, January 24, 1983.

objectives specified under Aim 10. With reference to the proposals made above for a geographical contribution to vocational studies, it raises questions about:

— the job attraction-power of different regions and sub-regions.

— regional and sub-regional disparity in job opportunities in new industries.

— trends in employment structure and the effects on employment levels.

— the role of financial incentives on new industry location.

— the impact of multi-national companies on regional economic health.

— the role of women in the workforce.

— the transferability of skills from one occupation to another.

In addition, coupled with supplementary data on population, employment and unemployment rates, grant aid, market location, etc., a number of data analysis exercises could be devised around the information presented on the map. These exercises might be designed to help with some of the objectives relevant to numeracy listed under ABC Aim 6.

In all cases resources should be selected and handled with care. Figure 9.10 is a demanding item, but one which has great potential for extending student understanding. Given the nature of the target group, few worthwhile resources are likely to be suitable for unguided student use, and most will require joint student-teacher exploration.

Conclusion

A publication by the FEU entitled *Common Core Teaching and Learning* (1984) draws attention to the fact that pre-vocational education, and the common core in particular, invariably presents problems to teachers whose teaching has been primarily related to conventional subject syllabuses. The emphasis of the publication is on ways in which an institution and its teachers may adopt an increasing commitment to an integrated core via the processes of matching, modification, extension and assimilation of existing single-subject syllabuses. Guidance is given on this important aspect of contributing to pre-vocational education, and on many other aspects of common core teaching within a pre-vocational framework. For teachers of geography who are becoming, or likely to become, involved with pre-vocational provision, this FEU document is strongly recommended.

Apart from adaptation to the demands of pre-vocational preparation, there are educational implications of a more fundamental nature to be coped with. Vocationally-orientated study inevitably represents a response to the labour market and the demands of employers. Constant vigilance will be required to prevent vocational training from becoming the dominant element in *pre*-vocational preparation, especially with the spread of vocationally-orientated initiatives into the last two years of compulsory education. This is an area of real concern because it weakens the contribution of education towards improving the quality of life, and undermines the role it can play in introducing necessary changes into society. It is to be hoped that a geographical contribution, particularly to vocational studies on the lines suggested here, will help to counter this potentially damaging development.

References

DES (1982) *17+ A New Qualification*, HMSO.
FEU (1979, 2nd edn. 1982) *A Basis for Choice* (The Mansell Report), HMSO.
FEU (1982) *Skills for Living*, Further Education Unit.
FEU (1984) *Common Core Teaching and Learning*, Further Education Unit.
Geographical Association (1982) *The Contribution of Geography to 17+ Courses*, Geographical Association.
Geography 16-19 Project (1981) *Geography and Pre-Employment Courses in the Sixth Form*, Geography 16-19 Project.
Geography 16-19 Project (1982) *The Geographical Component of 17+ Pre-Employment Courses*, Geography 16-19 Project.
Hart, C. R. (1983) 'Putting geography to work at 17+', *Teaching Geography*, Vol. 8, No. 3, pp. 112-3.
Hart, C. R. (1984) 'Common core — no cause for concern', *Teaching Geography*, Vol. 9, No. 4, pp. 148-50.

10. Framework Fieldwork

Clive Hart and Tony Thomas

In this chapter we propose a contextual framework for field study; a framework that strengthens and enhances the value of fieldwork in several respects, and makes it an essential, natural and active ingredient of all work in geography. We also show how a flexibility towards extending the scope of the subject is an important pre-requisite of meaningful fieldwork. By meaningful fieldwork we mean study that takes place largely in the field (but also including associated 'indoor' preparation and follow-up) and which seeks to find answers to pertinent questions about the many ways in which people interact with the various environments in which they live and work. This interaction is the key to the idea of a contextual framework for field study mentioned a few sentences earlier. The style of fieldwork it generates is termed *framework fieldwork* throughout the subsequent sections of this chapter, and discussion on its characteristics and demands is followed by a worked example of the approach.

The origins of framework fieldwork

Broadly speaking, fieldwork has evolved in such a way that the balance of activity has shifted in favour of increasing student involvement and participation. Again in broad terms, a transition from passive field teaching and eye-balling of the landscape (Figure 10.1) towards technique-orientated field research and measurement (Figure 10.2) can be recognised; this transition having accompanied the rise in popularity of an increasingly systematic and quantified form of geographical education. However, like quantitative techniques themselves, the purpose of research-mode fieldwork based on a divisive systematic treatment of the subject has come under close scrutiny and increasing criticism. In short, its long-term educational value is being questioned and, as it happens, as much by students as anyone else. Blunt but telling questions like 'What are we doing this for?' carry disturbing implications for the image and well-being of the subject and present a challenge we ignore at our peril. As Pocock (1983) has pointed out, 'the quest for a more scientific approach has meant that characteristic activities today concern the measurement of stream flow, delimiting zones of urban land use or eliciting customer orientation, the results then being subjected to a model-testing approach where goodness of fit is often the measure of success'. Much, therefore, hinges on the nature and quality of those models, and it does not need stressing here that many models perpetually featured in school geography bear little resemblance to the everyday observable world; a world which, in most instances, begs to be explored and examined at first hand 'in the field'.

In recent years valuable attempts have been made to introduce a more accurate human dimension into fieldwork, particularly in the realms of topics embraced by the human branch of the subject. These attempts have drawn on the increasing recognition that people are an inseparable part of geography, and that their responses to, and opinions about, the environments they encounter are an important (if not central) element of the observed world which they inhabit. Thus the de-humanising role of much systematic pseudo-scientific human geography has been reversed by a concern for matters humanistic and by a realisation of the significance of human subjective assessment.

Such developments are becoming the focal points of more and more fieldwork, often under the

Figure 10.1. Eye-balling of the landscape.

Figure 10.2. Technique-orientated field measurement.

guise of perception studies or, more recently, environmental engagement. Admirable and welcome as this humanistic trend may be, it is quite easy to envisage it becoming just another field activity employed largely in its own right and for its own sake. In other words, a student-centred activity to be tried out, appreciated, and then shelved with little chance of ever being used in a wider context applicable, say, to understanding the quality of life in a given area.

Physical geography, too, stands to benefit from a more people-related approach. With the possible exception of landscape appreciation, physical studies in the field have focused primarily on examining and appreciating the role of processes in the creation of form. The element of human concern over the functioning and management of the natural environment has not been a prominent feature of field studies in physical geography, and yet it provides an exciting opportunity for enhancing student awareness that natural systems do pose challenges for people, and that people themselves are frequently an important component of those systems. Few would deny that at school level detailed physical studies need a human interest context in order to make them meaningful and more accessible to student comprehension.

It should be stressed at this stage that the desirability of research-mode fieldwork *per se*, and the value of personal immersion in a particular topic, are not being challenged here. What is being questioned is whether or not any fieldwork organised solely on a systematic basis around separately identifiable components of the subject can ever provide learning experiences which are strikingly worthwhile, and which can clearly be seen to demonstrate the worth of geography as a contributor to environmental understanding and as a medium for detecting when and where action is needed to change the observed environmental conditions.

Hence, in their simplest form, the origins of framework fieldwork stem from an attempt to counter field study which has been over-directed by exclusive topic orientation and excessive technique enslavement. At a more sophisticated level, there lies behind framework fieldwork a quest for an approach to geography which seeks both to emphasise the significance to us all of people-environment interactions, and to explain the nature and relevance of geographical ideas and concepts by references to these interactions. Framework fieldwork is thus an integrative vehicle for geographical study.

The characteristics of framework fieldwork

The framework for the approach to field study that we are proposing here consists of the three basic parameters shown in Figure 10.3. Pre-field preparation and appropriate background data enter the frame as required, while the main outputs comprise the findings of the fieldwork, an element of personal experience and, after due evaluation, answers to the questions which initiated the field inquiry in the first place. Each of the fundamental sides of the frame are now described in detail.

People-environment interactions

The nature of interaction between people and the various environments in which they live tends not to be immediately apparent in many cases. In effect, environmental interaction is the essence of everyday life and is therefore easily taken for granted. However, when events occur which threaten to upset established patterns of activity and accepted ways of doing things, the character and strength of interaction is brought to the fore. This is as true of the natural environment as it is for the social environments created by human endeavour. For example, coastal erosion may destroy property and valuable amenity space, whereas by contrast, the proposed redevelopment of an inner city district may create concern among the residential population over future housing provision.

In all cases the changing relationships between people and their environmental circumstances will generate questions, issues, problems and challenges requiring resolution in one way or another. Accordingly, these questions, issues and problems become the focus of concern and require close examination if a solution is to be found. The fieldwork component of this examination is concerned with finding out the details of the topic under study through 'outdoor' contact with, and investigation of, all the relevant aspects. Such an investigation means formulating and posing secondary questions which are relevant to the central question or issue, but which primarily guide and fashion fieldwork at the practical level. These secondary questions must be pertinent to the central issue if they are to be worth trying to answer, and absorbing if they are to motivate the

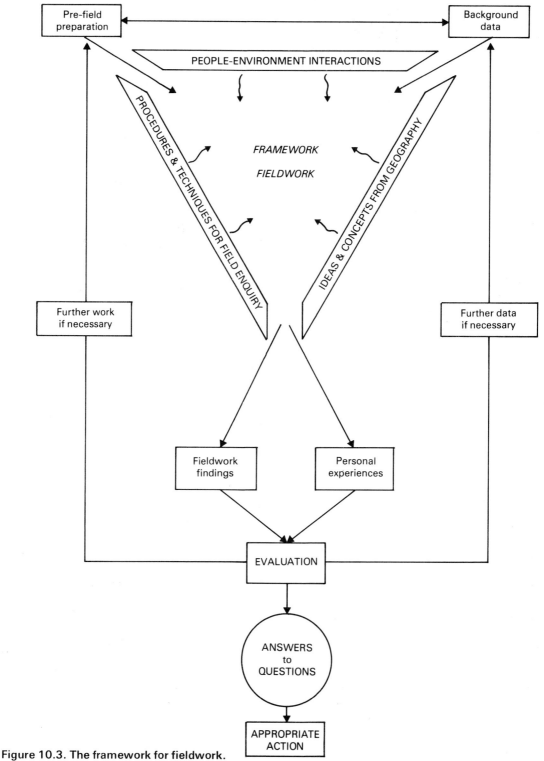

Figure 10.3. The framework for fieldwork.

students attempting to answer them. They must also be sufficiently rigorous if the answers to them are to make a valid contribution to knowledge and understanding of the central question or issue itself. Only in the field can the most effective efforts be made to examine and explain the active geography of everyday life; framework fieldwork seeks to do this through focusing attention on questions, issues and problems of a people-environment nature.

Figure 10.4 summarises how framework fieldwork differs from field study that is designed around a fundamentally systematic approach to the subject. Both parts of the diagram are concerned with coastal processes and morphology, but the framework fieldwork version on the right places immediate significance on the interest of people in the coast and its management.

Other features of framework fieldwork are also apparent from Figure 10.4. The most prominent of these is the recognition that environmental questions and issues have both factual, and opinion or values orientated facets. Both these facets require examination if some form of valid decision or conclusion is to be reached as a result of the work carried out in the field. For further guidance on implementing this dual approach the reader is referred to the route for geographical enquiry proposed by the Geography 16-19 Project and reproduced in Chapter 3 of this volume. The nature of this route is largely self-evident, as is its relevance for structuring framework fieldwork. The similarity between this route and the right-hand part of Figure 10.4 is also apparent. The practical implications of adopting a route of this type as a basis for field enquiry are discussed next.

Procedures and techniques for field enquiry

The context for field study proposed in the preceding section firmly casts all techniques of investigation, the left-hand side of the framework, in the role of servant rather than master. This has the advantage that all field techniques, however traditional or innovative, have a potentially significant part to play, but without becoming the dominant activity and obscuring the purpose of the work itself. They are not ends in themselves, but purely the means by which information is gathered. Their main purpose is to supply answers to questions; that is, to support and satisfy the process of enquiry and not to direct it.

Framework fieldwork does not rely solely on single-function techniques of a closely defined and highly specific nature, such as the levelling of a beach profile. Rather, framework fieldwork requires the use of procedures which are aimed at establishing the nature and extent of people-environment interactions. Specific techniques may be incorporated into these procedures, but the emphasis is on integrated information gathering which can lead directly to an overall result or conclusion.

It will be apparent from Figure 10.4 that framework fieldwork requires that data are collected not only on the factual aspects of the topic under study, but also for the human values dimension. Both these components of framework fieldwork require examination if some form of decision or conclusion is to be reached, however tentative it may be. In general terms, the collection of information relating to the factual side is likely to be more straightforward in that it is dependent on techniques and procedures which are inherently objective and require measurement along scientific lines.

The values dimension demands a different approach to information collection, namely, one that is organised but which recognises the significance of subjective human responses to given situations. In itself this is not an easy task and, to date, it has not formed a significant element in geographical fieldwork for students of school age. As Pocock (1983) has pointed out, 'after two decades when university and sixth form students have been largely taught to avoid the subjective — often using personal pronouns at their peril — students may need encouragement to respond and to express themselves'. This quotation is a useful reminder that values related work is not only concerned with the views and attitudes of people actually involved personally with a particular issue, ie, those caught up in the social or political ramifications of the issue, but also with the opportunities that framework fieldwork provides for students themselves to clarify where they stand over a particular matter. Procedures for use in the field are thus needed which can encourage individual assessment of the prevailing circumstances, and also provide insight on the values and attitudes of those people who are involved at first-hand.

At the simplest (but not the least effective) level, interviewing members of the public and examining letters in the press can provide valuable information on matters of human perception, attitude

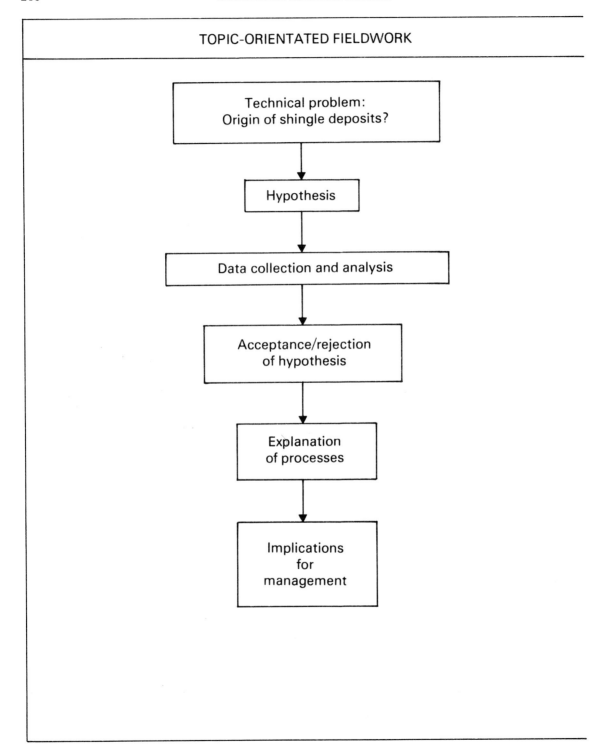

Figure 10.4. Two approaches to field study on coastal processes.

FRAMEWORK FIELDWORK

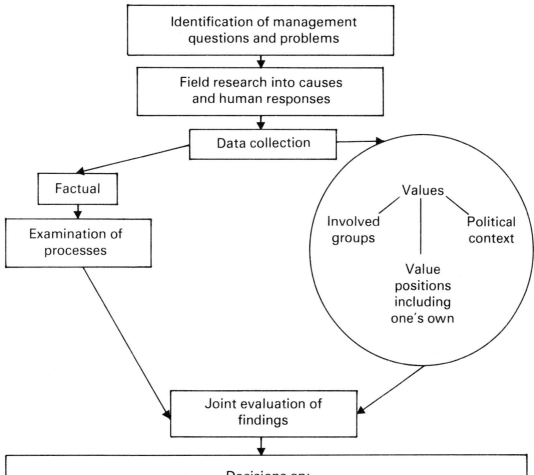

and opinion. The political context of a particular issue can also be discerned effectively in this way, and this approach might prove well suited to a clearly polarised issue where alternative courses of action are readily apparent. Alternatively, more complex approaches are possible, perhaps employing a technique designed to elicit responses from members of the public that do not overtly prejudice the line of enquiry itself. As an illustration, reactions to landscape change can be gauged using a semantic differential technique which enables students to assess their own responses to landscape attractiveness as well as those of others.

Another way in which field enquiry can take account of values-related aspects is for students to adopt the roles of people who are actually involved in a particular issue — whether actively or as interested onlookers. The purpose of this strategy is to help students develop an appreciation of the views and preferences of these people. The fieldwork may then perhaps culminate in a simulated public meeting with students speaking on the basis of their role-related field findings. For further details of this approach the reader is referred to 'Reservoir Site Evaluation' in Hart (1983). Further ideas on implementing this aspect of framework fieldwork may be found in recent volumes of *Geography, Teaching Geography,* and the *Bulletin of Environmental Education.*

Ideas and concepts from geography

Ideas and concepts from geography comprise the right-hand side of the framework. They should be seen as guides whereby the subject acts to direct enquiry as a whole, and to help in the exploration of particular lines of enquiry. The specific character of generally recognised processes may be investigated and reference to models of land use (both urban and agricultural) may help in developing an appreciation of how particular environmental issues and problems have arisen. Concepts of system response, dynamic equilibrium, distance decay, spatial justice, deprivation, the quality of life, spatial interaction, and so on, can all provide organising vehicles for posing questions about people-environment interactions. As all these concepts and ideas are well documented in school texts they are not explored in detail here, but the reader should also be aware of the potentially valuable contribution that other subjects can make to framework fieldwork. This is especially true of biology, ecology, economics amd sociology.

Who is framework fieldwork for?

We hope by now that the reader will appreciate that framework fieldwork provides a context for field study which is intended to validate fieldwork for all students of geography whatever their precise interest and level of study at the time.

Thus framework fieldwork can provide a common strategy for the whole of a department's fieldwork, with students participating according to age, ability, experience and personal competence. While older students may be able to cope more readily with all the aspects of a particular question or issue, the work of younger pupils may be orientated towards the collection of that data which is more accessible and more easily acquired through relatively straightforward measurement techniques. However, this 'lower level' work does not remain isolated and without apparent purpose and meaning. Under framework fieldwork, it would go forward to help answer the central question under study — the very question that initiated the work in the first instance.

For example, imagine that a by-pass has been proposed for a local village. For those likely to be affected adversely by this suggestion, two key questions arise. These are, do we really need a by-pass and, if so, where will it go? To answer the first question, one requirement would be a study of present traffic flow through the village. This is a comparatively straightforward task easily managed by younger pupils. In addition, the environmental impact of present traffic movements, and an assessment of likely future impacts on areas adjacent to the new by-pass would be required. This work, and that related to the values and political aspects of the by-pass proposal, are perhaps better undertaken by older students whose experience of the subject and ability to enquire in an orderly way are more highly developed. Combined with other lines of enquiry, work in geography for all ages can be given meaning and demonstrated to contribute to environmental understanding. All that is required is the exchange of each group's findings in order to provide the complete picture.

An important aspect of framework fieldwork which has not yet been mentioned is its ability to provide an appropriate vehicle for the achievement of those general educational objectives which underpin many of the new inter-disciplinary pre-vocational courses. Through the selection of questions and issues relevant to the particular interests of a given target group of students, the different facets of framework fieldwork can be used to enhance a wide range of personal skills and competencies. Opportunities for the application of the principles of framework fieldwork outside the single-subject geography context are on the increase and include courses run under the auspices of the MSC, B/TEC and the Joint Board for Pre-Vocational Education (CPVE). Further opportunities arise within other subject areas such as biology, economics and sociology. In all cases framework fieldwork provides opportunities for experiential learning which are not artificially contrived or unduly hypothetical.

Framework fieldwork in action

Implementing framework fieldwork depends essentially on the recognition of people-environment interactions in a variety of contexts and at differing scales. Most readily detected and investigated are those interactions which become the focus of controversy and may comprise an issue at the local or regional scale. Such issues receive considerable public exposure in the press and allow secondary questions to be posed and avenues of investigation to be formulated. However, not all questions for field study have to reach public issue proportions. Questions which might occur to geographers are equally valid even if they are not in the limelight. An example of this situation might occur where agricultural land use policy is changing in a particular district with implications for the appearance of the landscape. Recognising the signs of change and investigating the causes would represent the examination of an important geographical question, but one which had yet to attract local public indignation. For a detailed review of the opportunities provided by the study of local issues, the reader is referred to *Local Issues and Enquiry-Based Learning* (Rawling, 1981).

An important aspect of framework fieldwork is the need for supporting secondary data. It remains a weakness of much present field study that it is carried out in isolation from the general passage of events. The risk arises that conclusions may be based on a severely restricted sample of information and this is even more the case with framework fieldwork where the study of people-environment interactions often involves the investigation of impacts. Such impacts are rarely of a short term nature and any data collected over a necessarily limited period needs to be viewed against a longer time span. This requirement is even more essential for work involving landforms and natural systems, but it should not be regarded as a necessary evil. It can, in fact, be the means whereby work in the classroom is initially integrated with field studies yet to come; the fieldwork component being the natural extension of familiarisation with background data, as shown in Figure 10.3. It may also be the foundation for links with other subjects. Real questions and issues are never the sole province of one academic discipline, and as was claimed earlier, subjects such as biology, economics and sociology may have important contributions to make in raising the quality of explanation in geography.

Without doubt framework fieldwork has certain implications for residential field courses and for any centres which may provide them. Under a systematic approach to geography, locations for extended field visits are often chosen for their ability to provide specific opportunities for study, ie., where the raw material for study is instantly accessible the moment one steps out of doors. Framework fieldwork, by contrast, requires advance information on questions, issues and problems, as well as background information on them. These requirements present field centres with a new role — one which obliges them to keep up to date with the full range of local environmental questions and issues so that visiting parties may have the support of a sound data base on which to plan the detailed nature of their work. Centres may also turn this task to their own advantage as local questions, issues and problems can form the basis of very effective in-house courses on geography. Such courses impart both a sense of place and a sense of realism.

Finally, framework fieldwork must be followed up rigorously. Most issues are multi-dimensional and individually students cannot be expected to cover all facets. A degree of specialisation in the field will invariably be required and this makes it essential that all aspects are brought together effectively at the end so that serious attempts may be made to answer the central questions posed. It is also the time to review each student's experience of the issue and for them to move towards a clearer picture of where they stand over it. One way of achieving this de-briefing might be through an organised role-play simulating, for example, an appropriate form of public meeting. Whatever form the drawing together of threads may take, framework fieldwork can only achieve its full potential if the findings are used to come to a conclusion about the subject under study, and to decide on any action which may be necessary.

Framework fieldwork: an example

'Devon village battered by mountainous seas'

— Headline in *South Hams Gazette*, 3 January 1979.

The focus in this example is the problem which arises from the interaction of human activity at the marine-coastal interface. The storm which affected Start Bay between 31 December 1978 and 1 January 1979 had a major impact along sections of the coast and posed certain questions.

Why did this storm have such a devastating effect?

Which part of the Bay was worst affected?

Has it happened before?

Will it happen again?

In order to answer some of these key questions the relationship between human activity and the coastal system should be scrutinised. Every effort should be made to utilise the students' existing knowledge and experience and/or develop ideas and concepts from their geographical background.

Coastal advance and/or retreat only becomes a problem when it threatens human activity: settlement, roads, agriculture and recreational areas. The problems which arise can be investigated by concentrating on the linkages between individual components of the physical system and aspects of human activity. It can be useful to adopt a working hypothesis to give a focus to the investigations, for example:

> that the physical and the human activities within Start Bay render coastal engineering works unnecessary.

Such a hypothesis can be tested by establishing a set of objectives which provide the framework for field enquiry; this is shown in Figure 10.5.

Framework fieldwork is student orientated with the teacher cast in an array of roles which will vary with each investigation and depend on the students involved and the stage of the enquiry. Those roles may encompass those of archivist, technician, specialist, graphic designer and occasionally geographical pedant. In this example it is essential that students are familiar with the general geomorphic processes at work on the coast: longshore drift; the characteristics of waves; fetch; the influence of features responsible for wave refraction; beach profiles, surveying, graphical representation and component nomenclature; and the characteristics of emergent and submergent coastlines. Students also need a grounding in settlement location and patterns of settlement growth.

At the specific level a detailed knowledge of the study area will allow a greater appreciation of the interacting factors: coastal geology and landforms; past meteorological and tidal information; and settlement pattern and infrastructure. Sources must either be given or material supplied to create the historical context against which the present findings can be set: these may include paintings of the area; old postcards and photographs; old map editions; press reports of earlier storms; past research; evidence of past attempts at coastal protection; human interference in the coastal-marine zone; and District Council Structure Plans.

Data gathering activities

The data gathering activities for field enquiry are indicated in Figure 10.5. Those relating to objectives 1 and 2 include shingle sampling, pebble tracer experiments and beach profiles.

Shingle sampling involves the collection of samples for lithology, size and degree of sorting; roundness can also be incorporated. Samples should be collected at set metre intervals between low and high tide marks along the surveyed beach profile transects. Size and sorting is best determined by the use of a nest of phi sieves.

Pebble tracer experiments allow an assessment to be made of the direction and extent of beach material movement under the conditions prevailing at the time. Onshore feed and longshore drift can be investigated using differently coloured tracer pebbles collected from the site where they are to be used. The painted pebbles must conform with patterns of shingle size from the sampled beach. After one tidal cycle, ie. at the next low tide, a search for the tracers should be conducted. The new position of the tracers should be recorded in terms of direction of movement and distance travelled.

215

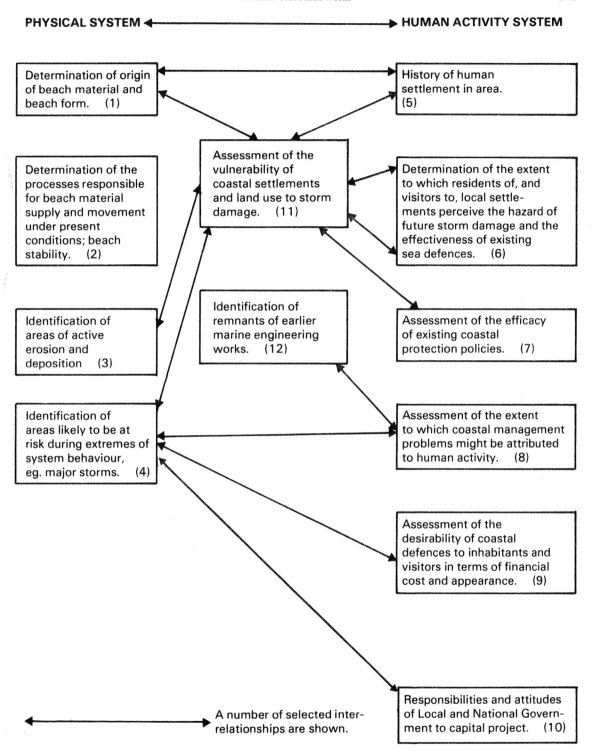

Figure 10.5. Field enquiry: data gathering activities.

Beach profiles should be surveyed at selected locations at 90° to the shoreline, preferably with a dumpy level and staff, though a gun clinometer and posts is adequate. Regular surveying of the profiles will allow seasonal and yearly changes to be deduced. The profiles should be evenly distributed over a wide area to allow the opportunity to sample sites which could possibly be experiencing depletion or addition of material. The beach should be surveyed after significant events, eg. storms, and before and after spring tide cycles.

Information about coastal hazard protection and the perceived attitudes of Local and National Government (objectives 6, 9 and 10) can be gathered from newspaper articles and letters regarding the response of people to specific coastal hazards and management problems. Certain aspects or issues may not have surfaced before the time of the field study and it may prove necessary to collect original data on the opinions of interested and involved people. Questionnaires are the usual format for sampling public opinion, although tape or video recorded interviews can give a more personal overall view. In this particular context the interviews concentrate on the perceived dangers imposed by the sea and level of public satisfaction with existing coastal management strategies and the related funding of such strategies.

Information relating to objectives 3, 4, 5, 7, 8, 11 and 12 (Figure 10.5) can be collected by field observation and mapping and/or the study of historical documents.

The physical processes involved in the coastal system must be understood before existing and proposed management strategies can be assessed and the attitudes and opinions of the people involved evaluated.

Shingle samples should be analysed for lithology, mean size and standard deviation as a surrogate for sorting. Beach lithology will be related to source areas and material availability. Trends of size and sorting can support ideas of longshore drift or suggest alternative ideas of beach development and maintenance. Permanent and consistent grading may indicate the continuation of transportational sorting but it may mean that events have yet to occur to upset beach characteristics which were determined at some time in the past, but are now cut off from their original lines of supply. Pebble tracer experiments should enable the distance and direction of movement to be mapped and provide an indication of movement trends.

Questions to answer

The analysis of the physical system data (including the general observations) should be employed to help provide answers to the following questions:

1. Where has the beach material come from and is the supply current or relict?

2. Can the beach be expected to regenerate after a storm, ie. is there a long-shore or offshore feed of material?

3. Are parts of the beach undergoing regular depletion or accretion, and is there any evidence to suggest that the beach and wave action are not in equilibrium?

4. Is there any evidence to suggest parts of the beach are more vulnerable than others to attack from the sea?

5. Is there any evidence to suggest that attempts at management or beach exploitation have altered the behaviour of the system?

6. Do present sea defences seem to be based on the assumption that the beach (or cliff) will retain its present characteristics of position and height?

7. To what extent are present sea defences in conflict with natural processes? (In the case of Start Bay, settlements are located behind areas of beach which are being reduced in mass as natural processes seek to establish a new beach form in response to changing conditions).

8. To what extent might the present sea defences create further management problems through alteration of natural processes?

The questionnaire on coastal hazard perception should be analysed to provide a picture of how respondents perceive the degree of marine hazard and their views on how it should be tackled through a policy of coastal landform management. This analysis should provide answers to the following questions:

1. How did past inhabitants of Start Bay coastal settlements perceive storm danger?

2. Did these perceptions change after the gravel abstraction of the 1890s?

3. What is the perceived level of hazard amongst residents and visitors?

4. Is the sea regarded as the main natural hazard of the location and to what extent?

5. Is present protection from the sea considered adequate?

6. What aspects of human activity (eg. roads, houses, farmland) should receive priority under a coastal protection scheme?

7. Should the Grade I Site of Special Scientific Interest, Slapton Ley Nature Reserve, be given priority in any coastal protection scheme?

8. What degree of importance do people attach to coastal protection measures?

The conclusion should achieve the dual function of evaluating coastal management strategies in the light of the findings of the analysis of the physical system and in respect of human preferences and activity. Students should be referred back to the working hypothesis and attempt to decide on its acceptance or rejection. In order to do this, they should try to provide answers to the following questions:

1. Is the present management policy working effectively?

2. Do involved people want sea defence works to be established and maintained?

3. Are involved people aware of the natural processes operating on the coast?

4. Does coastal protection seem possible in the long term?

5. Do the people in decision-making positions of authority appear aware of possible conflicts between the natural processes and existing management techniques?

6. Is there enough evidence to suggest that the natural processes pose a serious threat to the present pattern of coastal configuration and human activity?

7. What key relationships between human activity and the coastal system appear to be at risk from natural processes?

8. Who should pay for coastal protection?

9. If limited funds are available, which settlement(s) should be given priority?

The evaluation of the data collected, the answers to the questions, and the recommendations for appropriate action, all give rise to a number of related topics which would further enhance the study; for example:

1. What is the decision-making process leading to the acceptance or rejection of a defence scheme?

2. How does location of residence affect perception of responsibility and willingness to fund sea defence projects?

3. What are the implications for the local structure plans if no coastal protection work is undertaken?

Although the coastal study outline here is designed for older students it should be seen as an example of an approach which is applicable for all age groups and levels of ability.

Conclusion

Framework fieldwork does challenge the worth of systematic field study, but as the GA Sixth Form-University Working Group pointed out in the June 1984 edition of *Teaching Geography,* 'the wider purposes of fieldwork have tended to be at best implicit and at worst submerged under the operational nature of the work'. The nature of those 'wider purposes' is, of course, open to interpretation but five of them to which framework fieldwork contributes are:

— the opportunity to investigate the real geography that comprises the fabric of the everyday observable world

— increased awareness of the potential role of geography in resolving environmental questions and problems

— the recognition of the role of values and political considerations in environmental decision-making

— the development of an enquiring and interpretative ability

— the purposeful acquaintance with, and experience of, a range of techniques and enquiry procedures.

In his Presidential address to the 1984 GA Annual Conference, Rex Walford described some graffiti he had observed on a bridge in Bedworth, Warwickshire (Walford, 1984). The author of the graffiti had written 'Geography is everywhere', a timely suggestion indeed that we should not ignore any aspect of the web of environments that is responsible for the overall character of the real world around us. However, for fieldwork, and especially for framework fieldwork, we might do worse than to turn the slogan round and remind ourselves that 'everywhere is geography' — not the hypothetical geography of nowhere, but the particular geography of real places and real people. With a focus such as this, fieldwork can but advance the cause of geographical understanding.

References

Clark, M. L. (1982) 'Coastal geomorphology — pure and applied', *Geography*, vol 67, part 3, pp 235-43.

Clayton, K. (1979) *Coastal Geomorphology*, Macmillan.

GA Sixth Form-University Working Group (1984) 'The enduring purpose of fieldwork', *Teaching Geography*, vol 9, no 5, pp 209-11.

Geography 16-19 Project (1984) *Coastal Management*, Longman.

Gregory, K. and Walling, D. (eds) (1979) *Man and Environmental Processes*, Dawson Westview Press.

Hanwell, J. and Newson, M. (1973) *Techniques in Physical Geography*, Macmillan.

Hart, C. (ed) (1983) *Fieldwork the 16-19 Way*. Geography 16-19 Project Occasional Paper 4.

Pocock, D. (1983) 'Geographical fieldwork: an experiential perspective', *Geography*, vol 68, part 4, pp 319-25.

Rawling, E. (1981) *Local Issues and Enquiry-Based Learning*, Geography 16-19 Project Occasional Paper 2.

Walford, R. (1984) 'Geography and the future', *Geography*, vol 69, part 3, pp. 193-208. Reprinted in King, R. (ed) *Geographical Futures*, Geographical Association.

11. Combined Studies

11.1. Planning Combined Studies
Michael Williams

Introduction

The choice of the term 'combined studies' rather than the more common 'integrated studies' is deliberate. Experience suggests that the combination of subjects is much more frequently attempted than complete integration. Teachers find that they can accommodate their teaching more easily to courses which emphasise cross-referencing and sharing than to courses which threaten to de-emphasise their knowledge and skills. Indeed, discussions which seek to promote the integration of subjects bring out the Canute in many geography teachers: they see themselves overwhelmed by an irresistible tide which threatens to swamp them and geography. The pessimists can gain some succour from the robustness which geography consistently exhibits. Building defences to consolidate subject boundaries which will resist external contamination is energy consuming and time wasting at a time when teachers in all subjects are being encouraged to work more closely in harmony in defining a curriculum appropriate to the needs of all pupils in secondary schools.

Combined studies may be incorporated into the timetables of middle and secondary schools under a variety of titles: interdisciplinary studies, social studies, the humanities, general studies, and environmental studies as well as integrated studies. The term combined studies is an umbrella which covers course arrangements drawn from a continuum which has, on the one hand, informal links between different specialist teachers and, on the other hand, fully integrated courses.

There is no shortage of examples of schools in which combined courses, drawn from a position somewhere near the centre of the continuum, are taught, though there are marked variations from one local education authority to another. Finding examples of good practice, worthy of communication from one school to neighbouring schools, is much more difficult. Partly, this is a problem of defining good practice in a field which encompasses aims clarification, pupil learning experiences, teaching strategies, syllabus presentation, resources management and assessment and evaluation. Partly, it reflects the problems encountered in disentangling course arrangements from the context of a particular school.

In a very useful and cautious guide to interdisciplinarity in secondary education Wake (1979) uses the adjective 'interdisciplinary' rather than 'combined' or 'integrated'. To define 'interdisciplinary' he quotes from the definition adopted by a Council of Europe conference held in Strasbourg in 1976 which reads: 'an adjective describing the interaction and co-operation between two or more disciplines. This interaction and co-operation may range from the simple communication of ideas to a mutual integration over a wide field'. Such a definition is a useful starting point in examining the concept of interdisciplinarity, but even a peremptory analysis of the organisational possibilities and teaching implications of such terms as 'interaction', 'co-operation', 'simple communication' and 'mutual integration' soon exposes their lack of precision and the consequent difficulties of moving from ideas to action. Not surprisingly,

Wake warns, 'experience shows that interdisciplinary work can be a minefield for the optimistic and the unwary'. Such hidden dangers as the lack of sufficient attention to the nature of school geography, the managerial implications of curriculum revision and the adequacy of a resources infrastructure should not be underestimated.

Theory and practice

There is a gap between the way philosophers, sociologists, psychologists and curriculum theorists discuss combined studies and the way combined studies are taught and organised in schools. Geography teachers will find much that is stimulating and elucidating in the writings of Pring (1976), Bernstein (1971), Musgrove (1973) and Young (1971). Here they will be provoked by such questions as: Do pupils learn subjects or do they integrate knowledge? Are subject divisions merely conventional and arbitrary? How does a society select, classify, distribute, transmit and evaluate educational knowledge? Is geography a 'form' or a 'field' of knowledge? These are important questions but they generally seem remote from the daily concerns of geography teachers. They are discussions at the peaks of mountains, often shrouded in mists, distant from teachers struggling over the boulders which block the paths in the foothills.

In contemporary secondary schools geography teachers, while sometimes being familiar with the theoretical considerations relating to combined studies, are frequently obliged to implement course structures which are predetermined by school administrators. School curricula reflect the resources — time, personnel, teaching materials, accommodation — available and the perceived needs of pupils. These must be placed alongside the curriculum principles held to be valid by headteachers and senior teachers. Combined courses are often the product of particular administrative circumstances and it would be wrong to apply theoretical criteria in seeking to explain their origins and modes of operation. Falling rolls and financial stringency threaten to produce staff shortages and, in smaller schools, the end of some subject departments. The transfer of pupils from primary and middle schools to secondary schools creates problems which can sometimes be eased by simplifying timetables through, for example, reducing the numbers of lessons per day and increasing the contact of pupils with fewer teachers. Combined studies may be initiated by headteachers as solutions to problems of staff shortage, pupil transfer and timetabling. In these circumstances a theoretical rationale for the process of combination may be produced in retrospect. The bridge between theoretical principles derived from the educational disciplines may be ignored in favour of what may be seen as administrative principles underpinning effective school management. These assertions serve as warnings that the practice of organising and teaching combined studies in schools cannot be explained solely from the standpoint of blinkered theoretical positions: they are the product of many pressures, the strength of which vary from school to school.

Describing courses of combined studies

Syllabuses, written as statements of course aims and a synopsis of content, are frequently used as sufficient evidence on which to judge the quality of a course. They answer two questions: What are the teachers hoping to achieve? and: How is the course content perceived? Opportunities to visit schools where combined studies courses are taught yield answers to several other important questions including:

1. At the stage of course planning were the participating teachers given adequate time for designing the course?
2. Was provision made for the in-service education of the teachers?
3. Is the course taught to the most appropriate pupils? How are the pupils grouped for lessons?
4. Does the course content — knowledge, attitudes and skills — reflect accurately the statement of aims and objectives?
5. Have the strengths and particular contributions of the subjects combined in the course been retained?
6. How appropriate are the dominant teaching strategies for facilitating pupil learning?
7. Is there a carefully constructed timetable of pupil assessment, including classwork exercises, homework assignments, and regular testing?
8. Are the pupils encouraged to keep comprehensive records of work?

9. How suitable are the classrooms for teaching the course?
10. Is provision made for the maintenance and renewal of the stock of teaching resources, eg. textbooks, worksheets, audio-visual aids?
11. Do the teachers engage in formative and summative evaluation of the course?
12. Is time made available for the teachers engaged in teaching the course to meet on a regular, timetabled basis to plan and evaluate the course?
13. Are the senior staff and other teachers in the school kept fully informed about developments in the course?
14. Are parents informed about the rationale for the course?

Syllabuses are often cosmetic statements of intent, presented as lists of content disconnected from stated aims and providing little guidance, if any, about appropriate teaching methods and how and what the pupils are likely to learn. Much clearer information about a course of combined studies can usually be gleaned from pupils' notebooks and by the end-of-year examination paper together with pupils' marks.

Ideally statements of aims, content, teaching methods and modes of pupil assessment and course evaluation should be synthesised to give a coherent unity. However such idealism must be tempered by reality. Rarely in contemporary secondary schools are conditions conducive to such planning. Nevertheless careful and thorough curriculum planning is a goal at which to aim.

Aims and objectives

A review of statements of aims and objectives of courses of combined studies leads to the conclusion that there are three principal sources of aims. Firstly, there are those aims which derive from the organisational and administrative context of a particular school. Such aims are likely to figure prominently in courses which have been introduced to ease the transfer of pupils between schools, to simplify the timetable or to reduce the range of options for examination candidates. Secondly, there are aims which stem from the curriculum arrangements within a school. Discussions about the provision of a core curriculum have sometimes led to the decision to provide a humanities course for all pupils, thus avoiding the need for pupils to choose between history and geography. In a small number of schools teachers favour an open approach to teaching and learning in which subject boundaries are seen as barriers which limit study. Open plan classrooms, the availability of a wide range of resource materials, the utilisation of fieldwork and an emphasis on problem solving and enquiry-based learning are some of the features emphasised in such an open approach. Thirdly, there are aims related specifically to the subject matter which teachers wish to communicate to pupils. Topics or units of work in combined studies which are derived from conventional school subjects will have the same statements of aims as the topics when they are included in specialist courses.

The following statement of aims taken from a social studies course for 11-12-year-old pupils serves to illustrate the way statements of aims are commonly presented.

Social Studies: First Year

During the first year pupils follow an integrated scheme of work in history, geography and religious education. This is done so as to:

1. maximise our use of teaching time by eliminating overlap of content between subjects;
2. reduce the amount of fragmentation in the timetables of pupils used to having the same teacher all day in the junior schools;
3. enable the introduction of other social sciences in a co-ordinated way.
4. improve teacher knowledge of pupils by greater contact.

The criteria borne in mind when the course was formulated were that:

1. it should meet the objectives of the subjects involved in the course in terms of content, methods, skills and ideas;
2. it should meet general social expectations of the kinds of knowledge to be acquired;
3. it should deal with the more important aspects of general and long-lasting significance in preference to the ephemeral and unique;

4. it should meet the needs of the pupils in the way in which it is formulated and worked so that pupils benefit from a short-term sense of progress and a longer-term deepening of knowledge and awareness.

The broad aims of the course involve the development of:

1. knowledge and understanding of the origins and development of human civilisation;
2. knowledge and understanding of peoples of different cultures;
3. knowledge and understanding of the natural and human environment and the interrelationship between the two;
4. knowledge and understanding of the organisation of modern society;
5. attitudes towards, interest in and appreciation of the subject matter of the course;
6. general thinking skills, especially in relation to the subject matter of the course;
7. general practice skills necessary to learning progress;
8. the desire to learn which should be stimulated and fostered.

It is difficult in statements such as this to detect synthesised aims significantly different from those usually associated with specific specialist subjects. Statements of aims for combined studies generally reflect the way they have been written, that is, specialist teachers contribute items from their own subjects and these are added together to create a list. The same is generally true of statements of objectives. Lists of objectives associated with specific topics are not found as frequently as statements of aims, not least because they are difficult to construct. Geography teachers experience frustration and disappointment when they find themselves repeating the same objectives for a series of units. The repetition implies that the series lacks progression. When one topic follows another without apparent linkage the discontinuity is most obvious in statements of objectives. This discontinuity is a feature of courses in which geography is combined with one or two other subjects to create a course which is taught by one teacher to a class. The discontinuity in aims and objectives is a function of the discontinuity of content, and vice versa. It is a source of criticism for those who do not favour combined studies.

Content

The ways in which course content is written in syllabuses are directly related to the staffing arrangements for courses. The two most common arrangements are, firstly, a combined course taught to a class by one teacher, and, secondly, a course divided into subject components which are taught by subject specialists.

In the first arrangement the teacher is for some of the time a non-specialist teacher. The syllabus usually is presented as a list of topics: headings of units are followed by short statements of content. This is illustrated in the following example:

Year One

1. Mapwork
 a) direction
 b) distance
 c) scale
 d) contours

2. Village life
 a) historical development of the local village, the feudal system
 b) village life today
 c) the location of prominent buildings in the village, eg. the church

3. Religion in Britain
 a) survey of the major religions in Britain
 b) doctrines and practices of these religions

4. Peoples of the world
 Studies of contrasting lifestyles:
 a) Bushmen
 b) Japanese cities
 c) Chinese peasants

In the second arrangement, illustrated below, the syllabus is written as two related lists:

Year One: Population Problems

Geographical

1. Types of employment
2. What is 'unemployment'?
3. Industrial structure: primary, secondary and tertiary industries
4. Factors of production: land, labour and capital
5. Local jobs

Historical

1. The changing pattern of local employment
2. The Industrial Revolution
3. Industry in the twentieth century
4. Agricultural changes since the eighteenth century
5. Child labour in the nineteenth century

In this arrangement the geographical and historical items could be taught by the same teacher or by separate specialist geography and history teachers. In both arrangements there is a low level of integration of the content: there is no difficulty in detecting the contributions from the specialist subjects.

A greater movement towards integration can be seen in this syllabus statement:

Our Community

The school in the local community: its layout, organisation and staffing.

The local area: local services; housing and health; industry and commerce; law and order; leisure and recreation facilities; churches; the local newspaper; responsibilities of a young person in family and community; local shopping; the value of money; supply and demand for goods and services; inflation; jobs and unemployment.

In determining the sequence of topics in a course attention should be paid to the need for progression in the knowledge, attitudes and skills which must be included. In seeking compromises and consensus between subject specialists this can be overlooked. Teachers assert that syllabuses are rarely written to be slavishly followed item by item. There is usually scope for rearranging the sequence of topics and giving different emphases to topics with different classes. However, in combined studies courses teachers are often expected to teach topics with which they are unfamiliar. Geography teachers may well experience difficulty in teaching about the ancient Egyptians and history teachers may have a limited understanding of the intricacies of Ordnance Survey mapping. Selecting priorities in the content of combined studies courses may be very arbitrary and this threatens to disrupt progression in learning. In recent years geography teachers in many schools have become more sophisticated and successful in their attempts to structure their courses so that, for example, one kind of study skill leads to another and skills, once learned, are applied to new problems in later years. This structuring can easily be overlooked in the process of combining bodies of content from different subjects.

The content of combined studies courses is commonly drawn from history and geography, subjects which, in their aims, content, teaching methods and modes of assessment, are strikingly different. Geography is sometimes used to give a crude spatial context to historical events and historical factors are taken into account in geographical explanations, but these are unconvincing reasons for merging the two subjects into combined studies. Some of the least credible combined studies courses are those in which the Bible is used as an historical document and pupils are directed to compare and contrast the physical environment of Israel with that of the local community. In these courses the combination of subjects appears as an end in itself rather than a means to an end. Adding geography to history merely serves to devalue both subjects: functional cross-referencing can give both subjects a new level of interest.

Of the many contemporary pressures on the

curriculum which encourage the introduction of combined studies four are particularly influential. They are:
1. the need to give the curriculum in all schools a multi-ethnic dimension;
2. the introduction of world studies and peace studies;
3. the forging of links between schools and industrial organisations;
4. the need to protect and conserve the physical and human environment from small and large acts of vandalism.

Geography teachers are not alone in feeling these pressures. They surface in critiques of courses and resources in the popular media and in professional publications and meetings of teachers. They are the focal points of local, regional and national curriculum projects, such as the World Studies Project and the Keep Britain Tidy Campaign. These and other initiatives are of interest to geography teachers yet their messages are often best incorporated in curricula by adding units to combined studies courses.

Teachers engaged in teaching combined studies courses will find much that is directly useful and stimulating in the curriculum materials and publications of four curriculum development projects in combined studies: the Environmental Studies 5-13 Project, History, Geography and Social Science 8-13 Project, the Humanities Curriculum Project, and the Integrated Studies Project. Introductions to the work of the four projects can be read in the teacher's guides by Harris (1972), Blyth (1976), Stenhouse (1970) and Bolam (1972). The teacher's guides to the Geography for the Young School Leaver Project also take account of the possibility of incorporating its teaching strategies and curriculum materials into courses of combined studies.

The impact of the American project *Man: A Course of Study* (MACOS) has been felt in some localities in Britain. This carefully structured course is founded on the work of Bruner (1966). He pinpoints five humanising forces: tool-making, language, social organisation, the management of man's prolonged childhood, and man's urge to explain his world. Geography teachers have used some of the ideas and materials from this project to enrich their teaching about Third World countries.

The curriculum statement published by the Geographical Association (1981) identifies four special contributions of geography to the school curriculum: graphicacy, world knowledge, international understanding and environmental awareness. It is the spatial factor combined with the interaction of people with their natural and constructed environments that mark the distinctive role of the geography teacher. Multi-ethnic education, world studies, industry and environmental conservation are mainstream to the geography teacher's interest and much can be gained by cross-referencing the content of geography syllabuses with those in other subjects.

Teaching methods and assessment

Discussions about combined studies rarely proceed beyond concern with content and its organisation in syllabuses. There is far less attention paid to teaching methods and modes of assessment. To some extent this reflects the privacy with which teachers regard the way they teach their lessons. More importantly, methods are grounded in the traditions and conventions of the specialist subjects which have been combined. Topics with a strong geographical bias are taught in the same way as if they were in specialist courses. The same is true of homework assignments, class tests and terminal examinations. Some examination papers for combined studies courses are divided according to contributing school subject and pupils must answer questions drawn from each of the sections. The questions are set by specialist teachers and no attempt is made to combine the subjects in individual questions. This may be less true in those combined studies courses from which geography has been excluded, eg. in social studies courses which have been designed as social science courses where the contributing subjects are sociology, politics, economics and anthropology.

The teacher of a social studies course in which geography has been combined with history and religious education described his teaching methods in this way:

> Variety, both in methods and materials, is the keynote of our approach. Films, slides, television and radio recordings, talks by teachers with particular expertise in a particular subject, lectures by outside speakers, dramatic and artistic work, fieldwork, educational visits,

work based on enquiry and projects, have all been employed regularly. Selection and organisation of teaching method is decided in departmental or inter-departmental planning discussions. A range of project materials, text and reference books, and materials produced by staff, individually and in co-operation have been used. The school is fortunate to have an excellent reprographics department headed by one of the faculty staff. Materials produced by one teacher are usually made available for others, details being regularly circulated in departmental news sheets.

Mixed ability class groups are the rule in the first year only. To ensure that all children are learning purposefully teachers prepare worksheets suitable for children at all levels of ability. They also make use of Schools Council and other project materials, where they are appropriate for our courses. It is also a general rule that more advanced reading and research exercises are usually available for the more advanced pupils, while simplified reinforcing exercises are prepared to substitute for normal classwork and homework exercises where children have difficulty with these.

In this statement one finds reference to the principal methods employed by experienced geography teachers who are successful in arousing and maintaining the interest of their pupils in geographical topics. Methods regularly employed by geography teachers, such as mapwork exercises using maps drawn on a variety of scales, the study of photographs and the use of games and simulations, may be quite unfamiliar to teachers of other subjects. Unless teachers make deliberate attempts to help their colleagues in combined studies courses difficulties will certainly arise. Notice the reference to departmental and inter-departmental discussions in the quotation above. In some schools it would appear that the production of worksheets is both the beginning and the end of the process of inter-departmental co-operation. The assumption that teachers can distribute worksheets prepared by another subject specialist and then engage in worthwhile teaching is highly dubious. Doubtless it is a widespread coping strategy but such a strategy has very limited value. At worst it leads to the repetition of tasks which rarely go beyond copying text, diagrams and maps. In combined studies courses the avoidance of copying as a teaching strategy may be a principal aim and a prominent reason for seeking to reduce subject boundaries.

Course evaluation

Formative evaluation, the term used to describe the formal, regular monitoring of a course while it is being taught, is likely to be a prominent part of the departmental and interdepartmental planning discussions. Such evaluation, which should take account of the attainment of stated objectives, the use of resources, the teaching methods employed, the difficulties and successes of individual pupils and classes, and the maintenance of adequate records, needs to be carefully planned. Certainly pressure on teachers' time makes this difficult but if left to chance formative evaluation is likely to be neglected.

Summative evaluation should take place at the end of a course and it is the basis for course revision. At this stage the whole of a course should be evaluated, including all of the aims and objectives, the course content, teaching methods, resources and modes of assessment. It should be undertaken in sufficient time for any radical revisions in the course to be carefully planned prior to the changes being implemented in the next year. Summative evaluation should focus sharply on the nature of the combination in combined studies courses. It should seek to answer the key question: What is gained and what is lost in the process of combination? The gains and losses are best seen in pupil attainments and the extent to which the pupils' future studies may benefit or be impaired by studying topics in a combined studies course.

Combined studies courses are difficult to design and teach. To be successful they require a high level of co-operation and goodwill between teachers from various subjects who must make a concerted effort to learn about other subjects. The case as to whether pupils learn more or less through combined studies courses compared with specialist courses remains unproven. What is clear is that in many courses the addition of topics from various subjects appears to interrupt the progress of pupils' learning in the constituent subjects. If progression is threatened then geography teachers are well-advised to highlight this when they contemplate introducing a combined studies course and when they evaluate those courses in which they are participating. Teachers will find further guidance on designing and teaching combined studies courses in a book published by the Geographical Association (Williams, 1984).

References

Bernstein, B. (1971) 'On the classification and framing of educational knowledge', in Young, M. F. D. (ed) *Knowledge and Control*, Collier-Macmillan.

Blyth, W. A. L. et al (1976) *Place, Time and Society 8-13: Curriculum Planning in History, Geography and Social Science*, Collins/ESL for the Schools Council.

Bolam, D. et al, (1972) *Exploration Man: An Introduction to Integrated Studies*, Oxford University Press.

Bruner, J. S. (1966) *Towards a Theory of Instruction*, Harvard University Press.

Geographical Association (1981) *Geography in the School Curriculum 5-16*, The Geographical Association.

Harris, M. et al, (1972) *Teacher's Guide to the Environmental Studies 5-13 Project*, Rupert Hart-Davis for the Schools Council.

Musgrove, F. (1973) 'Power and the integrated curriculum', *Journal of Curriculum Studies*, Vol 5, No. 1, pp. 3-12.

Pring, R. (1976) 'Integrating the curriculum', in *Knowledge and Schooling*, Open Books.

Stenhouse, L. et al, (1970) *The Humanities Curriculum Project: An Introduction*, Heinemann Educational Books.

Wake, R. A. et al, (1979) *Innovation in Secondary Education in Europe*, Council of Europe.

Williams, M. (ed) (1976) *Geography and the Integrated Curriculum*, Heinemann Educational Books.

Williams, M. (1984) *Designing and Teaching Integrated Courses*, The Geographical Association.

Young, M. F. D. (1971) 'An approach to the study of curricula as socially organised knowledge', in Young, M. F. D. (ed) *Knowledge and Control*, Collier-Macmillan.

11.2. Humanities in Practice

Malcolm Brown

Course creation

Combined studies (henceforth referred to as 'Humanities') have been taught at Gordano School, Bristol, an outer suburban mixed 11-18 comprehensive school, for over 10 years. The decision to combine history and geography for pupils aged 11-13 in years 1 and 2 was taken at about the same time as that to move towards mixed ability teaching. Humanities, followed by history and geography in years 3, 4 and 5 are now all taught in mixed ability classes to 16+.

Although both integration and mixed ability teaching are closely identified with comprehensive education in many schools, they are not mutually dependent. Humanities occupies six 55-minute lessons over a 10-day timetable for all pupils in the first two years of the school. First-year pupils are taught in their tutor groups for the whole curriculum, thus reducing fragmentation.

The rationale for the Humanities course, from today's perspective, may reflect the full benefit of hindsight rather than the views of all concerned when the course commenced. The social and educational advantages of increasing contact time between teachers and pupils, and of reducing the curriculum fragmentation which resulted from the study of separate subjects, weighed heavily in favour of a Humanities course. The administrative convenience of timetabling and the opportunity to employ non-specialist staff probably helped its conception and certainly bolstered its continued existence.

A point worth stressing, raised by Michael Williams in the preceding section, is that as subject specialists, our resistance to non-specialist teaching and our demands for single-subject time may not always be in the best interests of our pupils. Much of this argument revolves around the importance attached to the subject content. Whenever content takes precedence over the *process* of learning then arguments about content will stand as barriers to subject combination.

Two further arguments were propounded by teachers at the inception of the course. One was that a combination of subjects might enable a fuller range of skills to be exploited in studying topics not covered in depth by either history or geography. It is possibly the case that pupils, particularly at the lower end of the secondary age range, integrate areas of knowledge rather than learn subjects. A second argument was that, in order to emphasise the process of learning, new teaching styles were being advocated. Team teaching, for instance, was considered a worthwhile idea, and this required blocking of history and geography on the timetable.

The aims of the Humanities course are as follows:

1. to develop a variety of map skills which form a basis for further work in Humanities, geography and history (including the use of scale, direction, bearings, references, conventional signs, keys etc.);

2. to teach basic historical skills and concepts through the use of evidence and clues to the past, and to develop a questioning approach to the use of clues from the past and the reliability of evidence;

3. to develop an appreciation of chronology and an understanding of evolution and adaptation through time;

4. to show the relationship between changing environmental, social factors and man's activities through time through specific projects, eg. farming, town and port development;

5. to use the crossover between history and geography when both subjects are needed to give a full picture of a topic eg. 'Exploration', or an 'issue' eg. 'Third World Development';

6. to develop study skills eg. the use of a variety of sources for project work;

7. to develop communication skills through group work eg. discussion, games, simulations, role playing exercises, etc.;
8. to develop the skills of reasoned judgement and empathy;
9. to help pupils appreciate how others in the world live and how all must work together for a secure future.

------The topics for each term are as follows:

Year 1

½ term   *Introductory mapwork skills.*

½ term   *Introductory history skills.*

1 term   *Farming:* resource based learning organised as a circus.

1 term   *Early life:* a wide range of resources, especially videotapes, plus archaeological and fossil evidence, again organised as a circus.

Year 2

1 term   *The development of towns:* concepts of siting, a case study of the local city and an analysis of port development.

1 term   *Rich world, poor world:* a critical view of development (see below).

1 term   *Exploration:* an attempt to explore the ubiquity of migration in both time and space.

For the purpose of illustration the aims and objectives of the term's work, 'Rich World, Poor World', are shown in Figure 11.1.

There are termly assessments and a full report requires regular checking or work, but no formal end-of-session examination is used. Sufficient contact with pupils and the mixed ability nature of groupings make formal examinations both less necessary and more invidious.

Course development

All curriculum development is difficult not least because it creates an additional burden on staff. At the outset two strategies were used. Individual staff produced various worksheets and one member of staff was 'internally' seconded for a term so that his teaching timetable with years 1 and 2 could be used for resource preparation. Later the special responsibility post of teacher in charge of Humanities in years 1 and 2 was created and this provided useful experience for post-holders before they became heads of department in other schools. Fortunately the incumbents tended to promote curriculum development in areas that were not within their subject specialisms so that 'empire building' was avoided. The present post-holder has ensured that there has been a steady improvement not only in

---

| Year 2 | Rich world, poor world | 1 Term |

1. **Rich or poor?**

   Aim: To allow pupils to explore their perspectives of poverty and wealth.

   Objectives: (i) To recognise that poverty and wealth are often relative and may exist within the same country.

   (ii) To recognise that poverty is not limited to one stereotype of country or people and can be eradicated over time.

   (iii) To introduce current ideas and terminology of the rich/poor divide.

2. **Measuring wealth in a country**

   Aim: To recognise the spatial variability in socio-economic indices.

   Objectives: (i) To gain a knowledge of terminology eg. GNP.

   (ii) To recognise that a number of indices taken together can help to distinguish between developed countries and less developed countries (LDCs).

(continued)

(continuation)

3. **Parameters of wealth**

    Aim: To consider, in detail, variations in life-chances and societies' reactions to this imbalance.

    (A) World Food Distribution

    Objectives:
    (i) To identify areas of the world where the basic right to a fair share of the sufficient world food supply is not achieved.
    (ii) To consider the crippling effects of hunger.
    (iii) To discuss viewpoints on food maldistribution from developed and LDC bases and to empathise with LDCs.

    (B) Disease

    Objectives:
    (i) To identify some of the more common diseases in LDCs and to assess why these are not common in Britain now.
    (ii) To consider leprosy and change perspectives about this much misunderstood disease.

4. **Ghana — a case study**

    Aims: To explore the problems and possibilities facing an LDC and to consider the value of a 'Western' route to development.

    Objectives:
    (i) To consider the term 'developed' in terms of the capability of self-sufficiency within local resource constraints.
    (ii) To recognise the problems associated with primary product export dependency.
    (iii) To consider the possibilities and problems, socio-economically, of tied aid and 'advanced-technology' capital development.
    (iv) To consider the possibilities and problems, socio-economically, of appropriate technology and local resource exploitation.
    (v) To evaluate the alternative strategies facing an LDC.
    (vi) To empathise, via role play, with the decisions facing leaders of LDCs in a world context of unequal resource distribution.

5. **Rich world**

    Aims: To contrast the 'two worlds'; to explore the range of socio-political organisations in the rich world and to recognise the interdependence of the whole world.

    Objectives:
    (i) To enable pupils to view the Industrial Revolution as part of a world-wide process and its spatial origins as only partly deterministic.
    (ii) To consider the past and present influences on industrial locations.
    (iii) To contrast the growth, successes and problems in USA, USSR and China.
    (iv) To conclude by generating (a) a list (b) a map of goods and services which show children their global interdependency.

Figure 11.1. Rich world, poor world: aims and objectives.

Weakness	Remedy
1. The syllabus aims and objectives were only implicit and whilst not being 'cosmetic' the syllabus was not sufficiently supportive of teachers in suggesting schemes of work that had proved to be effective.	A clearer statement of aims and objectives for the course has enabled more effective schemes of work to be suggested, supportive to both specialists and non-specialists.
2. Given the 'circus' nature of the topics studied, it was (and continues to be) difficult to ensure that progression is explicitly built in to each topic. It is not enough in a mixed ability classroom to assume that a non-specialist (or even a specialist) teacher will ensure that both the progression and 'stretching' of all pupils will occur. Something more explicit was needed to help staff identify areas of work that could be exploited to meet this criticism.	A 'staff guide' to each topic was produced to ensure that at whatever time of the academic year the topic appeared, progression could be effected. This includes: (a) A clear statement of objectives for each part of the topic (see example in Figure 11.1). This need not be as detailed as a complete 'scheme of work' but where a lead lesson is suggested, for example, then specific guidelines are issued. This challenges teachers to prepare lessons effectively in advance, for the range of resources and modes of learning are too extensive to be used without preparation. These staff guides quickly became effective tools to better and less eclectic classroom practices. (b) All the resources, both permanent and consumable, for pupils are listed and with copies inserted, so that a ready reference is available anytime, anywhere. (c) Supportive staff information plus additional reference texts are itemised and, where possible, 'staff copies' enclosed.
3. The reading age of many of the materials was too high. The measured reading ages of the first year intake range from 6 to 13+ years. A lack of awareness of reading age was initially evident.	Later materials generated by staff tried to overcome the readability problem. Suitable textbooks for less able pupils, however, are still difficult to find. It is important that the purchase of appropriate books takes place at an early stage to provide a professionally produced input to the course.
4. 'Death by worksheet'. Early curriculum development in schools was often based on idiosyncratic attempts to counter the dearth of commercially available material. We were no exception and many of the worksheets were poorly worded, poorly handwritten or typewritten, overcrowded and monotonous in their presentation. The tasks required of the pupils were also too narrow.	(a) Whilst it is important to utilise staff expertise and enthusiasms in the ideas/research periods of curriculum development, if *one* person is capable of the final production, then the materials will display a coherence of style, layout etc, which experience suggests is better than a range of idiosyncratic 'master-pieces'. (b) Layout and writing of materials has been improved by reducing the information on

(continued)

(continuation)		each page and breaking up the text with diagrams and other illustrations.
	(c)	Where possible, we have organised resource based learning along the lines of the classroom management packs developed in Bristol by the Avon LEA Resources for Learning Development Unit.
5.	Earlier course elements were marred by a narrow range of skills to be developed and tasks to be completed. This reflected the lack of variety of resources and the less well developed acceptance, initially, of the value of techniques such as role play, simulations, and small group discussions leading to plenary sessions.	The range of skills developed and tasks demanded has increased as the resources available have expanded. There is no need to rely on hastily prepared worksheets. Approximately 50 hours of videotaped material, efficiently stored and indexed, creates enormous scope for less didactic teaching. For instance, in the 'Rich World, Poor World' topic approximately 30-40 per cent of it is discussion based, using a range of stimulus material. Simulations have proved very successful and a very positive development has been the post-simulation generation by pupils of their own board game. The sophistication of some of the variables built into 'Development' games has been impressive.
6.	The course was, unintentionally, racist and sexist in some of its topic titles and content, eg. Rich *Man*, Poor *Man*. The 'Travelling Man' topic was felt by girls to be irrelevant as soon as project work on cars, trains etc., was initiated. The study of the Third World reflected the view, prevalent at the time, that it was a 'problem', the assumption being that the Rostow model and imitation of Western ways of 'development' would lead to 'take-off'.	Racism and sexism are being positively eradicated through a greater awareness of our values, as teachers and individuals. Images which challenge stereotypes are sought; for example, the archive photograph of London slum children in the 1890s (Figure 11.3.) has generated most effective discussions of change over time and challenged stereotypes of poverty. Similarly, the 'Rich or Poor' drawing (Figure 11.4.) has proved an excellent basis for discussing (a) the unequal distribution of wealth and (b) sex stereotyping.
7.	In-service support by colleagues in the separate disciplines was never really achieved at a satisfactory level, partly due to the lower status given to Humanities at the time. Early attempts at staff involvement were 'ad hoc' and often involved the circulation of memoranda rather than the meeting of groups.	One important step in generating a new wave of development was to get all the staff (11 at the time) to meet, often irregularly, to review systematically each topic. This identified dissatisfactions, the strengths and weaknesses of ideas and materials, and initiated change. Over time the move towards a Faculty structure has raised the status of Humanities and created expectations of change. A further move to hold timetabled meetings between Heads of History, Geography and Humanities with a formal agenda and minutes, has enabled changes to be implemented.

Figure 11.2. A Humanities course: weaknesses and remedies.

Figure 11.3. Slum children of London in the 1890s. Photo: BBC Hulton Picture Library.

materials and the range of classroom activities but in the *process* of curriculum development and course evaluation.

It needs to be borne in mind that course improvements, which require additional individual and team efforts, continue to be difficult to implement. This is partly due to the rigidity of scale-point responsibilities and the lack of motivation deriving from fewer promotion opportunities for those involved in the development of the course. Notwithstanding these constraints, complete revision of the topics is completed at the rate of one per academic year in close consultation with the heads of the history and geography departments.

The weaknesses identified in the first column of Figure 11.2 became apparent at different times in the development of the course. The criticisms are in no way meant to belittle the enthusiasm and professionalism of all involved in the course: rather, they reflect the acquisition of greater expertise, as will be seen from a study of the remedies shown in the second column.

Conclusion

Michael Williams poses the key question: what is lost and gained by combining courses? Most clearly lost is the single-subject, factual-knowledge content and the 'purist' viewpoint. I feel that as geographers we may still have some way to go in jettisoning some of the factual content of the subject, just as it seems strange to hear the cry from some historians on the loss of the chronology of British history. Geography has always found it difficult to view itself as a 'pure' discipline and at school level we should feel uneasy at disregarding the interfaces and duplications at the boundaries of our subject with those of other subjects. Many of these overlaps are the very ones which keep geography relevant to the lives of the pupils.

The gains can include better learning of skills and concepts valuable to all our pupils, a greater sensitivity to the pupils' needs and aspirations, and a realisation that the processes of learning and enquiry are superior to the acquisition of a narrower range of factual knowledge. Moving towards combined studies is probably a long journey, but a continually rewarding one for both pupils and teachers.

Note

A Handbook of Classroom Management in Humanities is available from the Resources for Learning Development Unit, Bishop Road, Bishopston, Bristol, BS7 8LS.

Figure 11.4. Rich or poor? Source Enquiries: Life in Developing Countries, Longman.

12. Evaluation and Assessment in Geography

12.1. Evaluation and Assessment 11-16

Melvyn Jones

Introduction

Evaluation is a greatly abused word. Its meaning is often misunderstood and the process of evaluation is sometimes seen as a threat. As a result, at one extreme the evaluation of what goes into and what comes out of the learning situation may be careless, incomplete and narrowly conceived, and at the other varied, structured and based on a well-defined rationale. It is the contention here that no curriculum is complete if insufficient thought has been given to evaluation.

But first some clear definitions are called for. There is often some confusion between the terms *evaluation* and *assessment,* and evaluation is often used when assessment would be more appropriate. Assessment may be defined as the means that are devised to describe and classify a pupil's performance. These usually include course assessments, classroom exercises and tests, homework assignments and formal examinations. The concern here, therefore, is on the pupil and on attainment. Assessments measure learning outcomes.

Evaluation, on the other hand, is defined by Marsden (1976) as 'the making of qualitative and quantitative judgements about the value of the various curriculum processes'. He goes on to emphasise that evaluation is 'critical *at each stage* of the curriculum process, from the first stage of evaluation of the broad aims of a course, to the late stage of evaluating which instruments of assessment to use and the final one of making judgements anew on the basis of the evidence which these instruments provide'. Evaluation is, therefore, a much wider term than assessment but usually includes assessment as one of its techniques. It is concerned as much with classroom processes as learning outcomes.

Two further terms need defining at this stage: formative and summative evaluation. Formative evaluation takes place while a course is in progress. In particular it provides 'corrective or confirmatory feedback' (Satterly, 1983) about the effectiveness of materials and procedures. Summative evaluation, on the other hand, takes place once a course or unit of work has been completed and seeks to find out whether goals have been attained.

Clearly, if evaluation is concerned with making judgements, it is very much concerned with collecting evidence. What should be collected, and how and when and under what circumstances it should be gathered, will obviously depend upon the use to which the information is intended to be put. In the context of the 11-16 geography curriculum there are many reasons for gathering evaluative evidence, for example:

1. to enable the teacher to reconsider course objectives, materials and procedures;
2. to enable the teacher to diagnose what groups or individuals do or do not know, do or do not understand, and can or cannot do, so that remedial action can be taken;
3. to motivate pupils;
4. to inform other teachers who may have to make decisions about pupils;
5. to make decisions about entering pupils for external examinations;
6. to inform parents about progress.

Assessment is clearly a useful way of gathering data for all these purposes. This is why it is often

Evaluation & Assessment in GEOGRAPHY

equated with evaluation and why the two terms are commonly used interchangeably. It needs to be stressed again, though, that evaluation is as much concerned with teacher performance as with pupil attainment and involves the gathering of opinions, impressions and judgements as well as quantitative data. But what all these data gathering activities have in common is their purpose in providing information for decision-making in one context or another.

With these considerations in mind the rest of this section will be concerned with brief discussions of classroom observation and pupil opinion surveys and a lengthier consideration of assessment techniques and the analysis of tests.

Structured observation in the classroom

Teachers, of course, observe in their classrooms all the time and most practitioners would agree that observation is an indispensible part of their repertoire of evaluative techniques. What is being advocated here is the use, periodically, of more formal approaches, to achieve particular evaluative goals. For example, structured observation may be a particularly effective way for a departmental team to analyse and reflect upon the consequences of classroom re-organisation (eg. the introduction of mixed ability groupings), the use of new teaching resources (eg. a curriculum project pack) or the introduction of new teaching/learning strategies (eg. the conduct of simulation exercises, games or individualised learning programmes) see Figure 12.1.

All teachers have undergone periods of close scrutiny by an observer during teaching practices and during the probationary year, and these will have been times of stress and anxiety. The suggestion that they should submit to further observation may not, therefore, be welcomed. In spite of possible objections, for busy teachers the most objective observation technique is for two colleagues to work together, observing each other in turn and discussing findings. Although the method is time-consuming and may prove organisationally difficult it should be seriously considered as a method of formative evaluation.

The alternative is self-observation. This involves the teacher in playing the roles of both practitioner and evaluator. Methods should obviously be closely related to objectives. If the objective is to investigate classroom organisation or pupil movement, then video-taping would seem to be the most appropriate method; if the interest is in dialogue, audio-recording may be sufficient. In both of these cases the equipment may intrude and the outcome may be atypical, though this is less likely in the case of the audio-recorder. Making a visual or oral record of a lesson may not in fact be the most difficult phase of the evaluation exercise; this is more likely to occur when the teacher replays the video- or audio-recording. At this stage the collected data needs to be transferred to written form from the video-recording or transcribed in some way from the audio-recording. After that, conclusions have to be drawn and decisions made.

A hypothetical example may show the value of this approach more clearly. Imagine a situation in which a teacher has recently been appointed to a geography post in a school where the pupils in the 4th and 5th years are preparing for examinations based on the Geography for the Young School Leaver (GYSL) Project and where, in consequence, the lower school geography curriculum is also broadly based on the approach of the GYSL project. He has had only limited experience of the GYSL approach (on his final teaching practice three years previously) and is concerned by the fact that whereas his new colleagues stress the 'pupil

Figure 12.1. Structured observation of classroom activity can be very illuminating.
Source: Cutts, Higginbottom and Jones (1981).

participation' and 'pupil thinking' aspects of the work, his experience in his previous school was with a rather traditional curriculum in which 'teacher talk' and closely supervised formal written work were the main characteristics of activity in geography classrooms. He wishes, therefore, to analyse, in the simplest possible way, his early lessons in order to gauge how much of the time is dominated by teacher talk and formal writing and how much and what kinds of pupil thinking and participation are taking place. He is primarily interested in dialogue and therefore an audio-recording is likely to be sufficient although supplementary 'field-notes' made during the lesson may provide useful additional data. Rather than transcribing the whole recorded lesson as a prelude to a formal analysis, he could make use of a schedule of the kind shown in Figure 12.2. The observation intervals used will depend on the length of the lesson and the detail of the analysis required and may vary from a few seconds to half a minute. This type of schedule is usually completed in the classroom by an observer while a lesson is in progress, but its value as a quick way of summarising audio-recordings of lessons should not be overlooked.

Obviously these approaches to evaluation are full of pitfalls, but as Harlen has pointed out '... evaluation ... has moved away from being the preserve of 'experts'. In many cases it uses methods which are accessible to, or can be devised by, teachers ... simple techniques, often built upon what teachers already do in some degree, can yield information which should be available when decisions of various kinds are made' (Harlen, 1978). A full discussion of observation techniques and methods of analysis is beyond the scope of this section, but a wide range of books is available, eg. Walker and Adelman (1975) and Boehm and Weinberg (1977). The ways in which teachers, departments and whole schools can use such techniques to advantage are also explained in the Open University continuing education course, *Curriculum in Action: An Approach to Evaluation* (1980 and 1981). Block 2 concentrates on the pupil and gives guidance on such matters as choosing an aspect of the curriculum for investigation, choosing a focus for the investigation, choosing a method of observation, choosing the method of recording the information, clarifying the collected information and, finally, considering the issues raised by the investigation. Block 3 focuses on the teacher. The course materials provide a useful focus for individual teachers, departments and local curriculum groups wishing to evaluate new procedures and materials.

Surveys of pupil opinion

Pupil opinions, reactions and perceptions may be surveyed in two quite separate ways. One way is to give pupils special tasks to do in order to gauge attitudinal change. It is usual to set such tasks before and after undergoing a course of study in the form of what are known as pre- and post-tests. The word test may not be the most appropriate word to use in this context, expecially if it is attitudinal change that is being measured. For example, if one of the objectives of a particular curriculum unit is to dispel stereotype images of a place all that may be necessary is to ask pupils to make lists, before the beginning of the unit and after its completion, of the mental images that they have of that place.

Another way of finding out what the pupils think is to conduct questionnaire surveys. Unfortunately surveys of pupil opinion of geography courses have not been widely reported. One of the few published surveys was concerned with the evaluation of the GYSL project with pupils of average and below average ability in four South Yorkshire schools (Hebden et al, 1977). The study was part of a larger research programme monitoring the impact of GYSL over a five-year period. Its main conclusions are worth repeating because they give some indication of the types of information that can only be obtained by asking the 'curriculum consumers'.

The survey involved nearly 400 pupils in 4th and 5th year classes. The 4th year pupils completed the survey twice, before embarking on GYSL and after completing one year of the project course. In the first section of the questionnaire a series of questions with a 'yes', 'no', 'don't know' format was designed to investigate the extent to which pupils liked or disliked geography. The responses suggested that pupils liked geography better after experiencing GYSL. 5th year pupils had generally more favourable attitudes towards geography than did 4th year pupils who had yet to begin work on the project. 4th year pupils recorded a greater liking for the subject at the end of one year of a course based on GYSL compared with their position a year earlier.

The second section of the survey consisted of a

Teacher activity

Teacher asking questions involving:

1. recalling facts and ideas
2. understanding ideas
3. offering hypotheses
4. making value judgements
5. pupil tasks
6. routine classroom matters

Teacher making statements about:

1. facts
2. ideas
3. values and attitudes
4. what to do (instructions)
5. routine classroom matters
6. small talk

Pupil activity

Pupil(s) answering solicited questions involving:

1. recalling facts and ideas
2. understanding ideas
3. offering hypotheses
4. making value judgements
5. pupil tasks
6. routine classroom matters

Pupils making unsolicited statements about:

1. facts
2. ideas
3. hypotheses
4. value judgements
5. pupil tasks
6. routine classroom matters

Silence during oral work

Transition between activities

Pupils engaged in written work

observation intervals

Figure 12.2. Observation schedule for recording teacher and pupil activity in a lesson.
Source: Boydell (1974, 1975).

list of resources and pupils were requested to signify whether or not they felt each resource appropriate for inclusion in a geography lesson. The responses indicated that experience of GYSL widened the pupils' views of the variety of resources appropriate to learning in geography.

The last section consisted of a number of invitations to comment freely on geography in general and their geography course and lessons in particular. It was interesting that a number of pupils indicated an appreciation that their views had been sought. The pupils made a large number of comments about fieldwork and visits, usually positive, and several pupils stressed the value of 'going and seeing'. Of the other learning activities to which reference was made, worksheets were rather more disliked than liked. Discussion was mentioned mostly in favourable terms, but the 4th year pupils showed much less support for it on the second occasion on which they completed the questionnaire.

The most significant response in the final section was that in terms of areal content, pupil interest was shown to be derived as much, if not more, from the wider world as from the local area. Though the local area obviously constitutes an indispensible resource in the geographical education of young people, the results of this survey showed that it should not be taken as marking the limits of interest and relevance for the less academic pupil. Yet views gathered independently from teachers in the surveyed schools at the time of the pupil survey indicated that they generally believed that places and issues not within the direct experience of average and less able pupils were of little interest to them. If for nothing else, the last finding showed the value of conducting pupil opinion surveys during the formative period of course development.

Assessment as an evaluative tool

Pupil assessment may be used to provide general information about the appropriateness and effectiveness of syllabus objectives, teaching materials, class organisation and teaching/learning styles as well as acting as a diagnostic tool for identifying individual pupil problems and as a means of classifying pupils. If assessment is used for a particular purpose then the fitness of the various assessment techniques for the particular task being undertaken needs to be carefully considered. When a particular technique or combination of techniques has been chosen, the test needs to be designed, administered, marked and analysed. In analysing tests use can be made of a well tried group of statistical techniques. Some of these involve tedious calculations but all are simple and straightforward. When calculations have been made, tables drawn up and graphs constructed, a second stage of analysis can begin: Why were certain questions found to be difficult? Was it the wording of the questions? Was it the ideas involved? Were the teaching/learning strategies used not appropriate for the range of abilities involved? Was insufficient time allowed to practise skills? These and other questions may lead the teacher to analyse in greater

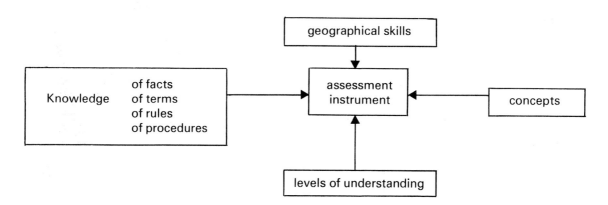

Figure 12.3. The four cognitive components of a geography assessment.

detail a selection of pupil scripts and perhaps discuss aspects of the test with the pupils. Then the crucial questions are reached, for example: What implications are there for my teaching style, the pace of the course, the nature of the resources, remedial work, test design? and so on. Clearly in all this test design and test analysis need very careful thought.

In terms of test design *validity* and *reliability* should be kept constantly in mind. There are different kinds of validity, the most important in this context being content validity. This is the extent to which a test or exercise samples the curriculum being tested. It means checking not only that a test fairly reflects the range of factual information in the curriculum being tested but also the main concepts, skills and levels of learning involved. Thus an assessment of cognitive abilities in geography may be seen in terms of four related components, as shown in Figure 12.3.

The levels of understanding component in geography assessments in the 11-16 age range have in recent years been based on Bloom's taxonomy of educational objectives (Bloom et al, 1956). Bloom identified six categories of ability within the cognitive domain: knowledge, comprehension, application, analysis, synthesis and evaluation. These were seen as being in an ascending order of difficulty, but this point has been a matter of contention. At a practical level it is clear that it is possible to set a difficult knowledge question and a rather easier comprehension question. Provided teachers do not follow the taxonomy too slavishly it is a useful tool in balancing the various abilities tested. A summary of Bloom's taxonomy as it may be applied to the 11-16 geography curriculum is given in Figure 12.4. For practical purposes the application and evaluation categories are often amalgamated; the analysis category is also often omitted on the grounds that it may in practice form

---

Knowledge: the pupil's ability to store information in his mind and to recall it later in substantially the same form. This may include knowledge of facts, terms, rules, procedures and theories.

Comprehension: the pupil's ability to know an abstraction (a concept or a skill, for example) well enough to be able to demonstrate it correctly when specifically asked to do so. This includes the separate abilities to translate data from one form to another (eg. to construct a climatic graph from temperature and rainfall figures), to interpret data and to predict trends from provided data.

Application: the pupil's ability, using accumulated knowledge and comprehension skills, to solve a new problem. Usually no mode of solution or procedure is specified.

Analysis: the pupil's ability to study a statement (which in the case of geography could be in the form of a map, diagram, graph or photograph as well as a prose passage) and separate it into its constituent parts, to distinguish between fact and opinion, between relevant and irrelevant and between cause and effect.

Synthesis: the pupil's ability to build up separate elements into a connected whole and so produce a unique communication on a topic, as in an investigation leading to an individual study or project.

Evaluation: the pupil's ability to make judgements about the value of various ways of solving problems or ways of tackling a problem.

Figure 12.4. Bloom's taxonomy applied to the 11-16 geography curriculum.

Question	Key idea 1	Key idea 2	Key idea 3	Knowl.	Comp.	Appl.
1a	1				1	
1b	1				1	
1c	1				1	
1d	1					1
1e	1					1
1f		3		3		
1g		2		2		
1	5	5	0	5	3	2
2a		1		1		
2b			6	3		3
2c			1	1		
2d		2		2		
2	0	3	7	7	0	3
3a			2	2		
3b			3		3	
3c		2		2		
3d		2		2		
3e		1		1		
3	0	5	5	7	3	0
4ai		1			1	
4aii		1				1
4aiii		1				1
4bi		1			1	
4bii		2			2	
4ci			1		1	
4cii			1		1	
4d		2		2		
4	0	8	2	2	6	2
Total	5	21	14	21	12	7

Figure 12.5. A specification grid for a test given to third year pupils.

part of the thinking involved in all the other abilities. A useful discussion of practical issues surrounding the use of Bloom's taxonomy by geography teachers will be found in Roe (1971) and Naish (1977).

One way of ensuring content validity is to compile a *specification grid* or *test blueprint*. This is simply a tabulated summary of the contents of a test. Omissions and unintended bias are thus easily identified and remedied. The construction of retrospective specification grids may reveal quite unintended weightings in exercises. The example shown in Figure 12.5 was constructed for a test on aspects of agriculture with third year classes (13-14-year-olds) in a large comprehensive school. It covered half a term's work. Note that besides indicating the balance between abilities that it also indicates the number of marks awarded to each of the three key ideas tested. The apparent imbalance in this respect in fact reflects the amount of lesson time devoted to the three key ideas.

Reliability, the second of the two important features of test design to which attention has been drawn, refers to the extent to which a test is dependable in measuring whatever it is that it sets out to measure. Measuring reliability is a rather technical matter (Satterly, 1983, Ch. 3) but in practical terms the main point to remember is that some assessment techniques are less reliable than others. The essay, for example, can be notoriously unreliable: not only may different examiners award markedly different marks for the same essay, but a marker may give widely different marks for the same essay when assessed at different times. Partly

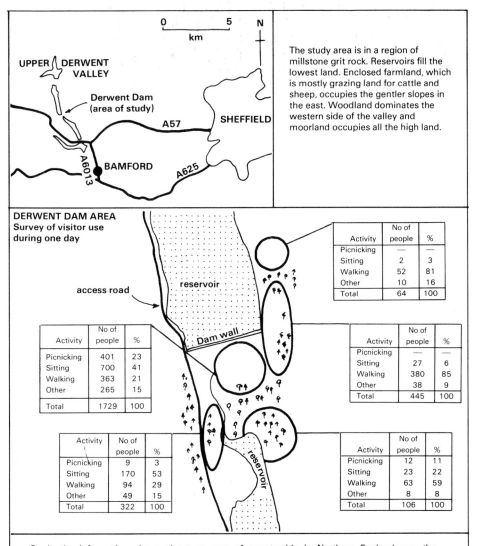

Study the information above about an area of countryside in Northern England near the Derwent Dam. It is an area which is frequently visited by motorists (there is no bus service). There is no planned provision for tourists and the planning board responsible for the area has made a study of the problems that have arisen from heavy visitor use and the improvements that visitors would like to see in the area.

(a) (i) Name the National Park in which the area is located.
    (ii) On what day of the week and in which season of the year do you think the visitor survey was undertaken?
(b) (i) Name the types of problems that are likely to have occurred in the area as a result of heavy and unplanned visitor use.
    (ii) Suggest ways in which the planning board responsible for the area can improve it as a tourist/recreational area. In making your suggestions you should bear in mind that it is necessary in such areas to conserve the landscape, cause as little disruption as possible to the economy and consider the interests of local people.
(c) For a *named* area of British countryside, describe its landscape and explain how this has influenced the types of leisure and recreational activity that are carried on there.

Figure 12.6. A structured question using a variety of data.

for this reason and partly because of the desire to test, explicitly, abilities other than recall, the essay is in relative decline as a means of assessing attainment in geography by pupils in the 11-16 age range. As structured questions and objective items have assumed greater significance in written assessments they are given further consideration below.

Structured questions

Structured questions, including data-response and stimulus-response elements, have become the most commonly used ways of assessment in teacher-set geography tests. Properly, a data-response question should be largely self-contained: the data provided should enable the pupil to draw conclusions without recourse to remembered facts, but of course such a question is set in the knowledge that a pupil has studied particular concepts and possesses certain skills. A stimulus-response question, as originally envisaged, is designed so that the data provided stimulates the pupil to write about a particular topic. For example, pupils may be provided with a photograph showing men and women working in a flooded paddy-field in the background; in the foreground, on a hillside women and young girls gather leaves from bushes on a tea plantation. The accompanying question invites pupils to describe and suggest reasons for the different kinds of agriculture in Monsoon Asia. The point is that having been reminded via the photograph of the existence of two types of Asian agriculture — subsistence farming and plantation agriculture — the pupils are not necessarily required to analyse the photograph in any detail; it is merely a stimulus. Invariably, structured questions used by geography teachers have data-response sub-sections and inevitably, because data are provided, include an element of data-stimulus, but they often go beyond these and include short essays and even objective items.

Structured questions usually begin with a statement which prescribes the limits of the question and directs the pupils' attention to the information that should be used in answering the various sub-questions. The information provided may be in the form of maps, photographs, prose passages, diagrams, graphs or tables or any combination of these. The internal structure of the question may vary. The data-response sections may be arranged in terms of increasing difficulty, the testing of different abilities (as in Bloom's taxonomy) or increasing generalisation. Sometimes all three changes occur more or less concurrently. Very often the last part of the question is in the form of a short essay related to the general theme of the question.

A number of advantages are associated with structured questions which account for their rise in popularity with geography teachers and examiners. First, they allow the teacher to test different abilities and skills within the same question and to quite consciously allot marks in varying proportions to the various abilities and skills. Secondly, because structured questions are prescriptive and sub-divided, the drawing-up of mark schemes is much easier than with essays and therefore tests made up of structured questions are likely to be more reliable than essay tests. Hudson (1973) has suggested that 'Perhaps the most important property of good structured questions is that they can test the ability of the pupil to deal with complex problem situations while at the same time defining clearly for him the nature of the answer required.'

Figure 12.6 is an example of a structured question which uses a variety of data: two maps, a short prose passage and statistical tables. As the question progresses the answers get longer and move, between (a)(i) and (b)(ii), from knowledge and comprehension to application and evaluation. The question ends with a short essay so that as a whole the question has tested at least four of Bloom's abilities.

Figure 12.7 shows part of a structured question which tests the ability to translate data from one form to another, in this case the basic geographical skill of compiling a map from written information. In doing so it involves the application of knowledge of points of the compass and the skill of using a linear scale. This of course raises an interesting point: application is, in Bloom's view, a higher order ability than translation. So is this a comprehension question (translation is a sub-category of comprehension) or an application question? In practical terms this may not be important; what is important is that geography teachers should decide what abilities they wish to test and then find the most appropriate means of testing them.

Objective items

The main type of objective items used in geography tests are multiple choice, multiple completion and matching pairs. A *multiple choice item* consists of

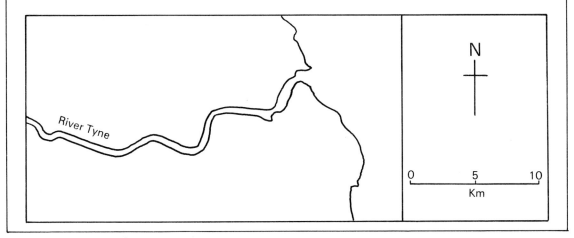

Figure 12.7. A question testing the ability to translate data from one form to another.

either an incomplete statement or a question (called the stem) followed by several, usually five, suggested answers or completions (called responses or options) one and only one of which is correct (called the key). The incorrect options (called distractors) should all be plausible. The options are usually lettered A, B, C, etc. and the pupil is required to choose the letter representing the correct response (Figure 12.8).

A multiple completion item, like a multiple choice item, consists of a stem followed by several responses. It differs from a multiple choice item in one important way: one or more of the responses may be correct (Figure 12.9.).

A matching pairs item consists of a stem and a series of responses followed by between two and four items. In this type of item a particular response may be used once, more than once or not at all. For example in Figure 12.10. the correct answers are: item 1, C; item 2, E; item 3, A; and item 4, C. Responses B and D are not used at all.

Advocates of objective tests have put forward some very persuasive reasons for their widespread adoption. Among these are their effectiveness in covering all aspects of a syllabus within one test (content validity), their reliability (in a good objective item there can be only one correct answer), their capacity to test a wide range of cognitive abilities and the ease with which they can be marked. Against this, though, must be set a number of disadvantages which have contributed to teacher wariness of their use in school tests. First, questions testing higher abilities are difficult to devise and involve the collection and preparation of many different sets of data. Secondly, a test of say 50 items takes a long time to prepare, especially if the items are previewed, pre-tested and then reviewed in the light of the pre-test before being used in an operational test. Thirdly, they cannot test the candidates' ability to marshall an argument and express it in writing. Finally, there is the problem of guessing, but this has been

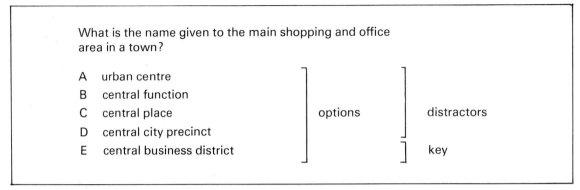

Figure 12.8. A multiple choice item.

over-emphasised. Provided that there is an adequate number of items (usually a minimum of 50) it is extremely unlikely that a pupil can obtain a good score through blind guesswork. For detailed discussions of objective testing in geography readers are referred to Hones (1973), Salmon and Masterton (1974) and Senathirajah and Weiss (1971); for a balanced discussion of the advantages and disadvantages of objective tests see Hudson (1973, Ch. 5).

As has already been pointed out, if tests are used as evaluative or diagnostic tools then their thorough analysis is critical. When essays and/or structured questions are used, analysis can take place in two main stages: a statistical analysis followed by a content analysis. In the case of objective items content analysis is less productive because there is little evidence in the completed tests; the only evidence is likely to be in the form of distractor choice where, if certain ones are chosen persistently, misconceptions can be investigated. This is the greatest weakness of objective items as evaluative and diagnostic tools: important questions that a teacher may wish to ask, for example, 'Why did that pupil think that distractor was the correct response?' cannot be answered from the internal evidence of the test. Interviews with pupils may remedy this deficiency but this is a time-consuming task and if done thoroughly amounts to reviewing the whole test with each pupil. In essays and structured questions evidence of failure to remember facts, of the way concepts are being grasped or models misapplied abounds and it must be remembered in what follows that content analysis and test review with the pupils is just as important as statistical analysis. What statistical analysis usually does is to point the direction in which further analysis should proceed.

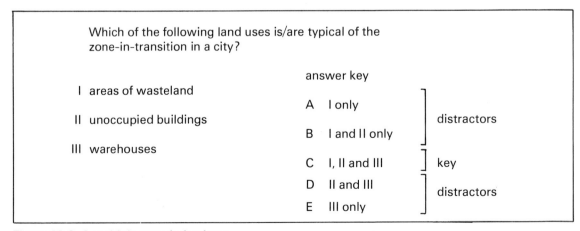

Figure 12.9. A multiple completion item.

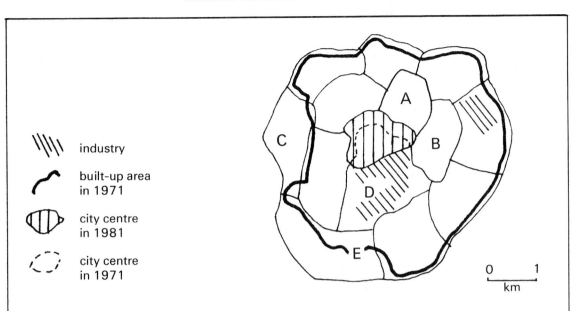

residential district	population		Persons living in:			Homes sharing or lacking:			hectares of open space per person
			rented accom-modation	council houses	privately owned homes	hot water	bath	inside w.c.	
	1971	1981	%	%	%	%	%	%	
A	20 000	17 000	61	9	30	32	69	80	0·70
B	15 000	17 500	43	8	49	2·2	48	50	1·65
C	420	9 000	6	93	1	0	0	0	3·90
D	23 000	21 000	46	8	46	15	28	15	0·75
E	4 500	20 000	1	0	99	4	8	5	4·35

The map above shows an industrial town somewhere in Britain. It is divided into 13 residential districts. Five of these are labelled A to E on the map. In the accompanying table information is provided about these five districts.

1. Which one of the districts (A to E) is an example of an outer suburban area made up entirely of local authority housing built in the last 10 years?

2. Which one of the districts (A to E) is best described as a well-planned suburb of modern owner-occupied houses?

3. Which one of the districts (A to E) has been depopulated in the last 10 years through the demolition of houses and the building of shops and offices in their place?

4. Which one of the districts (A to E) has had the greatest percentage increase in population in the last decade?

Figure 12.10. A matching pairs item.

Facility and discrimination

The two most important statistical measures that need to be derived are measures of *facility* and *discrimination*. Facility is a measure of the difficulty of an item or question, and discrimination is a measure of the effectiveness of an item or question in separating more able from less able pupils. In the case of objective items, facility is usually simply expressed as the percentage of all candidates attempting the item who selected the correct option and is called the *facility index*. A facility index can also be calculated for a whole test or for part of a test. In the case of structured questions the facility index is usually calculated by expressing the mean mark as a percentage. In a norm referenced examination a question at the appropriate level should have a facility index of 0.50 indicating that the mean mark is about half the total mark available.

To calculate a *discrimination index* it is usual to compare the performances of the top and bottom groups of pupils (as identified by the test as a whole) in each item or question. The top and bottom 27 per cent of candidates are usually used for this purpose. For example, in a group of 82 pupils who had completed an objective test, the top 22 and the bottom 22 would be compared item by item. If on an item 17 of the top 22 (77 per cent) and 6 of the bottom 22 (27 per cent) had correctly selected the correct option, the calculation would be 0.77–0.27, which would give a discrimination index of 0.50. Obviously the higher the discrimination index the more effectively an item has discriminated between the most and least able pupils and therefore reflects overall performance on the test. If the discrimination is very low the item has been unsuccessful in differentiating between pupils of different abilities. Of course if a test has been designed to test a homogeneous group of pupils then items will not have been designed with high levels of discrimination in mind and this type of analysis need not be carried out.

For teachers using structured question tests a useful way of measuring discrimination is to construct a series of bar graphs to show the mean scores of the upper, middle and lower bands of pupils (as identified by the test as a whole) on each question. A test consisting of three structured questions (each marked out of 15) has been analysed in Figure 12.11.

The total pupil group has been sub-divided into three groups, top (T), middle (M) and bottom (B) according to performance on the test as a whole. For each of the three questions a mean mark for each of the three groups has been calculated and converted into a facility index. The facility indices are also shown in the form of bar graphs. The

	T	M	B
Question 1			
mean mark	9·3	4·9	0·9
facility index	0·62	0·32	0·06
Question 2			
mean mark	5·75	5·25	1·15
facility index	0·38	0·35	0·10
Question 3			
mean mark	10·2	9·8	1·85
facility index	0·68	0·65	0·12

Figure 12.11. Question discrimination in a test.

variations in the way the questions have worked are very clear. Question 1 has discriminated well but in question 2 the performance of the top group is depressed, whereas in question 3 the performance of the middle group seems insufficiently differentiated from that of the top group. Clearly queries are raised about the content of the questions, their wording and their relationship with what was taught and how it was taught.

As Deale (1975) has reminded us, the assessment process is 'a rather complex jungle in which there are numerous false trails and pitfalls for the unwary.' Care must be used in employing the various evaluative tools discussed above. In particular, the discussion of test analysis here has been necessarily brief and teachers preparing to analyse their tests are advised to consult general texts on assessment and articles on specific topics. In this respect Marsden's (1974) article on objective test analysis and the general discussions on assessment techniques and methods of analysis in Hudson (1973), Deale (1975), Macintosh and Hale (1976) and Satterly (1983) will be found particularly useful. A bibliography on assessment in geography has been produced by Jones (1983).

Assessment mix and question choice

Inspection of public examination papers at 16+ reveals a great variety of practice in examining candidates. The mix of assessment techniques and procedures has evolved from the combined force of a number of factors, including the generally held view that geography examinations should test other abilities besides recall; the fact that some assessment methods are better than others at testing particular abilities; the knowledge that candidates are better at some tasks than others; the growing concern about the possible unfairness of testing attainment on a two-year course in one or two terminal examination papers; and the existence of more than one paradigm of geography so that courses are likely to have different organising principles.

One of the most striking features of examination papers is that whereas in some terminal papers there is wide choice, there is much less choice in others and no choice at all in some. There are supposed advantages in question choice, for example, they allow a certain amount of freedom in sampling the syllabus and allow candidates to show themselves in the best possible light by attempting those questions where they feel confident of producing good answers. Nevertheless there has been increasing disquiet over the possible negative effects of such choice on pupil performance. Even where only restricted choice has been retained there is some evidence that candidates are not always able to make suitable choices. This has been aggravated by the more widespread use of structured questions which has resulted in candidates having to make snap judgements when faced with complex data. In some examinations candidates may have to contend with a question and answer book between 25 and 40 pages in length. Little wonder that candidates are sometimes attracted to questions to which they do less than justice. The tendency for overwhelming numbers of candidates to answer the first question in each section has been a marked feature of a number of examinations.

Concern over the effects of question choice was expressed by Taylor and Nuttall (1974), who carried out an experiment with 16-year-old candidates sitting a public geography examination. The method used was to bring candidates back at the end of the two 'choice' papers and to ask them to attempt as many of the other questions as they could in a different coloured ink. 186 candidates took part in the experiment on paper 1 and 209 on paper 2. The results, concluded the authors, were encouraging in one respect but disquieting in another. Well over half did better in the real examination than in the experiment; these candidates seemed to have chosen wisely. What was worrying, however, was that 25 per cent of the candidates obtained better grades on the experimental paper than on the real paper — in spite of the fact that they had already chosen what they believed to be their best questions and had already been working for 2 hours before tackling the experimental paper.

Another reason for the move towards restriction of choice and compulsory questions is the content validity of the candidates' completed papers. A paper with 12 questions may be perfectly valid in syllabus content terms *before* the examination begins, but if the candidates are then asked to answer any five questions there are 792 combinations of questions and therefore *792 potentially different examinations,* each one of which would be highly suspect in content validity terms.

As far as reliability is concerned, it has already been pointed out that one of the supposed advantages of structured questions is their reliability and one of the supposed disadvantages of essays is their unreliability. Internal investigations carried out by examination boards have tended to confirm these generally held views. In one such piece of research an examination board (whose identity will not be divulged for reasons that will become clear) made a random selection of several hundred scripts from each of its examinations and had them re-marked by experienced examiners. Before re-marking took place all marks and annotations made by the original markers were removed. Geography emerged well from this re-marking; at the other extreme the results of the exercise in the case of English Literature were rather disturbing. The geography examination consisted of two terminal papers made up of structured questions; in English Literature the examination was in the form of one paper of essay questions. In geography the average mark change was ± 2; in English Literature it was ± 7. On looking at the different magnitudes of change the differences between the examinations were even more striking. Only 7 per cent of geography candidates had a mark change of more than 5 marks in contrast with 55 per cent in English Literature. In terms of grade changes, in geography less than 3 per cent of candidates would have had a different grade; in English Literature the figure was nearly 50 per cent. Obviously a combination of factors contributed to these differences, but question types (structured questions instead of essays), marking schemes (detailed schemes instead of impression marking) and assessment procedures (two papers instead of one paper) can be expected to have been very influential. How we measure attainment is as important a consideration as what we measure.

Norm- and criterion-referenced testing

Norm-referenced assessments are those in which levels of performances are related to the distribution of attainments in a population as a whole and which, when plotted in the form of a histogram, will tend towards the normal curve of distribution with the largest number of occurrences in the middle and a symmetrical tailing off at the top and bottom ends of the range. Although adjustments are made from year to year and boards use statutory criteria at some grade boundaries, this is, in general terms, the model traditionally used in most public examinations. Theoretically, the aim of examiners is to devise assessments in which pupils of average ability in a particular population will achieve about 50 per cent of the marks available, in which the very few brilliant candidates obtain 100 per cent and the least able 0 per cent. In practice there is much more bunching than this and a mark range of 15-85 on a 100-mark examination would be counted a great success. Clearly an examination in which all candidates obtained 45-55 marks would be reckoned to be very poor as there would be major problems in differentiating between candidates of different abilities.

Norm-referenced examination systems have been criticised because of the stifling effect that they may have on the curriculum, especially when combined with the practice of retaining question-setters for long periods. There is a sort of inertia in which the same types of questions on the same groups of topics become the norm. Under these circumstances critical re-appraisal and incremental improvement (as opposed to streamlining of existing procedures and techniques) tends to be inhibited. Tolley and Reynolds (1977) describe how this tends to lead to a 'vicious circle of curriculum underdevelopment'.

Another weakness of this approach to testing is that an aggregate mark conceals differences in the degree of mastery of particular cognitive abilities. For example, a mark of 45 per cent in an examination by a particular candidate may reflect a moderate ability to read and interpret Ordnance Survey maps and to interpret other types of data, the failure to recall appropriate factual information and a rather disorganised style of writing. For other candidates exactly the same mark may reflect a very different combination of strengths and weaknesses. A criticism of norm-referenced grading, therefore, is that it provides no useful information for teachers, parents or prospective employers about the meaning of particular grades in particular examinations. What, for example, can a candidate with a certain grade in geography at 16+ do, how well, and in what circumstances?

The move towards criterion-referenced grading at 16+ is, it must be said, much more related to central government pressure and the need to provide information for users than to concern over curriculum stagnation and undesirable backwash effects on teachers and pupils. In criterion-referenced testing the object is the mastery of previously stated objectives as in a driving test. There is no constraint on the number of candidates who may demonstrate competence at an appropriate level. How far it will be possible to introduce this type of grading into public examinations remains to be seen.

It may be putting the cart before the horse to offer grade descriptions before carefully considering whether a workable system can be devised that can identify attainment in particular abilities. Surely it needs to be demonstrated that examinations can be designed to consistently identify different degrees of competence in those abilities and skills that are thought to be appropriate to a course in geography for pupils in the 14-16 age range. This means that much more consideration needs to be given to the nature of assessment objectives, the fitness for purpose of the various assessment procedures and techniques (assessment mix) and question choice. It also suggests — if a certain grade is to have the same meaning over the whole country — a greater central control of curricula. The twin issues of criterion-referenced examinations and grade descriptions are likely to be debated hotly for many years.

The move towards examinations based on national criteria raises a number of important questions which have yet to be resolved. Firstly, to what extent will the general and subject-specific criteria be used as tools of curriculum control? Secondly, to what extent will the new administrative arrangements permit large-scale teacher involvement in the construction of syllabuses and control of assessment? And finally, what will be the effect of the new arrangements on curriculum review and renewal? Agencies other than examination boards, such as individual schools, local groups and curriculum developers, have been responsible for the initiation and subsequent development of some of the most exciting examination schemes. The loss of this vital developmental and entrepreneurial role by practitioners and the re-emergence of a closed, external and largely inert system will not serve the needs of pupils, teachers or the community at large.

References

Bloom, B. S. (ed) (1956) *Taxonomy of Educational Objectives: Handbook 1: Cognitive Domain*, Longman.

Boehm, A. E. and Weinberg, R. A. (1977) *The Classroom Observer*, Teachers College Press.

Boydell, D. (1974) 'Teacher pupil contact in junior classrooms', *British Journal of Educational Psychology*, vol 44, no 3, pp. 313-18.

Boydell, D. (1975) 'Pupil behaviour in junior classrooms', *British Journal of Educational Psychology*, vol 45, no 1, pp. 122-29.

Cutts, P. D., Higginbottom, T. and Jones, M. (eds) (1981) *The Geographical Curriculum 8-14: Planning and Practice*, City of Sheffield Education Committee.

Deale, R. N. (1975) *Assessment and Testing in the Secondary School*, Evans/Methuen.

Harlen, W. (ed) (1978), *Evaluation and the Teacher's Role*, Macmillan.

Hebden, R. E., Jones, M., Parsons, C. and Walsh, B. E. (1977) 'Changing the geography syllabus: what do the pupils think?', *Teaching Geography*, vol 3, no 1, pp. 30-33.

Hones, G. (1973) 'Objective tests in geography', *Geography*, vol 58, no 1, pp. 29-37.

Hudson, B. (ed) (1973) *Assessment Techniques: An Introduction*, Methuen.

Jones, M. (1983) *Assessment and the Geography Teacher*, Bibliographic Notes, no 21, Geographical Association.

Macintosh, H. G. and Hale, D. E. (1976) *Assessment and the Secondary School Teacher*, Routledge and Kegan Paul.

Marsden, W. E. (1974) 'Analysing classroom tests in geography', *Geography*, vol 59, no 1, pp. 55-64.

Marsden, W. E. (1976) *Evaluating the Geography Curriculum*, Oliver and Boyd.

Naish, M. (1977) 'The development of children's thinking in school geography', *Teaching Geography*, vol 3, no 2, pp. 81-84.

Open University, (1980 and 1981) *Curriculum in Action: An Approach to Evaluation.* Block 2, *The pupils and the curriculum* (1980); Block 3. *The teacher and the curriculum* (1980); Block 5, *Observing classroom processes* (1981) and Block 6, *Measuring learning outcomes* (1981).

Roe, P. E. (1971) 'Examining CSE geography', *Geography*, vol 56, no 2, pp. 105-111.

Salmon, R. B. and Masterton, T. H. (1974) *The Principles of Objective Testing in Geography*, Heinemann.

Satterly, D. (1983) *Assessment in Schools*, Basil Blackwell.

Senathirajah, N. and Weiss, J. (1971) *Evaluation in Geography: A Resource Book for Teachers*, The Ontario Institute for Studies in Education.

Taylor, E. G. and Nuttall, D. L. (1974) 'Question choice in examinations: an experiment in geography and science', *Educational Research*, vol 16, pp. 143-50.

Tolley, H. and Reynolds, J. B. (1977) *Geography 14-18: A Handbook for School-Based Curriculum Development*. Macmillan.

Walker, R. and Adelman, C. (1975) *A Guide to Classroom Observation*, Methuen.

12.2. Coursework Assessment 14-16
Sheila Jones

When the Geography 14-18 Project was initiated much of the planning was a joint venture between the project team and the staff of the pilot schools. On one planning issue there was unanimity — that of the inclusion of a continuous assessment element in the examination. The only disagreement concerned the proportion of the examination marks to be assessed in this way. The outcome was 50 per cent for the final examination and 50 per cent for continuous assessment (30 per cent for coursework and 20 per cent for an individual study).

All involved agreed that such a means of assessment offered three major advantages:

(a) the opportunity for pupils to produce in-depth studies without the pressure of time allocation as in a formal examination;

(b) the scope and incentive for teachers to incorporate innovative materials within these assessments;

(c) the encouragement of discussion between teachers and pupils so that the latter could evaluate and clarify their ideas and judgements, developing an increasing ability to monitor their own progress.

The 50/50 structure was retained until the 1981-83 examination cycle. The weighting given to continuous assessment in the 1982-84 and subsequent cycles, however, has been reduced to 40 per cent (24 per cent for coursework and 16 per cent for the individual study). At the same time the number of coursework units has been reduced from 5 to 3. Each unit now represents 8 per cent of the marks for the final examination compared with 6 per cent when there were 5 units.

This change has reduced the workload of both pupil and teacher. As the number of schools participating in the examination has increased, so some teachers who were new to the course found the workload of the 5 unit course too great. It was also argued that 3 units are more realistic for the average pupil. The timing of units in the latter half of the Spring term in the 4th year and in the Autumn and Spring terms of the 5th year allows for the increasing maturity of pupils during a two-year course.

The three units of coursework assessment are designed, implemented and marked by teachers. Each unit takes up the normal geography teaching time for about two weeks and constitutes a study which is not easily assessed in a final examination paper. It may involve, for example, skills which require perseverance and initiative or creativity and originality on the part of the pupil. The study may be specific to a school's local area or it may utilise a teacher's special interest or expertise.

One coursework unit is devised for each of the three categories:

1. Regional or synoptic studies;
2. Studies of the physical environment;
3. Planning problems.

In designing coursework units there is as much concern with the testing of educational objectives as with assessing geographical knowledge. The objectives for the planning and assessment of individual studies are also relevant to coursework:

(i) knowledge of relevant concepts, generalisations and facts;

(ii) reflective or thinking abilities;

(iii) attitudes, feelings and sensibilities;

(iv) facilitative skills.

At the start of each examination cycle the geography staff at Colston's Girls' School consider the possibilities of coursework units in relation to the current course. Changes occur in the school's course organisation and content and, true to the Project's aim of encouraging curriculum renewal, coursework units are both modified and replaced as the department attempts to innovate and experiment. Some units prove more successful than others and are used regularly, others are used once and then abandoned.

The physical unit currently completed by fourth year pupils at the school will be used here for the purposes of illustration. It is concerned with the study of soils and incorporates practical soil study, an analysis of statistical information on a soil investigation, and consideration of problems of soil erosion with particular reference to the Tennessee Valley Authority. A testing piece of work, it has proved a good discriminator.

The coursework unit on soils requires the pupils to consider the following ideas:

1. Soil and vegetation result from a variety of *processes* which vary in their influence, eg. local geology and climate. Land use is the result of an even more complex relationship between a variety of physical and human factors.
2. Soil and vegetation and land use form a *system* in which soil is a complex body that differs from rock and is constantly changing. Vegetation and land use are varied in type and are also constantly changing.
3. Soil and vegetation are important in many varied *environments,* eg, primitive economies, advanced economies, recreation, conservation. Soil is the basis of plant life, natural or cultivated, and it is upon this plant life that all animals (including man) depend.

For the first part of the unit the pupils have to collect and analyse two soil samples of contrasting origins, taken from different vegetation or land use areas. The pupils are given worksheets containing a series of exercises. The first two are compulsory and *one* other exercise has to be chosen.

1. Soil collection and soil acidity.
2. Soil depth and stoniness.
3. Soil profile.
4. Moisture content.
5. Moisture retention.
6. Air content.
7. Soil texture.
8. Infiltration.

The first six exercises of the coursework are shown in Figure 12.12. It will be obvious that the practical work in these exercises could not be done as part of a timed, written examination. The pupils have to conclude this part of the unit with a concise statement (not more than two sides of file paper) outlining their main conclusions from their analysis of the two soil samples. Small samples of each soil have to be handed in with the written work. Each exercise and the conclusion carry 4 marks (total 16 marks).

The second part of this coursework unit consists of a consideration of the problems of soil erosion and its consequences together with methods of prevention and cure. The pupils are given an account of the work of the Tennessee Valley Authority and answer a number of questions on it (5 marks). In conclusion they comment on the following quotation: 'Soil erosion should be made an offence like careless driving. After all it is careless farming' (4 marks).

The marking scheme is devised at the same time as each new coursework unit is prepared. Each unit is always marked out of a total of 25. This has the advantage of continuity in marking schemes, comparability with examination question design and ease of cross-reference from one unit to another. The unit design has the same plan as the final examination questions, ie.

i   A section which involves some form of data response. This carries the smallest proportion of marks.

ii   A section containing some element of investigation. This carries the greatest proportion of marks.

iii   A section devoted to analysis and original comment. The mark allocation is similar to that for (i).

Thus each unit aims to give all pupils opportunity for success in (i) and (ii), whilst the most able

Exercise 1. Soil collection and soil acidity.

Aim. To collect two contrasting samples of soil use for analysis and measure their acidity.

Equipment. Polythene bags; spade or trowel; labels; clinometer made from protractor; compass; notebook; OS and geology maps.

A   In the field: collecting soil.

Choose two contrasting sites with different vegetation or land use, eg. coniferous or deciduous woodland, arable or pasture land, garden or heath.

1. At first site note location, geology, altitude and land use; write these on label.
2. Measure angle of slope with clinometer; angle will normally be a small one.
3. Decide on aspect, ie. direction which slope faces, or which *you* face looking *away* from slope.
4. Take as deep a soil sample as possible, preferably at 1-2 metres. You will need about 1-2 kg, especially if doing Exercises 3 or 4. Place sample in bag and label it.
5. Note colour (dry and moist).
6. Note texture (plastic, sticky, sandy, peaty).
7. Use above information and maps to complete results sheet (not reproduced here).
8. Repeat instructions 2-8 at second site.

B   In the classroom: measuring acidity.

Use the BDH Soil Testing Kit to measure the acidity of the soil.
Note. Do not let any of the materials touch your hands as this will distort your results.

1. Place 3 cm sandy/2 cm loamy/1 cm clayey soil in the test tube.
2. Make up to 5 cm with barium sulphate powder.
3. Add distilled water up to lower mark.
4. Top up with BDH Indicator to upper mark.
5. Replace stopper and shake well.
6. Leave to settle and compare colour of liquid with colour chart (hold test tube against chart).
7. Note the pH value.
8. Repeat the experiment with second soil sample.

Conclusions. Compare the pH values for the two soils and try to explain the similarity or difference in terms of the factors recorded in the field, eg. land use, geology, slope.

Exercise 2. Soil depth and stoniness.

Aim. To examine the relationship between soil depth and stoniness.

The fieldwork for this exercise has already been undertaken. 40 sites were visited in the southern Cotswolds (between Bath and Cirencester). The sites were chosen on a random basis after making sure that they were all on Jurassic Limestone. At each site a measurement was made of soil depth and the percentage of stones by weight. The figures are therefore from one soil type and show the variation that might be expected within that type. The results for the 40 sites are presented in the table (*not reproduced here*).

A   Examine the table and then:
  1. Calculate:
     (a) the average soil depth for all sites.
     (b) the average percentage of stones.
  2. Draw histograms to show:
     (a) the frequencies of soil depth.
     (b) the frequencies of soil stones.
  3. Draw a scatter graph, plotting soil depth against percentage of stones. Plot all of the 40 sites and then draw a line of best fit through the points.

B   Consider these calculations and graphs and answer the following questions:
  1. Using the information from 1(a) and 2(a) comment on soil depth.
  2. Using the information from 1(b) and 2(b) comment on soil stoniness.
  3. What is the relationship between depth and percentage of stones? Use the graph from A3. Try to give an explanation for any relationship that you think exists. Remember that limestone is a rock that tends to break up easily.
  4. Write a report to a farmer, giving details of your conclusions. Remember that he will know nothing of your work or any technical details — so you will have to explain things carefully and clearly.

Figure 12.12. A fourth year coursework unit: the study of soils (continued overleaf).

Exercise 3. Soil Profile.

Aim. To make a record of the soil profile of two sites, and to measure the depth of the soil and the depth of each horizon within the profile.

Equipment. Trowel or spade; ruler (metric) or tape; notebook; some sticks; plain paper; double-sided sellotape; clingfilm.

Instructions.

1. Note details of site and of soil exposure accurately as in Exercise 1.
2. Scrape away the surface which might be weathered to get a *clean new* 'face'.
3. Find the different layers in the soil and mark the junctions between each with sticks. Measure the depth of each horizon and record carefully in notebook. If there are no distinct horizons mark out the profile with sticks at 10 cm intervals and describe the charactertistics of each 10 cm.
4. Describe the colour of each layer, eg. Is the top layer very dark or only light brown? Can you see any sign of roots? Is the second layer red or grey? Is it sandy? Are there stones in it? Is there any evidence of worms? In the lowest horizon, what size are the stones? Do they change in size? What sort of rock? (Bring some to school for identification.).
5. Using a suitable scale make a coloured diagram of the profile, marking in all the characteristics you have noted.
6. Make a sealed diagram to scale using soil and sellotape. Cover it with clingfilm.
7. Repeat for the second site.
8. Write your observations and comments.

Exercise 5. Moisture Retention.

Aim. To discover the potential of a soil to absorb water.

Equipment. Medium sized can (baked bean tin); enough soil to fill the tin; a piece of butter muslin or gauze or an old tight leg; a large bowl; a measuring jug; a litre of water.

Instructions.

1. Remove lid and base from tin; secure tightly over one end a piece of muslin, etc. to act as a sieve.
2. Place a 'core' sample of soil in the tin. Note whether it is wet or dry.
3. Hold the tin carefully, with the sieve at the lower end, over the bowl.
4. Pour 1 litre of water *slowly* into the tin of soil and allow it to pass through the soil and drip into the bowl.
5. Wait until the dripping stops.
6. Measure the amount of water which has passed through the soil using a measuring jug. Record the result.
7. Subtract the amount of water in the bowl (A) from 1 litre to obtain the amount (B) left in the soil.
8. Calculate $\frac{B}{1}$ x 100 to obtain the percentage moisture retention.
9. Repeat for the second sample using the same amount of soil.
10. Write your observations and comments.

Exercise 4. Moisture Content.

Aim. To investigate wet and dry samples of the same soil, and to discover its moisture content.

Equipment. Medium sized can (baked bean tin); enough soil to fill the tin; foil to 'cap' the tin; scales; the use of an oven (warn the family that there may be a smell).

Instructions.

1. In the field push soil firmly into tin to collect a 'core' in the tin. Seal to prevent moisture loss.
2. At home remove seal and weigh tin and soil together. Record weight (A).
3. Put the open tin containing the moist soil into the oven at a very low heat (100°F; 38°C; Gas Regulo ½). Leave for 24 hours. If it is not possible to leave the tin in the oven for this length of time, put it in the oven after a roast meal has been cooked and leave it in the oven until about 24 hours later.
4. Take the tin of dry soil out of the oven and weigh it again. Record weight (B).
5. Subtract (B) from (A) to obtain the weight of moisture (C) in the original soil.
6. Calculate $\frac{C}{A}$ x 100 to obtain the percentage moisture content.
7. Repeat for the second sample. The 'cooking' can be done at the same time but a similar size tin and weight of soil should be used.
8. Write your observations and comments.

Exercise 6. Air Content.

Aim. To find out whether the air held in the pore spaces of the soil amounts to about 25 per cent of the total soil matter.

Equipment. Medium sized can (baked bean tin) with top removed; newspaper; rolling pin.

Instructions.

1. Push the tin into the soil to collect a core of soil in the tin.
2. Dry the tin of soil at very low heat (100°F; 38°C; Gas Regulo ½) for a period of 24 hours. Keep soil in the tin.
3. Calculate the volume of soil in the tin using $\pi r^2 h$, where h is the height to which the soil reaches in the tin. Call this volume A.
4. Shake the soil out of the tin on to a sheet of newspaper. Cover with another sheet and crush the soil very finely with a rolling pin. In this way you get rid of air spaces but you will not be able to remove all of them.
5. Put the soil back into the tin and again calculate the volume of soil in the tin using $\pi r^2 h$, where h is the new (lower) height to which the soil reaches in the tin. Call this volume B.
6. Subtract B from A to find the air content (C).
7. Calculate $\frac{C}{A}$ x 100 to obtain the percentage air content.
8. How does your result compare with the average of about 25 per cent?
9. Repeat for the second sample.
10. Write your observations and comments.

have the chance to demonstrate their potential in (iii). Mark schemes are agreed within the department before marking begins, but if it is discovered that the scheme for a new unit is in reality either over-generous, over-ambitious or imbalanced, then modifications are made at the internal moderation meeting.

Internal moderation always take place as soon as possible after coursework units have been completed and normally involves the checking of about a quarter of the pupils' studies. These sessions are invaluable in the department because they ensure a degree of comparability of standards at all levels of teaching. New colleagues can immediately become aware of departmental policies and can make their criticisms and suggestions. Thus the department's marking system is reinvigorated and stagnation of standards prevented.

The school has been lucky in its external moderators. All have had a close accord with the department's aims and objectives, and criticisms have usually reinforced existing doubts or have recognised flaws in structure. One slight problem is that of timing, for not surprisingly moderators may go weeks without receiving work and then may receive several batches from different schools in a short period. This means that there may be delays in the return of moderated material.

The new procedure in which the moderator gives more time to working with schools to develop new units, rather than to moderate old ones with defined marking schemes, is to be welcomed. The development of new coursework units is particularly time consuming for spontaneous innovation is the gift of few people. Most new units are team efforts developed and enlarged from a 'bright idea' arising from departmental expertise, a specific interest or an outside influence. Thus the presence of the external moderator at this stage can provide the necessary stimulus for curriculum development.

Finally, what of pupil reaction and evaluation? This is most difficult to establish. Most pupils understandably want to be successful, so the main interest is always in what grade has been achieved. The more able pupils take full advantage of the discussion following the return of work and some ask personally what went wrong. The less successful probably 'write off' each study and hope for better things next time, but do not always take note of the relevant comments on the returned studies. Geography is not considered an easy option but many worthy, hardworking pupils realise that continuous industry is rewarded and their confidence is enhanced for the final examination. It is often the visiting former pupil who comments on the way in which this type of study trained her in good work practice.

It has been suggested by Tolley (1984) that school-based assessment is integral to good teaching, learning and examining, whilst Orrell (1984) has emphasised that coursework units should measure the pupils' active learning — the process, not the product. There is no doubt that devising and assessing coursework units is worthwhile and rewarding for teacher and pupil, but it is not a method of assessment to be undertaken without the realisation that it increases the workload of both participants.

References

Graves, N. J. (1979) *Curriculum Planning in Geography,* Heinemann.

Hickman, G., Reynolds, J. and Tolley, H. (1973) *A New Professionalism for a Changing Geography,* Schools Council.

Orrell, K. (1984) 'Designing coursework assessment units', in Orrell, K. and Wiegand, P. (eds) *Evaluation and Assessment in Geography.* Geographical Association.

Tolley, H. (1984) 'A fresh look at the contribution of school-based assessment to examinations', in Orrell, K. and Wiegand, P. (eds) *Evaluation and Assessment in Geography.* Geographical Association.

Tolley, H. and Reynolds, J. B. (1977) *Geography 14-18: A Handbook for School-Based Curriculum Development,* Macmillan Education.

University of Cambridge Local Examinations Syndicate (1982) *Schools Council 14-18 Geography Project,* Subject No 2202. Syllabus Booklet.

12.3. Advanced Level Examinations
David Hall

An overview

A notable feature of A-level syllabuses is differentiation of outlook and of methods of examining. There are marked differences in the approach to content, to the availability or not of choice in papers, the presence of core questions, of short answer or data response and stimulus response questions, either as separate papers, separate sections, or just mixed in with essay questions. Some boards have practical papers or technique papers and the stress on quantitative methods is uneven. Fieldwork may be encouraged, but not compulsory, and individual projects may be optional and either marked externally, sometimes including an oral, or be internally assessed. Recently decision-making papers have been introduced where values analysis is required in response to unseen stimulus material provided in the examination.

Different viewpoints as indicated by syllabus design are shown in Figure 12.13. The Oxford and Cambridge Board and the Associated Examining Board have undertaken comprehensive revision from first principles, starting with explicit statements of the aims and objectives of the examination *as a whole,* and designing question styles in support of such statements. Thus, for Oxford and Cambridge, there is a lucid statement of the systems approach to be applied in both the human and the physical papers. The AEB specifies an approach through man and his behaviour in an ecological and spatial context. The AEB candidate does not study the human and physical parts of geography separately: man is placed centrally in the seven topics listed (from population to current problems) to merge the physical and the human into a total man/environment approach. Oxford and Cambridge use a techniques paper to cement a connection between the study of human and physical processes. London has been equally ruthless in the use of structured and essay questions to test human and physical geography as separate sections in single papers, and applying the ideas gained from such study in an integrated form in a third paper of area studies cast in a modern rather than descriptive vein. The Joint Matriculation Board sets separate papers for physical and human geography in Syllabus B, together with a practical paper which is an alternative to project work.

Other boards have favoured the 'additive response' to syllabus revision rather than tackled the problem of subject change root and branch. Cambridge offers five papers, with only the first paper compulsory, where the questions are essay choice. The concept of core content is weak. Oxford simply offers three papers across the three conventional divisions of physical, human and regional geography, but considers three questions per paper a more sensible requirement than the four required by Cambridge under similar time-limit conditions. The 'core' is in this case the obligation of all candidates to study three divisions — a common rather than core curriculum.

The issue of A-level syllabuses remains open following the brief put by the Secretary of State for Education and Science that Inter-Board Working Parties should undertake 'the clarification and rationalisation of syllabuses' and 'identify where possible an appropriate common core'. The draft for geography was completed in 1982 and published in 1983 (GCE Boards, 1983). As Daugherty (1982) observes 'It could have a marked influence on future developments, or it may gather dust on the shelves alongside the many excellent papers which the N and F debate gave rise to'.

Syllabuses vary, however, not just because

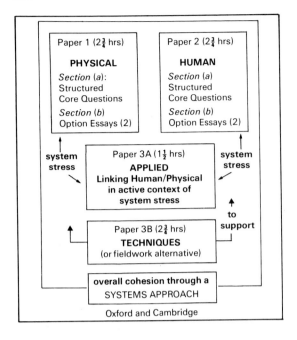
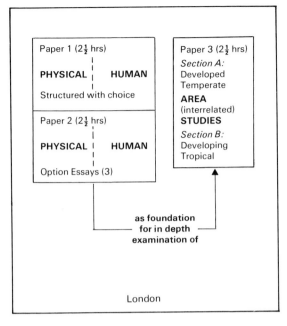
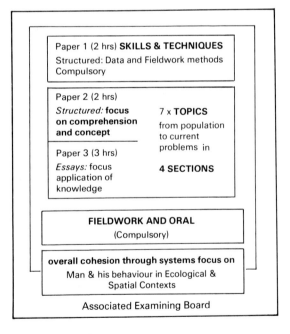
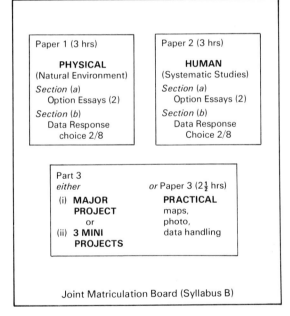

Figure 12.13. Different viewpoints about A-level geography as indicated by syllabus design.

geography is a broad discipline capable of many interpretations, but because each board has evolved its own way of working and set of beliefs about the principles of examining. Thus syllabuses are the product of two distinct forces inside the system: (i) the impress of strongly-minded chief examiners who may have forceful views about the nature of the subject for this age group but whose classroom contact with them is usually remote, and (ii) the charisma of the secretaries and their subject officers who will be concerned with efficiency and cost, as well as the reliability and validity at the statistical level, of the examination. It is, as Becher and Maclure (1978) have commented, the formidable strength of this combined professionalism whose main aim is with a merciless routine under pressure of tight time schedules, which exerts a powerful control over the A-level curriculum.

The mechanics of examining

Figure 12.14 indicates the general processes operated by the boards, under the five staged headings of *syllabus, examination papers, administration, the award,* and *feedback* to the schools.

New *syllabuses,* or revisions to syllabuses, will be initiated or approved by subject panels, although the substantial work will be undertaken by a subcommittee. Whilst some teachers, courageous and diligent panel members, and thick-skinned subject officers may push for change rather than merely obstruct it, external forces also influence the process. In particular, until 1983 the Schools Council had the remit to approve any new or substantially revised A-level syllabus through its own subject committee and also scrutinised every year two of the GCE Board's examinations in geography A-level on a rotating basis. About every five years, therefore, two assessors scrutinised the syllabus, mark schemes, the marking and the awarding procedures of a particular board. Since 1984 the Secondary Examinations Council has inherited this function, and it may be that a more directive role is envisaged than has been the case in recent years, when it has not been unknown for the criticisms of the assessors to be discounted, and merely noted rather than acted upon.

The chief examiner drafts the *examination paper.* It is the reviser's responsibility to check that the wording of each question is intelligible, and also to point out areas of overlap between questions which would allow content to be repeated and reduce the validity of the examination paper. The reviser will also check that the spread of questions fairly reflect the breadth of the syllabus, and are pitched inside the tenets of the syllabus aims and objectives. Much time can be taken at this stage, particularly where differences of view occur. For example, short questions are easy to read but may be more ambiguous than a question which is subdivided. But sectionalised questions, even when sub-marks are indicated on the paper, often produce unbalanced responses and create their own inflexibilities at the marking stage. Such issues often reflect the lack of consistency in the syllabus itself, and ambiguities in the aims of the examination. But decisions have to be taken, usually at a moderation meeting, with the chief examiner having the ultimate power to direct.

Candidates rightly expect examination papers to be of good reprographic standard and logically arranged. Typesetting, origination and graphic layout are vital in geography, and although excellence can be achieved, some boards lack the resources or the returns to scale (candidate numbers) which will finance this work by subcontracting or direct labour to a desirable standard. In this sense of publishing, geography is seen by board secretaries as a costly examination which is always seeking to cut its cloth far in excess of its fee income, and many uneasy compromises have to be accepted by chief examiners. If other subjects can be reliably and validly assessed by a simply worded set of essay questions, how can complicated systems with oblique photographs, Ordnance Survey maps, diagrams, and original material be justified? Even schools have been known to complain at the amount of information and complex instructions required by a geography examination!

A year has usually elapsed between the first draft of the papers and their printing in their final form. At Easter, the chief examiner will prepare a mark scheme, which will be circulated to all assistant examiners with the papers near the date of the examination. The *administration* of the examining process begins as assistant examiners mark a range of questions from 20 to 30 scripts from randomly selected candidates and centres. Within a week of the examination sitting, the chief examiner will chair a standardisation meeting at which mark schemes are discussed and adjusted. Many boards require a full day's 'workshop' meeting in which

Stage	Task and Timing		Completion of Work
SYLLABUS	Syllabus is published under aims/objectives of the examination paper headings and time allowed Content of the papers and marks for each paper Mode of examining by papers Degree of choice by (i) papers, (ii) sections, (iii) questions		By Subcommittee of the GCE Subject Panel (i) usually lecturers in higher education and nominated teachers; (ii) subject officers of the GCE Board, advising on cost, feasibility of production and technicalities (iii) Board's examiner(s), reviser(s)
EXAMINATION PAPERS	January Year I	Papers drafted	By Chief Examiner (sometimes Assistant Chief Examiners for each paper)
	Spring Year I	Papers moderated	By Reviser(s): sometimes 2-stage process involving higher education and school teachers
	Autumn Year I	Approved papers typeset	By Board staff or outside contract printer. Maps, etc. obtained
	Winter Year I	Drafts approved	By Chief Examiner, Reviser and Subject Officer
	Spring Year II	Mark scheme prepared	By Chief Examiner
ADMINISTRATION	June Year II	(i) Examination worked in schools	'The administration' of the examination (the long hot summer)
	July Year II	(ii) Examiners' Meeting	Discussion of mark scheme. Samples of marked papers discussed as cases. Samples of unmarked papers marked by examining team
		(iii) Scripts marked	By examiners with mid-July deadlines. Samples remarked by Chief Examiner
AWARD	Summer Year II	(i) candidates placed in rank order (ii) grades allocated to candidates	Raw scores of individual examiners converted to marks by (i) statistical adjustments and (ii) inspection of referrals by Chief Examiner Grade boundaries drawn, borderlines checked
	August		Results published
FEED-BACK	September Late Autumn		Appeal procedure Report of examiners in all subjects issued to all centres

Figure 12.14. A typical A-level examining process.

photostats of unmarked scripts are trial marked by all the examining team, and scores compared and discussed. Even the boards with large entries have teams which are sufficiently small for effective control without recourse to hierarchies of team leaders.

The main concern of the chief examiner is that assistant examiners are consistent in the marks they give one day as with the next, and that the judgements they make about a response to one question conforms to syllabus aims and objectives at a general level as well as with the particularities of the actual question set. Throughout the marking period, samples from each assistant examiner are remarked by the chief examiner to monitor both these points.

By the end of July, the chief examiner with a small team of *awarders* will spend two or three days deciding the allocation of grades. Computer print-outs will be available of the distribution of marks given by each assistant examiner and statistical measures of spread (mean, standard deviation, quartiles etc.) used for comparison with those of the chief examiner or of the total population as standard. The raw scores of each individual examiner can be adjusted at various points across his distribution if necessary, and the final scores of candidates calculated and printed out.

With three examination papers to consider, it is likely that these scores will cover a possible mark range of 0-300. Theoretically the scores should be normally distributed. A-level is no exception to the problem that examiners are parsimonious on using the upper range of marks available for individual questions, so the actual range of scores is compressed. But the main concern at the award is where to draw the grade boundaries. The award is a matter of selecting an appropriate cut-off point by reference to precedent, interpreting reports about standards from assistant examiners, and taking note of the chief examiner's assertion that overall standards have either slightly risen, fallen, or been sustained. Thus a score of 220, for example, is selected as a probable cut-off point for grade A, and is adjusted upwards or downwards after re-reading a number of selected scripts. If the final decision is then 223/222 all scripts between 225 and 220 are re-examined and graded by inspection. The process is repeated at other critical grade boundaries.

There has been considerable discussion about the differences in the percentages of candidates who are awarded particular grades. These vary from 10 per cent (grades A and C) to 15 per cent (grades B and D) and 20 per cent (grade E). Thus 70 per cent of the candidates entered for A-level can expect to pass and half achieve grade D or better. The narrow grade C band, however, lies towards the centre of the distribution of marks where candidates tend to be bunched together. It is common for only a few percentage points to separate a grade B from a grade D. Yet the increasingly competitive entry into higher education depends crucially upon a points count based on grades.

Commendable efforts to bring these defects into public debate were made by the JMB (1983). In 1984 the Secondary Examinations Council set up a working party, with members from all boards, to look into the A-level grading system. The working party recommended that the five pass grades should be retained but should be related to mark ranges rather than defined in terms of percentile ranges of candidates. Examiners would award A and E grades according to the standards which earned those grades under the existing system. Each grade between A and E would then cover the same number of marks, thus removing the narrow band of marks for grade C.

In *feedback* all boards have appeal procedures, but they are expensive and can unearth errors only of serious magnitude. These are rare, because arithmetic is checked, and in the writer's experience, variations in marks awarded to questions by the assistant examiner have already been statistically adjusted after conscientious marking, and outright misjudgement is unusual. A mismatch of two grades between a school estimate and the board's award is due to the unstable elements of the examination structure: the grade band percentages, candidates who have been unable to answer four questions in the allotted time, the misunderstanding of key operating concepts in more than one answer, and/or personal difficulties in physical health or mental stress which have produced an incoherent script (or even set of scripts) where answers bear little relation to the questions set. In a system of external public examinations, students are as patients in a field operating theatre, the clients of a system in which the capital invested in teaching and learning, the skill of the surgeon as examiner, and the sharpness of the examining instrument, all contribute to the outcome as well as the physical state of the patient on the day of the examination.

Finally a formal report is written by the chief examiner, helped by comments from assistant examiners, and circulated to centres. Too often this lists the shortcomings of candidates, sweetened by a few encouraging asides. Indeed, as the mechanics of examining have been discussed throughout this section, it has been difficult not to pick up the general issues which constrain and underpin the whole process. In these remote post hoc reports we witness the consequences of the separation of examining from teaching, and the fact that time, manpower and money limit what is possible.

The reader is invited to read all the published reports in a given year as the writer has just done, and ask him/herself whether they convey a feeling of hope for the next generation of young adults, and a feeling that for them, two years of study in depth has been rich in its rewards. They might also see illustrated the industry with which some boards write bland documents about these processes, designed principally to reassure its clients that the workings of the machine are excellent, although not open to inspection.

The criteria of quality in assessment

What features distinguish a good candidate from a weak one at A-level? In practical terms, examiners, as they mark particular papers, concentrate on responses to particular questions and are not asked to trouble themselves over problems of the effects on the order of merit of the overall design of the examination syllabus, and the problems of grade validity in the norm-referenced bands. The good candidate is therefore one who repeatedly scores high marks on a question by question basis, and who neither falters nor is side-tracked into a misjudgement of time or meaning. The issue of quality is therefore argued out at the level of a debate on the wording of questions set (the teeth of the system), and hidden criteria of excellence written up in the mark scheme by the chief examiner.

Whereas question papers become public knowledge, mark schemes are confidential — hard evidence indeed of the separation of powers at A-level between the activity of teaching and that of assessing. Only once has an inter-board cross moderation exercise (in 1979 on the 1978 papers) allowed even chief examiners the opportunity to read the mark schemes of the other boards: indeed the one characteristic common to all boards is their paranoia about confidentiality in matters of mark schemes and assessment procedures.

It is certainly true that chief examiners differ in the quality of information and direction given to their assistant examiners in marking individual questions, ranging from three or four terse sentences to handbooks of content in miniature with detailed instructions about sub-marking procedures. More explicit operational instructions may be necessary in marking practical or techniques papers. Elsewhere, mark schemes reflect differing viewpoints about the needs of examiners for guidance in marking, and also differing viewpoints about the principles of assessment. For example, should a point by point system operate, or would this merely reward the gradgrind pedant at the expense of the divergent thinker? If the syllabus has explicit overall aims and objectives, should these not be the criteria of quality at all times irrespective of the particular question being assessed? Or if little is said about viewpoints of the subject in the preface, and even less about content under each of the three papers, what is the value placed upon accurate factual knowledge against a grasp of concepts or ideas?

A-level schemes of assessment fall into two broad categories. At one end of the spectrum is the *broad* view: this supposes that all assistant examiners (and teachers in schools) possess a common shared understanding about quality. It contends that an over-elaborate syllabus and tightly structured mark scheme not only inhibits teaching and marking, but penalises rather than rewards centres or individuals who offer answers which differ strongly from the detailed guidelines. If essay questions are set in speculative and divergent mode, an inflexible and overburdened mark scheme inhibits the examiner. At the other end of the spectrum is the *narrow* view, which is suspicious of assumptions that the examiners enjoy a consensus about quality, and argues for greater control over the definition of quality by the use of tighter questions, often with sub-sections, which supposedly help the majority of candidates to chain their response by more explicit directives, which are then mirrored by sub-mark routines in the mark scheme. Structured questions were discussed by Melvyn Jones earlier in this chapter; the examples used here are questions which require essay-type answers.

Impression marking of essays reflects the

academic tradition that a good candidate will offer in response to any question (i) an understanding of issues and concepts; (ii) appropriate use of accurate data and information; (iii) a method of analysis which provides for internal consistency in discussion or treatment, thus linking (i) and (ii) in an organised structure; and (iv) the opportunity to include some evaluation, judgement and imaginative insight. This is a personal draft for formulation of 'quality' which is responsive to changes in outlook in the subject in higher education and in recent writings for teacher audiences (HMI, 1978). But even if the above criteria are in a sense content free, considerable variance exists about the relationship and weighting of (i) and (ii), reflecting differences of view in the subject about the nature of the conceptual revolution: the emphasis on information content has been reversed and treated as a dependent variable of issues and concepts, but the relationship is ambiguous.

Figure 12.15 translates these generalised criteria of quality into a scheme of assessment which operates across all the questions in an A-level choice examination paper in an attempt to improve reliability and validity of these particular variables. In fact it is virtually impossible to set equivalent questions in syllabuses where content is used as the basic building block in its construction. Thus a syllabus which includes a study of world population may specify trends, problems, distribution, density, growth and pressure on resources. Population is a vast topic, full of possibilities in terms of issues, ideas, concepts, knowledge of rates of change, ratios and indices. In conditions of subject overload, it may be argued that it is fairer to all candidates to set broad questions, so that all have equal opportunity to respond according to their learning experiences under very different classroom conditions. Consider, for example, the question:

> With reference to contrasting situations, indicate why an examination of the age structure of a population is of importance in calculating the resource base of a country.

The mark scheme would refer the marker to the general criteria as given in Figure 12.15 and add specific comments along the following lines:

> Contrast supposes reference to a minimum of two examples drawn from geographical case studies with age structure represented graphically by population pyramids, differentiated by age cohorts and sex; which will be deployed to illustrate the linkage between the structure and resources. Expect shape to be related to the concept of dependency — the burden of both young and old with reference to pyramids of (say) a developing and a developed country. Credit an appreciation of the concepts of birth rates, death rates, sex ratios, fertility, and negative and positive changes if linked with a discussion on the dynamics of any analysis. Do not penalise comments on the limits of the relationship between the labour availability and the resource base, particularly in relation to gainful employment, available technology and labour quality, etc, where it is integrated with the analysis of a particular structure. Similar comment would apply to the 'S' and 'J' curve patterns.

The same syllabus might also include a section on agricultural systems, in which types of farming and a study of elementary location theory are specified. This part of the syllabus is effectively much narrower than that on population; textbooks are more limiting, and it is not easy to set broad questions which are effectively different from one year to the next. Von Thunen is the only model studied, and many candidates appear to learn the classic diagram and some of the theory with varying degrees of accuracy and success. The predictable question runs:

> Discuss the main elements of *one* theory of agricultural location and assess its (a) variety and (b) value in any one specific system of farming.

The assumption appears to be that the iron necessity of friction of distance applies equally to any of the commodities named in the syllabus. The framing of both the syllabus, and in consequence the question, are strongly positivistic, however, and exposed to criticism in the absence of understanding the historical circumstances under which the

Qualitative Description	Marks out of 25	Criteria in Assessment
Excellent	25 24 23	Where a candidate of 18 could be expected to achieve no more in 35/40 minutes, on the criteria for scores of 20+. Give 25 where merited; there is no such thing as a perfect answer.
Very Good	22 21 20	Ability to analyse, permutate, recombine. Discriminating in selection of construct to generate economical description. Information logically sequenced and argument organised with style and balance which comes from a thorough understanding of concepts and principles. Creative and purposeful in argument; appreciation of techniques and their proper application in appropriate context.
Good	19 18 17	Evidence of *both* structural *and* process/dynamic aspects of a theme. Ability to represent an idea in the most appropriate mode (verbal, graph, diagram). Use of relevant and accurate information/data linked with generalisation; sensible comment, lateral comparison.
Sound	16 15	Accurate in the use of quantities, locations and terms; but evidence of some deficiencies (i) in organisation (ii) inadequate appreciation of structural/dynamic/conceptual dimension of the theme, or in lateral comparison.
Fair	14 13	Mundane, even where thorough, linear regurgitation of notes; *or* evidence of partial grasp of argument/idea, but impoverished by lack of development or illustration.
Pass Plus	12	Covers basic textbook material, but unable to reconstruct directly to the question.

Figure 12.15. Criteria of quality in A-level assessment linked to a mark scheme.

theory was elaborated (Hall, 1976; Gregory, 1984).

The less predictable question runs:

> Analyse the differences of land tenure on rural land use patterns.

This is likely to produce many answers which show that the concept of tenure is unfamiliar, many candidates filtering out the word 'tenure' and reinstating 'distance from the market' which allows standard Von Thunen material to be reported back from revised notebooks. If Von Thunen's arguments about the bid rent made to the landlord by a prospective tenant had been studied, the question is amenable to some discussion along this dimension. The mark scheme, however, would not expect anything more than analysis drawn from local examples and regional studies within the syllabus.

This last example illustrates a vital point. The quality and flexibility of what is learnt is linked to the pressures of syllabus overload. Teachers resort to expository methods and dictated notes to meet such pressures. Students reproduce in their answers the content learnt under such conditions, rather than what is planned as desirable in the abstract by a subject panel. For the agenda of the panel too readily sees its task as negotiating the diverse claims of content for a place in the syllabus core, losing sight of the client group in its deliberations. What is missing is the awareness of the need for

Qualitative Description	Marks out of 25	Criteria in Assessment
Pass	11	Oblique sections but some evidence of information processing; *or* short on subject content but strong evidence of skill in argument or ideas.
Bare Pass	10	Genuine attempt at question, showing some profit from study at this level, but lacking the ability to develop either ideas or reorganise information.
Marginal Fail	9	Sporadic attempt at relevance, or the enunciation of an idea, principle or technique; weak in the use of information; ignorance of ideas or material which function as the key to an acceptable response.
Clear Fail	8 7 6	Descriptive linear treatment with content and illustration no improvement on 16+ exam; often oblique; irrelevant material, little evidence of thought or understanding; *or* evidence of scanty attention to the coursework. Thus an assemblage of half truths, misunderstood argument, etc.
Very weak	5 4 3 2	Random assemblage of information; insubstantial in the time available.
Abysmal	1 0	Carrots not just grown in Corsica but everywhere.

explicit syllabus planning using criteria based upon a diagnosis of student need.

From such experience we can better appreciate the crucial significance of the conclusions of the working party on common cores in A-level geography syllabuses (GCE Boards, 1983) which was unable to support the content view of the core as nucleus of essential material surrounded by a series of options particular to one board or another, but saw the core as 'a dimension of study and as a mode of inquiry about aspects of the world rather than as a substantive body of factual material'. The implications of this statement have not yet been absorbed by many of the boards, although its recommendations have generally been accepted.

Some major contemporary issues

Previous sections have attempted to look at A-level as a dynamic process striking a balance between information and argument. Attention has been drawn to the increasing differentiation of syllabuses and question styles, the lack of consistency between the aims of the examination and the amount of content specified which affects both teaching and learning; additive and root and branch attempts at syllabus modernisation, and differences of view about whether assessment should be broad or narrow in its objectives. Besides technical problems over the proportions of the existing grade bands, it was argued that the separation of teaching from examining and the closed systems of administration themselves create problems of teacher alienation and perpetuate what Tolley and Reynolds (1977) call 'a vicious cycle of curriculum under-development' and an obstacle to the development of 'enhanced professionalism' in the classroom.

If an effective guide and companion is needed to help teacher, let alone student, through a course to A-level in contemporary terms, Lines and Bolwell (1982) have analysed in depth the thirty most frequently recurring topics. But the purpose of this closing section is to identify the major contemporary issues which will affect A-level towards the end of the century.

There is a continuing debate on the nature of the 'client group' for whom A-level is intended. Thus if we accept the client group classifications of the Macfarlane Report (DES, 1981), only 6 per cent of all candidates taking geography at A-level in any one year will continue the subject to degree standard and less than 1 in 5 will take the subject in any form in higher education. Yet the system still reflects the influence of the 'University connection' (Reid, 1972) — the concept of criteria which starts *from the top* with the ideal entrant for an honours degree in mind. Its main purpose is that of a selection device to separate out the best candidates at the upper end of the distribution. Although some undergraduate departments prefer the style and emphasis of one board rather than another, the control of success by norm-referencing the grade bands while allowing performance to float remains a sufficient mechanism for admission combined with interview and school report.

But in recent years, teachers of students in the 16-19 age group have become increasingly aware of the wide ranging needs of a client group many of whom will not reach the level required for entry to higher education. Pupils in the open-access sixth form are seeking syllabuses which they see as relevant to their interests — present and future. These need to be based upon the principle of knowing 'how' rather than knowing 'that', with an emphasis upon skills and the acquisition of knowledge based upon application in situations which are seen as challenging and relevant. Even for the minority who aspire and have the ability to respond to degree courses, the criteria of quality should be linked to content which is neither overloaded nor scholastic 'shelf knowledge', but mindful of the research in the psychology of cognitive development (eg. Bruner, 1966, 1974) and the curriculum (Stenhouse, 1975) which emphasises the instrumental and transactive, process-dominated nature of learning that emanates from participation in the modes of search.

The survey undertaken by the Geography 16-19 Project showed that teachers regretted the weight of existing syllabuses which forced them into the expository mode of teaching for the greater part of contact time. There was a strong desire for greater participation by the classroom teacher in assessment based on experience at 16+ with mixed mode examinations. At the same time the Mansell Report (FEU, 1979) has encouraged courses which are more concerned with the 'how' rather than the 'that' of things. With the selection of content delegated to the teacher within a controlling framework specifying processes, it has created alternative models for syllabus planning. A-level needs to be defined in terms of content sampling within a coherent philosophy and an operating framework. Syllabuses which begin from arguments over the representation of content sectors will be abandoned.

In many schools, looking outwards from the conventional assumption of course recruitment for a University minded client group may be a matter of survival. Teachers in schools with falling rolls will not have the bonus available from the lower school to run non-viable small conventional A-level groups, but will need to plan much more flexible courses for their students. These issues are cultural: we need to consider as a single problem (i) content overload, (ii) teacher participation in syllabus and assessment, (iii) pupil needs, rights and expectations and (iv) a geography which is in harmony with contemporary issues and which is conceptually open-ended, exciting and integrative with feeling and emotion as well as with the intellect.

In this context, a modular syllabus such as that devised by the Geography 16-19 Project has overwhelming advantages. Firstly, it places a limit on content, by defining modules as units of *contact time,* such as half-a-term, giving a greater sense of realism over what can be accomplished. Secondly, modules can be designated as either compulsory or optional, and an element of flexibility can if necessary be introduced even into the core components. Thirdly, option modules can be 'colour coded' as either enrichment modules which support the core (as by a comparative or regionally grounded study), or 'self-directed' in which the teacher proposes a completely independent theme.

Indeed with, say, a double module allocated to pupil-conceived project work, a complete continuum is possible between board-based prescriptive content and that autonomously chosen by teacher and pupil. Fourthly, the modules can be modified and changed as the subject and the culture change without sudden dislocation, if the framework remains stable. It would allow a school, for example, to study a module in humanistic developments (Huckle, 1983) or in contemporary issues in the local region. Modules can also make a contribution to B/TEC courses and to the CPVE; important work which has already been done in this area is discussed by Clive Hart in Chapter 9 of this volume. Core and option modules are also extremely promising as a means of facilitating both complementary and contrasting courses proposed at AS level (DES, 1984).

The Geography 16-19 Project has adopted the concept of a continuum between culturally dominated systems at one end (ie. urban environments) and physically dominated systems at the other. With the syllabus aims stressing the inter-relationship between people and their environments, modules planned at the physically dominant pole are linked with man in terms of management, systems equilibrium and hazard. Clearly stated aims not only reduce problems of the inclusion or exclusion of content, but also assist in the design of the scheme of assessment. The enquiry-based approach of the Geography 16-19 Project, which in itself supports both the learning needs of the client group and our knowledge of how new learning is integrated into existing thresholds of individual understanding, is translated into an examination which includes a decision making paper and a resource-based stimulus-response paper testing applied skills and grasp of general principles, ideas and concepts.

In their new A-level examination for the late eighties, the Welsh Joint Education Committee has also endorsed the advantages of a modular system. The content of the core modules is more traditional than that of the Geography 16-19 Project, reflecting the diversity of viewpoint of teachers in Wales and the importance of existing resources in a regionally based board offering one examination. The systems approach is emphasised, but doubts expressed in the working party about the feasibility of an enquiry-based approach in all centres modified this to 'investigative approach'. Two columns describe the syllabus, one with statements of generalisation, the other illustrative content. The key here is that questions will be set from the generalisations in the sense used by HMI (1978) and not upon specific content. Learning, assessment and aims are cohesive.

The writer has taught A-level students for three boards, examined for five and had stints as chief examiner for two. Recent work as an examiner for Geography 16-19 Project centres in both the individual study and in the final examination has provided convincing evidence at the grass roots of the arguments expressed in this section. These are that syllabus aims and objectives should be defined with a coherent set of principles indicating the dimensions of experience to be studied. A framework should control content, and modules should function flexibly in the contract agreed between the board and its clients. An appropriate methodology based upon diagnosis of student need should be expressed both in the statement of aims and in a scheme of assessment explicitly designed to nurture it. In this lies the hope that scripts consisting of dictated notes and the regurgitation of inert content will be but echoes of a past era.

Mr. Examiner

> Today I am 18 — what a birthday! Geog. this morning, Econ. this afternoon and I couldn't vote yesterday (June 9th). So why don't you have a nice cup of coffee before marking my last essay, relax and, *PLEASE BE GENEROUS* — a 'D' *will* do.
>
> Thank you.

Figure 12.16. Postscript — from an A-level script.

References

Becher, T. and Maclure, S. (1978) *The Politics of Curriculum Change*, Hutchinson.

Bruner, J. (1966) *Towards a Theory of Instruction*, Belknap.

Bruner, J. (1974) *Beyond the Information Given*, Allen and Unwin.

Daugherty, R. (1982) 'A common core for Advanced Level?' *Teaching Geography*, Vol 8, No 2., pp. 77-9.

DES (1981) *Education for 16-19 year olds*, (The Macfarlane Report), HMSO.

DES (1984) *AS Levels*, DES.

FEU (1979) *A Basis for Choice* (The Mansell Report), HMSO.

GCE Boards (1983) *Common Cores at Advanced Level*, GCE Boards.

Gregory, D. (1984) 'The future of human geography' in King, R. (ed) *Geographical Futures*, Geographical Association.

Hall, D. (1976) *Geography and the Geography Teacher*, Allen and Unwin.

HMI (1978) *The Teaching of Ideas in Geography*, HMSO.

Huckle, J. (ed) (1983) *Geographical Education: Reflection and Action*, Oxford University Press.

JMB (1983) *Problems of the GCE Advanced Level Grading Scheme*, Joint Matriculation Board.

Lines, C. and Bolwell, L. (1982) *A-level Geography*, Letts.

Reid, W. A. (1972) *The Universities and the Sixth Form Curriculum*, Macmillan.

Stenhouse, L. (1975) *An Introduction to Curriculum Research and Development*, Heinemann.

Tolley, H. and Reynolds, J. B. (1977) *Geography 14-18: A Handbook for School-Based Curriculum Development*, Macmillan.

Appendix 1

Guide to the Selection of an Atlas for Secondary School Pupils

Herbert Sandford

This guide is mainly concerned with pupils' desk atlases, that is, those atlases that are usually purchased in sets and issued to individual pupils for classwork and homework. Whatever the level of their desk atlas, however, pupils will need to refer to more advanced ones in the classroom or school libraries, and so a short selection of the more scholastic of these is also included.

Not all teachers have an opportunity to see a full range of atlases, and so it is hoped that this guide will help them to make an initial selection before requesting inspection copies. Atlases vary greatly in their objectives and so no attempt is made here to suggest that any one is better than any other: it is up to individual teachers to judge the suitability of the atlases for their pupils and courses.

Table 1. Atlases for Years 1-5

Column 3 gives the date of the latest available edition or impression.

Column 4 gives the page size in cm.

Columns 5 and 6 indicate atlases with soft or flexible covers (rather than hard or stiff ones), and those with the pages sewn in by thread (rather than being fastened by wire staples).

Columns 7 to 12. Column 7 gives the number of pages (including cover but excluding dust jacket), which is then broken down into sections of guidance, British Isles, continents, the world as a whole, and index — with a small residue of pages not separately counted but including title page, credits, content list, blank pages and so forth. By guidance is meant any explanation about using the atlas, with or without questions and exercises.

Columns 13 to 18. Column 13 shows the number of entries in the index; this may not include all the names on the maps. Then follow a number of aspects of an index that are relevant to its use: the spacing of entries down the page, the use of geographical co-ordinates of latitude and longitude (rather than letters and numbers); whether the co-ordinates are ranged in columns; whether the entries are type coded or otherwise printed as they appear on the maps; whether the page number comes just before the co-ordinates.

Columns 19 to 27. Column 19 gives the number of individual or separable graphic items regardless of their size, which total is then broken down into several categories. Column 20, text, refers mostly to written passages providing information to supplement the maps, but there may also be explanation of questions detached from the main guidance section of Column 8. The figures in this column are not very reliable owing to the difficulty of defining 'individual or separable' with the extensive textual passages of some atlases. Column 21 gives the number of 'vocabulary' pictures, perhaps a convenient name to use for pictures that help to define common nouns like coconut, oasis, fiord and ranch, or proper nouns like Fujiyama, Matterhorn, Grand Canyon and Brasilia. The number and chief kinds of diagrams and the like are shown in Columns 22 and 23. Most maps are clearly distinguishable as being 'of atlas scale' or larger than that (Columns 24 and 25): by and large the scale of 1:1M separates them. Globes (Column 26) include space views of the Earth as well as pictures of globes and hemisphere maps. Satellite pictures of parts of the Earth are put with all other landscape pictures into Column 27.

Table 1. Descriptive analysis of atlases for pupils in years 1-5.

		BIBLIOGRAPHIC DETAILS					CONTENT (In pages)						INDEX	
	Publisher	Title b	Date	Format	Soft cover	Sewn pages	Total	Guidance	British Isles	Continents	World	Index	Total	Lines per cm.
	1	2	3	4	5	6	7	8	9	10	11	12	13	14
1	Arnold-Wheaton	At. for the Mid. Sch.	'79	27 x 22	S	S	60	6.6	18	15	2	9	2,000	3.2
2	,, ,,	Secondary Sch. At.	'80	27 x 22	S	S	112	0.0	20	40	10	33 g	17,300	5.0
3	Bartholomew	First At. of the Envirt.	'81	30 x 21	S		44	16.2	5	13	5	1	130	2.0
4	,,	Second At. of the Envirt.	'82	30 x 21	S		44	1.8	12	16	7	5	2,100	3.5
5	,,	Third At. of the Envirt.	'80	33 x 24		S	128	4.2	17	46	18	26	8,300	3.8
6	,,	Jnr. At. of the World	'82	30 x 21	S		76	17.1	19	13	10	5	1,300	3.6
7	,, /Holmes McDoug.	Problems of Our Planet	'77	30 x 21	S	S	72	0.0	0	16	48	0	0	—
8	,, / ,, ,,	World Envirtl. Problems	'81	30 x 21	S	S	72	0.0	1	24	39	0	0	—
9	Cassell	World Study At.	'78	27 x 21	S	S	68	0.0	16	10	20	4	1,800	4.8
10	Collins-Longman	Atlas Two	'83	26 x 21	S	S	88	2.1	21	24	14	17	5,500	4.7
11	,, ,,	Atlas Three	'83	26 x 21	S	S	144	4.2	50	20	25	22	6,700	4.7
12	,, ,,	Atlas Four	'83	26 x 21	S	S	200	2.2	21	86	21	57	15,900	4.0
13	Educl. Co. of Ireland	2nd At. for Irish Schs. c.	'80	26 x 21	S	S	68	2.1	19 e	25	12	6	2,200	4.0
14	,, ,, ,, ,,	Cert. At. for Irish Schs. c.	'82	26 x 21	S	S	200	2.2	27 e	86	19	57	15,900	4.0
15	Nelson	Atlas 80	'81	28 x 21	S	S	68	1.0	13	28	8	12	3,300	4.0
16	,,	Atlas Scotland d	'81	28 x 21	S	S	72	1.0	17 f	28	8	12	3,300	4.0
17	,,	Secondary Sch. At.	'83	30 x 21		S	88	0.1	12	47	4	13	10,500	5.0
18	Oxford UP	Ox. Sch. At.	'81	25 x 19	S	S	148	1.0	20	74	21	24 g	14,900	5.7
19	,,	Shorter Ox. Sch. At.	'78	25 x 19	S	S	76	0.4	18	36	12	3 g	2,200	5.7
20	Philip	Certificate At.	'83	28 x 23		S	160	2.2	23	72	25	22	10,500	5.0
21	,,	Elementary At.	'83	28 x 23	S		68	1.0	6	33	10	12	5,600	4.7
22	,,	First Venture At.	'79	23 x 19	S		36	3.0	15	2	13	0	0	—
23	,,	Middle Sch. At.	'82	28 x 23	S		36	3.4	10	12	8	1	450	4.3
24	,,	Modern Sch. At.	'83	28 x 23		S	160	0.0	19	82	19	32	17,900	4.7
25	,,	Modern Sch. Econ. At.	'83	28 x 23			192	0.0	21	91	39	32	17,900	4.7
26	,,	New Sch. At.	'83	28 x 23	S	S	100	0.0	10	52	9	24	13,400	4.7
27	,,	Secondary Sch. At.	'83	28 x 23		S	76	0.0	7	33	14	12	5,600	4.7
28	,,	Venture At.	'79	28 x 23	S		52	0.4	6	25	8	7	3,200	4.7
29	,,	Visual At.	'78	23 x 19	S		60	2.0	10	28	7	9	4,400	5.3
30	P.F.E.S. a	Atlas	'75	32 x 24		S	268	2.0	16	163	13	42	6,700	2.7
31	Schofield & Sims	Our World	'76	31 x 23	S	S	52	2.0	21	3	19	0	0	—
32	,, ,,	The Whole World	'81	31 x 23	S	S	52	0.0	5	12	27	0	0	—

Notes

a   Publications Fund of the European Schools. A special atlas produced by the cartographers Hölzel of Vienna and obtainable from the Internationales Landkartenhaus Zweigniederlassung der GeoCenter Verlagsvertrieb GmbH, D-7000 Stuttgart 80, Schockenriedstr. 40a, West Germany. It is in six Community languages, and follows the frequent continental practice of pairing lexical maps with economic ones at the same scale.
b   For associated atlas-based workbooks, etc., see Chapter 6.
c   These two are Irish editions of Nos. 10 and 12.
d   This is the Scottish edition of No. 13.
e   These two include 13 and 9 pages on Ireland.

APPENDIX 1

ENTRIES				GRAPHIC ITEMS									LEXICAL MAPS						THEMES			
													NAMES		SINGLE MAPS		MAP PAIRS					
Geogl. coords.	Ranged	Type coded	Page before coords.	Total	Text	Vocab. pics.	Diagrams, etc.	Predom. diag. etc. i	Small scale maps	Large scale maps	Globes, etc.	Landscape pics.	Names per cm²	Use of Eng. names: None (0) to (9) all	Pages	Type k	Pages	Ease of comparison: Hard (0) to (9) easy	Biblical	Historical	Social	Environmental
15	16	17	18	19	20	21	22	23	24	25	26	27	28	29	30	31	32	33	34	35	36	37
	R		P	131	42	19	16	B	50	2	0	2	6.8	7.7	12	H	2.0	9	0	0	0	0
G	R		P	128	6	0	11	BT	110	1	0	0	2.4	0.5	37	H	2.8	9	0	0	1	0
	R	T	P	159	30 h	88	0		24	0	9	8	3.9	7.7	18	L			0	0	0	0
			P	66	1	0	0		57	0	0	8	5.0	7.2	21	L			0	0	0	0
G				190	9	3	14	C	153	2	1	8	1.4	8.1	35	L			0	0	11	2
			P	324	58 h	181	14		51	0	10	10	3.9	7.7	18	L			0	0	0	0
—	—	—	—	521	114	142	150	B	87	8	1	19	—	—	0	—			0	0	18	14
—	—	—	—	542	138	126	156	B	78	19	1	24	—	—	0	—			0	0	19	18
G	R			313	63	177	23	BDS	41	0	1	8	3.9	7.7	13	R			3	9	2	2
G			P	128	11	40	5	LT	61	1	2	8	2.3	5.4	14	R	15.1	8	2	8	0	1
G			P	102	6	0	6	BP	86	4	0	0	6.8	5.4	26	R	4.0	8	0	0	22	2
G			P	218	1	0	77	AC	139	1	0	0	2.4	5.4	55	R	4.0	8	0	0	3	1
G			P	124	6	33	2	C	79	0	0	4	2.3	5.9	9	R	16.1	8	0	5	2	0
G			P	236	1	0	79	AC	156	0	0	0	2.4	5.4	57	R	4.0	8	0	0	3	2
		T	P	121	7	23	9	C	70	0	3	9	4.9	7.7	17	R	6.0	9	0	6	6	2
		T	P	126	7	23	9	C	73	0	3	11	4.9	7.7	19	R	6.0	9	0	7	6	2
			P	95	0	0	10	PT	82	0	0	0	1.2	5.4	45	R	4.8	8	0	3	0	0
G	R		P	274	13	0	141	C	118	2	0	0	3.4	7.2	67	R			0	2	0	1
G	R		P	87	8	0	20	C	59	0	0	0	3.4	7.7	47	R			0	0	0	0
G	R		P	765	90	82	256	BT	257	10	1	69 j	3.4	6.3	41	H	1.0	7	0	0	19	7
G				169	2	11	74	C	73	0	4	5	2.2	7.7	22	H	11.7	9	0	0	0	0
—	—	—	—	215	32 h	104	4	D	29	4	5	37	3.0	7.2	12	R	6.0	9	0	0	0	0
G				47	0	0	1	B	39	5	1	0	3.0	7.7	13	R	4.0	9	2	10	3	0
G				417	15	22	188	C	178	1	4	9	3.4	7.7	57	H	12.1	9	0	0	1	4
G				567	15	11	318	CP	223	0	0	0	3.4	7.7	57	H	12.1	9	0	0	2	5
G				140	2	11	2	D	114	2	4	5	3.4	7.7	38	R	13.2	9	0	0	0	0
G				185	2	11	78	C	85	0	4	5	2.2	7.7	22	H	11.7	9	0	0	1	1
G				65	0	3	6	B	55	0	1	0	1.9	7.7	26	R			0	0	3	0
G				57	0	0	2	BT	51	0	4	0	2.8	7.7	17	H	20.6	9	0	0	0	0
	R		P	346	0	0	29	C	244	70	0	3	2.0	5.5	72	R	13.3	8	0	0	0	0
—	—	—	—	627	89	353	113	3D	62	0	2	8	1.7	8.1	0	—	6.0	8	0	0	4	2
—	—	—	—	716	129	204	292	D	65	0	4	17	4.4	4.5	0	—	12.7	8	0	0	4	3

f   This includes 5 pages on Scotland.
g   These three include 15, 7 and 1 page of separate G.B. index with National Grid.
h   Mostly exercises and questions.
i   Age pyramids, Climate graphs, Diagrammatic pictures, Line diagrams, Pie graphs, Tables, Sections.
j   Includes 11 satellite images.
k   Height (Hypsometric); Landscape (land cover and slopes); true Relief (Hypsographic). See text.

Footnote
The Bartholomew atlases Nos. 3, 4 and 5 are now published by Heinemann and No. 6 is discontinued.

Columns 28 to 33 refer to 'lexical' maps, those intended to provide the locations of places mentioned in the index and the names of places found on the maps. Column 28 gives the density of names on the main lexical map for Europe, expressed as so many names for every square centimetre; density is commonly greater on British Isles maps and less elsewhere. Generally, the greater the density, the smaller the lettering.

For many places all over the world there are standard English names that are different from the names those places have in the foreign language. Twenty English names were examined to see how far they had been replaced by foreign language names: Amoy (Xiamen), Archangel (Arkhangel'sk), Athens (Athínai), Bangkok (Krung Thep), Cairo (Al-Qāhirah), Canton (Guangzhou), Chungking (Chongqing), Cracow (Kraków), Foochow (Fuzhou), Gothenburg (Göteborg), Jerusalem (Yerushalayim — Israeli; Al-Quds esh Sherif — Jordanian), Marseilles (Marseille), Munich (München), Oporto (Porto), Peking (Beijing), Prague (Praha), Rome (Roma), Swatow (Shantou), The Hague (s' Gravenhage), Wenchow (Wenzhou). Only the commonest foreign alternative names are given here, and all Chinese-language tonal marks are omitted as they are so difficult to reproduce. Column 29 provides an index of the extent to which English names for these places have been *retained*, using a scale from 0, none, to 9, all.

Almost all lexical maps combine both physical features, like rivers and mountains, with cultural features, like routeways and towns (Columns 30 and 31), but some atlases have maps in pairs, one showing and naming the physical features and the other showing and naming the cultural features, these last often being called political maps (Columns 32 and 33). The base map upon which the names are printed is stated in Column 31, the alternatives being the Height (or Hypsometric) map, coloured for the altitudes; the true Relief (or Hypsographic) map, which adds hill shading to the altitude colours; and the Landscape map, on which there are colours for forest and farm, city and desert, together with hill shading. The ease of comparing the two maps in a pair is indexed in Column 33, difficulty arising when they differ in scale, projection or orientation, or are on different openings of the atlas.

Columns 34 to 37. Many atlases contain clearly identifiable thematic presentations of global topics that are not traditional for school atlases: pollution, wildlife, famine, literacy, recreation, natural hazards, trade blocs and so forth. These are totalled in Columns 36 and 37, while Columns 34 and 35 refer to some more traditional but not strictly geographical global thematic presentations.

It may help to give an example of the use of the table. Suppose one wanted to purchase a class set of atlases with a strong statistical and diagrammatic emphasis combined with a generous provision of satellite or other landscape and panoramic pictures. A glance down Columns 22/23 and 26/27 reveals a choice of three atlases: those numbered 7, 8 and 20. Perhaps the class is to make a detailed study of the British Isles: Column 9 reveals that out of these three atlases only atlas 20 has a British section and so would be worthwhile inspecting. On the other hand, if this syllabus requirement were the overriding consideration, then one might look down Column 9 first and discover the very thorough treatment of the British Isles in atlas 11.

Table 2. Sixth Form Atlases

Column 5 gives the date of the latest available edition or impression.

Columns 6 and 7 give the page size in cm and the total number of pages.

Columns 8 to 19 refer to the main map section in the atlas. Total pages (Column 8) is broken down by area covered (Columns 9 to 11) and by type of graphic image (Columns 12 to 14): lexical maps (the maps to which the index refers, whether general, physical or political), thematic maps (maps of climate, population, products and so on), and non-map material like text, diagrams and pictures. Some special features are described in Columns 15 to 19: the largest scale map for all the British Isles and for almost all of the continents; pages of landscape maps; number of themes treated by maps (plate tectonics, precipitation, oil supplies, etc.); the kind of lexical map (classified by background colours).

Columns 20 to 26 refer to the atlas's introductory and concluding material. The total (Column 20) is broken down by type of graphic image (Columns 21 to 24), any remainder being title page, credits and so on, and some special features are described in Columns 25 and 26. Column 25 describes the kind of guidance, while Column 26 gives the number of themes, whether treated by maps alone (total pages in Column 22) or by maps with accompanying illustrative text, diagram and picture (total pages in Column 23).

Columns 27 to 33 provide information about the index. Word-by-word indexing (Column 29) is best explained by an example. Where names have two words, like those beginning with El (= The), they are all put together, but in the alphabetical order of the second word. The alternative system of indexing is to mix them up with all the other words that happen to begin with El like Elbe or Ely.

Columns 30 and 31 pay special attention to the names chosen by the map compiler for China, both as his first choice and as supplementary (generally printed lighter, smaller and bracketed). Column 30 gives the compiler's preference: Wade-Giles and pinyin being respectively older and newer ways of transcribing Chinese characters into the Latin alphabet; both of course represent Chinese language names. Column 31 provides an index of the provisions of English language names, on the scale 0 (none) to 9 (all), for major features, whether the English language name is first choice, eg. Canton, or second choice, eg. (Canton); foreign language forms of this geographical name (such as the pinyin Guangzhou and Wade-Giles variants like Kuangchou and Kwangchow) are ignored for the purpose of compiling this index.

Columns 32 and 33 relate to the names of places outside China and the British Isles that are significant and frequent in standard English speech and literature, such as The Hague (known to tourists and traders), Bear Island (the title and story location of Alistair Maclean's famous novel), Bangkok (repeatedly in the news), and so on. Column 32 provides a scale of the ease of locating such places through the index, while Column 33 provides a scale for the ease of locating these same places through any one of their different foreign language forms, eg. if one has come across The Hague written as Den Haag or as 's Gravenhage (or one of its variants).

An example might help. If one wanted an atlas small enough to carry around easily and with a substantial illustrated thematic section, a glance down Columns 6 and 23 might suggest atlases numbered 2 and 9. A final choice of one to inspect might be made from Columns 30 and 31 where it is seen that one has and the other has not largely replaced English language names of Chinese places by Chinese language names (or by transcriptions imitative of the Chinese language names). On the other hand, if one wants a comprehensive atlas with individual conurbation of Metropolitan Area maps, a perusal of Columns 15, 16 and 20 might suggest atlas number 3.

Table 2. Descriptive analysis of atlases for sixth form students.

		BIBLIOGRAPHIC DETAILS						MAP SECTION						
									CONTENT (in pages)					
										BY AREA			BY GRAPHIC IMAGE	
	Publisher	Title	Cartographers e	Suppliers (if not the publisher) e	Date	Format	Pages	Total map section pages	British Isles	Continents	World	Lexical maps	Thematic maps	Not maps
	1	2	3	4	5	6	7	8	9	10	11	12	13	14
1	Bartholomew	World Atlas	B		'83	38 x 26	176	100	9	77	14	76	24	0
2	Collins	Atlas of the World	C-L f		'83	30 x 23	288	126	6	117	5	115	0	13
3	Encyclo. Brit.	Britannica Atlas	R McN		'82	38 x 29	576	320	10	271	39	254	29	37
4	Hammond	World Atlas	H	B g	'81	32 x 24	208	192	8	181	3	82	45	65
5	Heinemann	World Atlas b	R McN		'79	38 x 29	136	74	10	58	6	54	16	4
6	Philip	Illust. At. of the Wd.	Ph		'81	37 x 26	215	162	10	152	0	60	0	102
7	,,	Library Atlas c	Ph		'81	31 x 24	384	208	21	157	30	150	55	2
8	,,	New World Atlas c	Ph		'83	28 x 23	292	128	5	120	3	127	h	h
9	,,	Universal Atlas c	Ph		'83	31 x 24	392	176	19	148	9	150	25	1
10	,,	University Atlas c	Ph		'83	31 x 24	336	176	19	148	9	150	25	1
11	,,	World Atlas	Ph		'83	28 x 23	160	80	4	73	3	80	0	h
12	P.F.E.S. a	Atlas	a	a	'75	32 x 24	268	192	16	163	13	85	81	26
13	Rand McNally	Goode's Wd. At.	R McN		'80	28 x 23	388	220	3	167	50	172	45	3
14	,,	Worldmaster Wd. At.	R McN		'83	28 x 23	260	89	1	86	2	89	0	0
15	Times Books	Concise At. of the Wd.	B		'82	38 x 28	284	147	10	134	3	127	1	19
16	Ward Lock	Atlas of the World d	Ph		'79	33 x 24	232	100	19	71	10	75	22	3

Notes

a   See Table 1, footnote a.
b   The lexical maps are a selection out of those in atlas number 3.
c   These three atlases have their lexical maps in common: atlas 7 has additional climatic graphs and economic maps: atlas 9 has an extensive illustrated thematic introduction.
d   Built around and contains Philips' Modern School Atlas pages 1 to 5 and 10 to 100 (see Table 1).
e   These are the initials of firms given in Column 1 unless otherwise stated.

APPENDIX 1

SPECIAL FEATURES				REST OF ATLAS														
				CONTENT BY GRAPHIC IMAGE (in pages)				SPECIAL FEATURES		INDEXING								
													GEOGRAPHICAL NAMES					
													CHINA			REST OF WORLD		
Largest scale for all B.I. (10^6)	Largest scale for conts. (10^6)	Pages of landscape maps	Number of themes	Main kind of lexical map j	Total rest-of-atlas pages	Guidance	Thematic, not illustrated	Thematic, illustrated	Index	Kind of guidance k	Number of map-based themes	Entries (,000)	Co-ordinate type l	Word by word indexing	First language of names	Use of Eng. Lang. (0 to 9)	Indexing by Eng. Lang. (0 to 9)	Indexing by foreign lang. (0 to 9)
15	16	17	18	19	20	21	22	23	24	25	26	27	28	29	30	31	32	33
$1\frac{1}{4}$	$12\frac{1}{2}$	0	19	H	76	8	6	0	50	cgmp	3	24	H		W	9	5	7
2	$12\frac{1}{2}$	0	0	R	160	8	5	48	84	gmo	15	39	G		P	3	3	8
1	6	16	33	R	256	16	20	0	205	blmo	0	197	G		P	8	4	9
$2\frac{3}{4}$	$17\frac{1}{3}$	6	1	N	16	2	0	0	3	o	0	43	L		E	8	7	5
1	12	14	1	G	62	4	15	13	19	lmp	9	18	G		P	5	2	7
$2\frac{1}{2}$	16	0	0	R	54	h	7	6	24	l	4	18	G	W	P	1	5	7
$1\frac{1}{4}$	16	0	8	R	176	2	18	0	142	g	37	64	G	W	W	8	9	8
2	16	0	0	R	164	2	0	47	96	o	47	46	G	W	P	1	5	7
$1\frac{1}{4}$	16	0	8	R	216	6	10	45	142	gm	43	64	G	W	W	8	9	8
$1\frac{1}{4}$	16	0	8	R	160	6	1	0	142	gm	0	64	G	W	W	8	7	8
2	16	0	0	R	80	3	0	0	53	c	0	22	G	W	W	8	5	8
$1\frac{1}{4}$	25	0	7	H	76	2	2	0	42	m	1	7	L	W	E	8	5	4
4	16	16	39	R	168	4	6	2	41	lop	1	11	G		E	9	4	7
$5\frac{1}{4}$	21	0 i	0	P	171	1	1	1	75	l	1 i	28	L		E	9	5	3
$1\frac{1}{2}$	12		2	H	137	4	0	38	85	m	25	102	L		P	8	6	8
1	$18\frac{3}{4}$		3	H	132	0	0	82	35		6	19	G	W	E	7	7	4

f   Collins-Longman.
g   Bartholomew.
h   Trace.
i   None in the map section but the introduction has a 28-page theme on landscape, with maps.
j   Gradient shading (hill shading); Height (Hypsometric); Neutral (background colours plain and without meaning); Political; true Relief (Hypsographic). See text.
k   Co-ordinates; Geographical terms; Legend; Map key to maps; Other matters; Projections.
l   Geographical (latitude and longitude); Hour co-ordinates (latitude and relative time); Letter-number.

Appendix 2

Satellite Imagery: Sources and Publications

1. Atlases

a) Sheffield, C. (1981) *Earthwatch: A Survey of the World from Space,* Sidgwick and Jackson.

b) Sheffield, C. (1983) *Man on Earth — the Marks of Man: A Survey from Space,* Sidgwick and Jackson.

c) *Landsat Views the World* (1976) NASA, Washington DC, US Government Printing Office. Nearly 400 colour composite images.

d) *Images of the World: An Atlas of Satellite Imagery and Maps* (1983) Collins-Longman. A concluding chapter reviews the applicability of satellite remote sensing to the teaching of geography at secondary school level.

e) Bullard, R. K. and Dixon-Gough, R. W. (1985) *Britain from Space: An Atlas of Landsat Images,* Taylor and Francis.

2. Slides, Cassette Tapes and Overhead Transparencies

a) *The Earth From Space*. 14 sets of 35mm slides of selected colour Landsat images, arranged regionally; the series is prefaced by a 50 minute slide/tape 'Interpretation of Remotely-Sensed Images' written specifically for A-level geographers. Focal Point Audiovisual, 251 Copnor Road, Portsmouth PO3 5EE.

b) *Remote Sensing*. A well selected set of 18 slides covering the scope of remote sensing and accompanied by a detailed explanatory booklet. National Remote Sensing Centre, Royal Aircraft Establishment, Q134 Building, Farnborough, Hampshire GU14 6TD.

c) *Satellite Geography*. A series of overhead transparencies of Landsat colour composites, with interpretative overlays. Nelson Audio-Visual, Walton-on-Thames, Surrey.

d) *Satpack I and II*. Two sets of interpretative data and exercises from meteorological satellites as described in the text. Sponsored by the Royal Meteorological Society; available from Focal Point Audiovisual, 251 Copnor Road, Portsmouth, PO3 5EE.

e) *Weather Study with Satellites*. (1983) 3 packs of slides; tape cassette and text. Diana Wyllie, 1 Park Road, Baker Street, London NW 6TX.

3. Texts

All those available to date are written for students of advanced courses. The most useful to the busy teacher requiring relevant background and a range of ideas are:

Barrett, E. C. and Curtis, L. F. (1982) *Environmental Remote Sensing,* Chapman and Hall.

Curran, P. J. (1985) *Principles of Remote Sensing,* Longman.

Harper, D. (1983, 2nd edn) *Eye in the Sky,* Multiscience Publications, Montreal, Canada.

Lillesand, T. M. and Kiefer, R. W. (1979) *Remote Sensing and Image Interpretation,* John Wiley.

4. Original Imagery

a) *Air Photographs*. Vertical and oblique air photographs of all parts of England and

Wales for various dates. The main two sources are the Ordnance Survey, Romsey Road, Maybush, Southampton, SO9 4DH, and Aerofilms, Elstree Way, Boreham Wood, Hertfordshire, WD6 1SB.

b) *National Remote Sensing Centre.* The main institution for research and development in the UK. Complete collection of satellite image cover for the UK with sophisticated image analysis facilities. Library open to public (by arrangement). Visits from schools and colleges can be arranged. Full details from Education Officer, National Remote Sensing Centre, R190 Building, Royal Aircraft Establishment, Farnborough, Hampshire, tel. 0252 541464. The NRSC also publishes a range of posters of true colour and false colour mosaics of the British Isles.

c) *The British Library.* Science Reference Section, 9-13 Kean Street, London WC2 4DE maintains a microfiche archive of 'quick look' Landsat images together with a collection of reference material of value.

d) *Films* may be hired from The British Interplanetary Society, 12 Bessborough Gardens, London SW1V 2JJ, which acts as UK distributor for the many titles made by NASA. Also recommended is a 30 minute film titled 'Remote Sensing: A New Angle on the World' produced by the French Scientific Films Service (1980), available from the Higher Education and Video Library, 74 Victoria Crescent Road, Glasgow G12 9JN.

Appendix 3

Books and Journals for Teachers

Books

Books written for geography teachers during the last ten years are listed below. They might form the basis of a geography department's staff reference library. Textbooks for pupils and students are too numerous to list here for reasons of space. Teachers are recommended to consult publishers' catalogues and send for inspection copies. New books are announced and reviewed in *Geography* and *Teaching Geography*.

Bailey, P. (1974) *Teaching Geography*, David and Charles.
Boardman, D. (1983) *Graphicacy and Geography Teaching*, Croom Helm.
Boardman, D. (ed) (1985) *New Directions in Geographical Education*, Falmer Press.
Corney, G. (ed) (1985) *Geography, Schools and Industry*, Geographical Association.
Corney, G. and Rawling, E. (eds) (1985) *Teaching Slow Learners through Geography*, Geographical Association.
Fien, J., Wilson, P. and Gerber, R. (eds) (1984) *The Geography Teacher's Guide to the Classroom*, Macmillan.
Graves, N. J. (1975) *Geography in Education*, Heinemann.
Graves, N. J. (1979) *Curriculum Planning in Geography*, Heinemann.
Graves, N. J. (ed) (1982) *New UNESCO Source Book for Geography Teaching*, Longman/UNESCO.
Hall, D. (1976) *Geography and the Geography Teacher*, George Allen and Unwin.
HMI (1978) *The Teaching of Ideas in Geography: Some Suggestions for the Middle and Secondary Years*, HMI Series: Matters for Discussion, 5, HMSO.
Huckle, J. (ed) (1983) *Geographical Education: Reflection and Action*, Oxford University Press.
Jay, L. J. (1981) *Geography Teaching with a Little Latitude*, George Allen and Unwin.
Marsden, W. (1976) *Evaluating the Geography Curriculum*, Oliver and Boyd.
Mills, D. (ed) (1981) *Geographical Work in Primary and Middle Schools*, Geographical Association.
Slater, F. (1982) *Learning through Geography*, Heinemann.
Walford, R. (ed) (1981) *Signposts for Geography Teaching*, Longman.
Walford, R. (ed) (1985) *Geographical Education for a Multi-cultural Society*, Geographical Association.
Williams, M. (ed) (1976) *Geography and the Integrated Curriculum*, Heinemann.
Williams, M. (1984) *Designing and Teaching Integrated Courses*, Geographical Association.

Journals

Subscription details for the journals and periodicals listed below are available from the addresses indicated. In some cases specimen copies may be supplied on request.

Geography. Quarterly. The Geographical Association, 343 Fulwood Road, Sheffield S10 3BP.
Teaching Geography. Quarterly. The Geographical Association, 343 Fulwood Road, Sheffield S10 3BP.
The Geographical Journal. 3 issues per year. The Royal Geographical Society, 1 Kensington Gore, London SW7 2AR.
The Geographical Magazine. Monthly. Magnum Distribution Ltd, Watling Street, Milton Keynes MK2 2BW.

Bulletin of Environmental Education. Monthly. Streetwork, Notting Dale Urban Studies Centre, 189 Freston Road, London W10 6TH.

Contemporary Issues in Geography and Education. 3 issues per year. Comedia, Farndon Road, Market Harborough, Leicestershire LE1 9NR.

The New Internationalist. Monthly. 42 The Hythe, Bridge Street, Oxford OX1 2EP.

Journal of Geography. 6 issues per year. National Council for Geographic Education, Department of Geography, University of Wisconsin — Eau Claire, Eau Claire, WI 54701, USA.

National Geographic. Monthly. National Geographic Society, PO Box 2895, Washington, DC 20013, USA.

Geographical Education. Annually. Australian Geography Teachers Association, School of Geography, University of New South Wales, PO Box 1, Kensington, NSW 2033, Australia.

Projects

Teachers' books and guides produced by the three national curriculum development projects in geography are listed below. Further details about the projects are available from the addresses shown.

Geography for the Young School Leaver. The project, based at Avery Hill College, London, published a teacher's guide for each of the three themes *Man, Land and Leisure, Cities and People,* and *People, Place and Work* (Nelson, 1974-75). A teacher's guide also accompanied the GYSL Development Education Project books *Geography and Change* (Nelson, 1981-82). CSE syllabuses and examinations based on GYSL were submitted and approved throughout the country. The Avery Hill O-level and joint 16-plus syllabuses and examinations were subsequently developed into a GCSE syllabus and examination. Details are available from the GYSL Project Secretary, Centre for Geography and Environmental Study, Sheffield City Polytechnic, 51 Broomgrove Road, Sheffield S10 2NA.

Geography 14-18. The project, based at the University of Bristol School of Education and later at Leeds Polytechnic, published *Geography 14-18: A Handbook for School-Based Curriculum Development,* by H. Tolley and J. B. Reynolds (Macmillan, 1977). A teacher's guide is included in each of the project's classroom units: *Population, Urban Geography, Transport Networks, Industry,* and *Water and Rivers* (Macmillan, 1978-80). The project's O-level and joint 16-plus syllabuses and examinations were subsequently developed into a GCSE syllabus and examination. Details are available from Keith Orrell, School of Education, Leeds Polytechnic, Cavendish Hall, Beckett Park, Leeds LS6 3QS.

Geography 16-19. The project, based at the University of London Institute of Education, has produced a handbook *Geography 16-19: The Contribution of a Curriculum Project to 16-19 Education,* by M. Naish, E. Rawling and C. Hart (Longman, forthcoming). A series of student booklets is published by the Longman Resources Unit. The project developed a new modular A-level syllabus and examination, and also modules for other courses such as the CEE and CPVE. Details are available from Michael Naish, University of London Institute of Education, 20 Bedford Way, London WC1H 0AL.

Appendix 4

Geography in Challenging Times

The following is the main text of a leaflet published by the Geographical Association and distributed to all members in 1984.

Ten guidelines to successful geography teaching: A checklist for every teacher

The first priority must be to ensure that all the geography taught in schools is both *useful* and *interesting*. Many teachers today apply the following ideas but the more *you* can do, the stronger will be your position.

1. Ensure that your topics provide a reasonable coverage of different parts of the world: work at a range of scales from very local to global.
2. Vary your teaching approaches: decision-making exercises, debates, discussions, games, fieldwork and role-play all have value as well as chalk and talk.
3. Use the visual appeal of the subject (slides, television, film and posters) to illustrate any analytical or statistical work in class.
4. Don't avoid controversial topics or duck issues relevant to your subject. An issue-based approach adds point and spice as long as it is not propagandist.
5. Emphasise the usefulness of geography by including practical and applied work.
6. Avoid turning the subject into a long list of world problems: school geography should reflect the wonder and enjoyment of the world as well as its difficulties.
7. Have supplementary materials ready so that those who finish quickly have other rewarding things to do.
8. Remember that whatever the multicultural mix of your class, you, as a geographer, have a key role to play in developing understanding of a multicultural society.
9. Help other teachers by sharing your ideas, materials and successful lessons especially with non-specialists.
10. Above all, if you have an enthusiasm or special interest in your subject, show it and share it with your pupils: it can be infectious!

Ten things a head of department can do

1. Encourage and foster contacts with feeder schools: offer to run fieldwork for them, to give talks and/or show films. Your school is a feeder school too; do you follow up what happens to your students when they leave you?
2. Identify new opportunities for geography to meet your students' needs. Offer courses and activities for less able as well as brighter pupils.
3. Make sure that your department makes contributions to general or interdisciplinary studies. Suggest ideas with a geographical base, eg. the growth of towns, knowing your local environment.
4. Find out about your school's involvement with the Certificate of Pre-Vocational Education or other similar courses and offer a geography component.

5. Make sure your best teachers take classes at all levels. If students are given good early experience in the subject, they will opt for it.
6. Start a geography club. It can provide the opportunity for visits, videos, competitions and computer work which your timetable may not allow.
7. Make sure your careers department has plenty of information about geography's value for career prospects.
8. Invite parents to exhibitions, club activities and field trips. Use your school magazine to spread news about your department's activities.
9. Encourage in-service training for your staff (and yourself). Use your GA branch and keep in contact with your LEA adviser.
10. Meet regularly with heads of department in other subjects and demonstrate the usefulness of your subject to their needs.

Now check: Are there any gaps that you could usefully fill?

What are you going to do *now* **about those gaps?**

Index

A-level examinations, 23–5, 257–68
  enquiry-based approach, 24, 63, *64*, 267
  map reading skills for, 135, 138
  syllabus diversity, 23–5, 257–9
ability
  disadvantages of separation based on, 172
  wide range of, common syllabus for, 20–2
  see also gifted children; mixed ability *etc.*
abstract concepts
  learning difficulties related to, 162–3, 176
  understanding of, 13–14, 161
administration
  A-level examinations, 259
  combined studies course structures and, 220
Advanced Supplementary Examination, 25
aerial photographs, 133–4, 145–6
aid, international, teaching strategies on, 185–91
aims
  A-level syllabus, 267
  combined studies courses, 221–2, 227–9, 230
  curriculum planning, 150–4; content selection in relation to, 154–7
  educational, 14
  humanities course, 227–9, 230
  in lesson planning, 29, 52
  pre-vocational education, *199*, 200
  16+ courses, 21
  see also objectives
Amazon Forest, lesson planning on, 46–52, 54
appeal procedures, examination boards', 261–2
area, grouping of topics based on, 158, 159–60
assessment
  combined studies courses, 222–4, 227, 232
  coursework, 250–6
  criterion-referenced, 248–9
  definition of, 32, 234–5
  evaluation through, 238–49
  mixed ability groups, 164
  norm-referenced, 248
  objective tests for, 242–4
  quality criteria in at A-level, 262–5
  structured questions for, 242, 246–7, 248
  techniques: mixture of at 16+, 247
  see also examinations; tests
atlases
  guide to selection of, 269–75
  purpose and interpretation of, 139–44
  of satellite imagery, 276
attitudes, *see* values and attitudes
audio-visual resources
  mixed-ability groups, 53–4
  management problems and, 43
  see also specific resources

banding, 164, *165*
BASIC, 113–14, 121
Basis for Choice, A (Mansell Report), 22, 198, 199, 200, 266, 270

blackboards, 86–7, 121
books
  management of stock, 87–8
  for teachers, 278
  see also textbooks
boredom, avoidance of, 166–7
Brazil, videotape on, 110

CSE examination boards: Joint Council with GCE boards, 20, 21
CSE examinations, 20, 22
calculations, computers and calculators for, 120–1
case studies: in multi-cultural education, 183
Centre for World Development Education, 191, 192
Certificate of Extended Education (CEE), 22
Certificate of Pre-Vocational Education (CPVE), 22–3, 149, 198, 199, 213
charity: included in work on aid, 187
city, spatial variation in, objectives in planning study unit on, 37–40
classroom management, *see* management
coastal processes and morphology, approaches to fieldwork on, 208, 210–11, 214–17
combined studies, 149
  example of, 227–33
  planning, 219–26
communication skills, 15
  slow learners, 178–9
  see also graphicacy; language
compass points, 124–5
complexity, learning difficulties related to, 162
computers, 112–22
conceptual development, 13–14
content
  avoidance of sex bias in, 195
  combined studies courses, 222–4, 227, 233
  objectives relating to in lesson planning, 30, 32
  organisation, 158–63
  selection: in curriculum planning, 154–7; in lesson planning, 45–52
  validity: in tests, 239, 247
continuous assessment, 250–6
contours, teaching concept of, 130–3
control, *see management*
copyright laws, 100, 111
core, common
  A-level syllabus, 257, 265
  in pre-vocational education, 199, 200, 204
core curriculum, 11
  combined studies courses and, 221
costs, teaching resources', 85
  books, 92
  satellite imagery, 148
coursework assessment, 250–6
creative activities: in environmental studies 60–2
criterion-referenced assessment, 248–9
cross-section construction, 133
culture, *see* multi-cultural education

curriculum
　common, 9, 10, 11; skills and, 14–15
　core, 11, 221
　DES and HMI publications on, 9–11, 14, 22, 25, 150, 181
　evaluation, *see* evaluation
　geography's contribution to, 9–19, 150–4, 199–204, 224
　hidden, sex bias in, 194–5
　public examinations influence on, 19–25, 149
curriculum development
　criterion-referenced testing and, 248–9
　Geography 16–19 Project, 24
　hindered by examination system, 265–6
　humanities course, 230–3
　microcomputers in, 112
　projects: books and guides from, 279; combined studies, 224; contribution to development of joint 16+ examination, 20–1
　16+ examinations and, 19–22
curriculum planning, 149–63
　combined studies, 219–26
　objectives model, 29–32

DARTS, 74–5
data-response questions, 242
Department of Education and Science, curriculum publications, 9–11, 14, 22, 25, 150, 181
Development Education Centres, 191–2
development studies, 185–92, 228–9
direction, learning concept of, 124, 136
discrimination, question, measurement of, 246–7
discrimination, racial, *see* racism
discrimination, sexual, 193
discussion
　learning through, 70–1
　with slow learners, 178–9
　in Third World Studies, 189–91
drama: in environmental studies course, 60–2

education, aims of, 14
employment, courses preceding, 22–3, 198–204
enquiry-based learning, 58
　in fieldwork, 208, 212, 216–217
　in Geography 16–19 Project, 24, 62–3, *64, 65, 66*, 200, 201, 208, 267
　in pre-vocational education, 200, 201
environment
　attitudes and values concerning, 15–18, 63–5
　experience of, sex differentiation in, 193–4
　man's interaction with: fieldwork on, 207, *211*, 212, 213–217; in Geography 16–19 Project, 267
　pupils' mental maps of, 123–4
　residential, humanistic approach to study of, 35–7
Environmental Studies 5–13 Project, 224
equal opportunities, 193
essays, assessment by, 242, 248, 262–3
evaluation
　through assessment, 238–49
　combined studies courses, 225–6
　in humanistic geography, 37
　lesson planning and, 28–9, *31*, 32, *33*, 52–3
　meaning of, 234–5

mixed ability teaching, 164
　through pupil opinion surveys, 236–8
　through structured observation, 235–6
examinations, public
　A-level, 23–5, 149, 257–68
　assessment techniques and question choice in, 247–8
　curriculum influenced by, 19–25, 149
　map reading skills for, 135, 138
　mechanics of, 259–62
　norm- and criterion-referenced grading, 248–9
　resources policy influenced by, 86
　at 17+, 22–3, 149, 198, 199
　sex differentiation in, 194
　at 16+, 19–22, 156, 247–9
　see also assessment
experience, pupils'
　environmental, sex differentiation in, 193–4
　influence on lesson planning, 27–8
　learning influenced by, 13
　learning through, *see* simulation
　thinking related to, 161

facility, measurement of, 246
fiction, advantages of using, 74
fieldwork
　contextual framework for, 205–7; characteristics, 207–12; examples, 214–17; implementation, 213–24, wider purposes, 217–18
　evolution, 205–7
　humanistic approach to, 35
　language and literary skills and, 18
　mapping skills and, 129, 138
　in pre-vocational education, 200
　sex bias in, 195
films: for more able pupils, 175
filmstrips, 90
　with more able pupils, 173, 174
formative evaluation, 225, 234, 235

GA
　anti-racist policy statement, 182
　on geography's contribution to curriculum, 9, 11, 224
　guideline for success, 280–1
　on multi-cultural education, 184
　Package Exchange (GAPE), 114
　Worldwise Quiz, 12, 143
GCE boards: Joint Council with CSE boards, 20, 21
GCSE, 19–22
games, 80–1
　see also simulations
Geographical Association, *see* GA
geography
　contribution to curriculum, 9–19, 150–4, 199–204, 224
　links with other subjects, 11, 19; *see also* combined studies
Geography for the Young School Leaver Project (GYSL)
　books and guides, 279
　contribution to development 16+ exam, 20–1
　curriculum units, objectives of, 154
　relevance to combined studies, 224
　resources suitable for more able pupils, 173

survey of pupil opinion on, 236–8
writing skills developed by, 76
Geography 14–18 Project
  books and guides, 279
  contribution to development 16+ exam, 20–1
  coursework assessment, 250–6
  resources suitable for more able pupils, 173
  syllabus goals, 154
  writing skills developed by, 76
Geography in the School Curriculum (GA), 11
Geography 16–19 Project
  books and guides, 279
  contribution to pre-vocational courses, 23, 200, 201–2
  enquiry-based approach, 24, 62–3, *64*, 65, *66*, 200, 201, 208, 267
  modular syllabus, advantages of, 266–7
  resources for more able pupils, 173
  survey on A-level syllabuses, 266
  values education strategy, 65, *66*, 202
Ghana, videotape on, 109–10
gifted pupils, 171–5
globes, 88–9
graded exercises: for mixed ability groups, 170
graphic display, microcomputer for, 119, 121
graphic techniques: in resource sheets, 100
graphicacy, 15, 19
  see also map reading
grid references, 125–6
group work: in mixed ability teaching, 168, 170
grouping of topics, 159–60

hardware, computer, 113–14, 120
hardware simulations, 81
Her Majesty's Inspectorate
  on aims of geography, 151–4
  curriculum publications, 9–11, 154
  on teaching of ideas, 14
history, geography and, 11, 223, 227–33
History, Geography and Social Sciences 8–13 Project, 224
humanistic geography
  in fieldwork, 205–7
  objectives, 35–7
  simulation in, 84
humanities course, example of, 227–33
Humanities Curriculum Project, 224
hydrology, microcomputer simulation in, 118–19

ice, drama and creative work in study of, 60–2
ideas, development of
  geography's contribution to, 13–14
  in mapwork lesson, 30
  in urban spatial variation study unit, 39, 40
inner city
  residential environments in, objectives in lesson planning on, 33–7
  videotape on, 110–11
  in-service training (INSET): on microcomputers, 112
integrated studies, 219
  see also combined studies
Integrated Studies Project, 224
intellectual development, 13–14

curriculum planning and, 161–3
  lesson planning and, 27–8
interdisciplinary studies, 19, 219–20
  see also combined studies
interest, gaining of, 166–7

Joseph, Sir Keith: address to GA, 11
journals, 278–9

key concepts, content selection using, 156
knowledge, acquisition of
  contribution of 16+ courses to, 21
  geography's contribution to, 12–13

labelling techniques: in Third World studies, 187
Lake District: use in values education, 15–18
Landsat imagery, 146–7
landscape model, 135
language
  increasing precision in use of, 163
  mixed ability groups, 167
  role in learning, 54–5, 68
  in textbooks, 92
  understanding influenced by, 68–9, 72–3
language skills
  geography's contribution to, 18, 68–78
  sex differentiation in, 193
  slow learners, 176–9
  see also reading; talking; writing
learning
  enquiry-based, *see* enquiry-based learning
  gifted children, 171–2
  influence of age and experience on, 13
  language's role in, 54–5, 68
  measurement of, *see* assessment
  progression in, 160–3, 223, 230
  slow, *see* slow learners
  through simulation, 79
learning activities
  organisation of, 158–60
  selection of, 157–8
  teaching approaches in relation to, 56–63
learning episodes, lesson planning and, 40
less able pupils, *see* slow learners
lesson planning
  learning experience and, 40
  management problems and, 42, 43
  for microcomputer use, 120
  mixed ability groups, 53–4, 60–2
  objectives, 27–37, 52, 60–2
  performance in relation to, 42, 43
  role of language in relation to, 54–5
  steps involved in, 45–53
  for values education, 53, 54, 55
lesson structure: mixed ability teaching, 167
literary skills, geography's contribution to, 18
location, learning concept of, 125–6, 136

management, classroom
  lesson planning and, 42, 43
  microcomputers and, 120
  of simulations, 82–3
Mansell Report (*A Basis for Choice*), 22, 198, 199, 200,

266, 270
map collection management, 88–90
map reading
  lesson planning and objectives for, 27–32
  skills, 15, 19, 123–34; progression in, 134–8, 140–1; sex differentiation in, 193–4; using atlases, 139–44
maps, textbook, 93
marking
  A-level exams, 259–61, 262–3
  types of, writing skills in relation to, 77
matching pairs questions, 243, *245*
mathematical simulations, 81
mathematical skills
  in geography, 18–19
  slow learners, 179–81
memory, testing of, 69–70
mental maps, 123–4
meteorological satellite, teaching resources from, 147
microcomputers, 112–22
Microelectronics Education Programme, 112, 122
mixed ability groups, 164–70
  audio-visual resources for, 53–4
  drama and creative activity with, 60–2
  games and simulations with, 172
  humanities course for, 227–33
  lesson planning for, 53–4, 164–5
models
  fieldwork testing of, 205
  hardware: in teaching contour concept, 130–3
  landscape: in GCSE course, 135
  planning, 29–32, *33*, 45–8, 52
moderation: coursework assessment, 256
motivation
  bright pupils, 171
  mixed ability groups, 167
  slow learners, 176
multi-cultural education, geography's contribution to, 181–4
multiple choice questions, 242–3, *244*
multiple completion test, 243, *244*

news items: as resources, 189, *205*
norm-referenced assessment, 248
nuclear power industry, values concerning, 16–18
numerical techniques, 18–19
  with slow learners, 179–80

O-level examinations, 20, 22
objective tests, 242–4
objectives
  A-level syllabus, 267
  combined studies course, 221, 222
  coursework assessment, 250
  in curriculum planning, 150, 154
  formulation of, 33–7
  humanities course, 228–9, 230, 231
  in lesson planning, 27–37, 52, 60–2
  pre-vocational education, 200
  in study unit planning, 36–40
  teaching methods in relation to, 32, 33–5, 65
objectivity: in multi-cultural education, 183
observation, structured, evaluation by, 235–6
operational thinking, concrete and formal, 13

Ordnance Survey maps
  collection management, 89–90
  interpretation, 27–9, 126–7, 128–9, 131–3, 135
  public examination questions on, 135
overhead projector, 87, 121

personal relationships, lesson planning and, 41–2
photographs
  aerial, 133–4, 145–6
  contrasting uses of, 33–5
  guidelines for taking successfully, 90
  interpretation, 15, 19, 103–7
  in textbooks, 94
place names: on atlas maps, 142–3, 272, 273
planning
  mixed ability teaching, 53–4, 164–5
  models of, 29–32, *33*, 45–6, 52
  objectives in, 27–40
  see also curriculum planning; lesson planning
precision, learning difficulties related to, 163
pre-vocational education, 22–3, 149, 198–204, 212–13
programs, computer, 113–14
  lessons using, 114–19
  lesson planning and, 120
  modification of, 121
progression in learning, 160–3, 223, 230
projection facilities, 90–1
pupils
  mixed ability, interaction between, 167–8, 170
  opinions, course evaluation through, 236–8
  participation, 70–1
  reaction to coursework assessment, 256
  see also particular pupils, e.g. slow learners

question choice: in public examinations, 247–8
question facility and discrimination, 246–7
questioning
  analysis of, 69–70
  more able pupils, 173–4
  in Third World studies, 187
  see also enquiry-based learning
questions, key: in lesson planning, 48
questions, structured, assessment by, 242, 246–7

racism, opposition to, policies on, 181–2
readability, 73, 92, 177, 230, 231
reading, 71–5
  slow learners, 177–8
"Red Book", 9–10, 14
relationships diagram, 174
reliability, tests, 239, 240–2
relief interpretation, 130–4
remote sensing, 145–8, 276–7
reprographic equipment, 100, 101
resource sheets
  Amazon Forest lesson, 48–52
  design and production, 96–102
  for slow learners, 177–8
  see also worksheets
resources
  for combined studies, 231
  guides to acquisition of, 268–77
  management of, 85–91

for more able pupils, 173–5
for multi-cultural education, 183–4
for pre-vocational education, 202–4
sex bias in and avoidance of, 194–5, 196
for Third World studies, 191–2
role playing, 79–80
   fieldwork and, 212, 213–14
   microcomputers for, 115, 118
   sex bias in, 195
   with slow learners, 179
routes, progression in programme on, 161–2

satellite imagery, 145–8, 276–7
scale, map, 127–9, 134, 136
School Curriculum, The (DES), 10–11, 14, 150, 181
Schools Council
   16+ common exam recommendations, 19
   projects: contribution to development joint exam system, 20–1; books and guides, 279; *see also particular projects, e.g.* Geography for the Young School Leaver
Secondary Examinations Council, 20, 259
settlement, microcomputer for role play on, 115, 118
17+ examinations, 22–3, 149, 198, 199
sex bias, 193, 194–6
sex differentiation, 193–4, 195
simulations, 79–84
   microcomputers in, 118–19
   in mixed ability groups, 172
   sex bias in, 195
   with slow learners, 179
16+ examinations
   assessment mix and question choice, 247–9
   development of common system for, 19–22
   key concept planning grid, 156
skills, matching ability with, 172, *173*
skills acquisition and development, 14–15, 18–19, 153
   A-level, 266
   contribution of 16+ courses to, 21
   on humanities course, 227, 228, 230
   in mapwork lesson, 30
   in mixed ability groups, 168
   in urban spatial variation study unit, 39, 40
   see also particular skills, e.g. language, map reading
slides, 90
   for more able pupils, 173, 174
   of satellite imagery, 276
slopes, microcomputer in lesson on, 115
slow learners, 176–80
   in mixed ability groups, 165
social skills, geography's contribution to, 19
social studies course, aims of, 221–2
soil study, coursework assessment in, 251–6
software, computer, 114, 120, 121
spatial ability, sex differentiation in, 193
spatial literacy, 153
special education, 176
specification grid, 240
staffing arrangements, combined studies, 222–3
Start Bay, fieldwork in, 214–17
stereotypes, contribution of videotapes to countering, 110
stimulus-response questions, 242

streaming, 164, *165*
summative evaluation, 225, 234
syllabuses
   A-level: diversity, 23–5, 257–9; enquiry-based approach, 24, 63, *64*, 267; overload in, 26–5; revision, 259
   combined studies courses, 220–1, 22–3
   common, contribution of Schools Council Projects to development of, 20–1
   modular: Geography 16–19 Project, 266–7
symbols, map, 127, 136

talking, 68–71
   learning through, 54–5
   slow learners, 178–9
   see also discussion; groupwork
teachers
   coursework assessment by, 250
   as most valuable resource, 85
   observation of, 235–6
   organisation of for combined studies, 222–3
teaching approaches
   analysis, 56, 58, 59
   choice of, 65–7
   learning activities and, 56–63
   personalised, 67
   resources policy influenced by, 85–6
   values education in relation to, 63–5
teaching as performance, 41–2, 43
teaching methods and strategies
   combined studies, 224–5
   influence of management problems on, 43
   influence of pupil age and experience on, 28
   in mapwork lesson planning, 30–1
   in mixed ability groups, 168–70
   for more able pupils, 172, 173–5
   for multi-cultural education, 184
   objectives in relation to, 32, 33–5, 65
   in pre-vocational education, 200
   selection of in curriculum planning, 158
   sex bias in and avoidance of, 195, 196
   for Third World studies, 185–91
Technical and Vocational Educational Initiative (TVEI), 198
techniques
   fieldwork, 208–12
   planning teaching of, 160
   see also map reading; mathematical skills
tests
   reliability, 239, 240–2, 247–8
   results, statistical analysis of, 238, 244–7
   validity, 239–40, 247
textbooks
   alternatives to, 73–4; *see also* worksheets
   evaluation of, 92–5
   for more able pupils, 173
   for multi-cultural education, 183
   reading difficulties with, 72–3
   sex bias in, 194–5, 196
   for slow learners, 177
   for Third World studies, 192
thinking
   influence of age and experience on, 13, 27–8, 161

operational, concrete and formal, 13
questioning in relation to, 69–70
Third World studies, 185–92
topic web, *186,* 187
transactional writing, 75, 76
typing and typefaces: on resource sheets, 99–100
under-achievement, 171
understanding
   assessment of, 239–40
   development of, geography's contribution to, 151–3
   influence of language use on, 68–9, 72–3
   slow learners, 177
university entrance, A-level geared to needs of, 266

validity, test, 239–40, 247
values and attitudes
   contribution of 16+ courses to, 21
   difference between teacher's and pupils', 183
   included in fieldwork, 208, 212
   geography's contribution to, 15–18
   in inner city residential environment lesson, 35
   in mapwork lesson, 30
   in pre-vocational education, 202
   in urban spatial variation study unit, 39, 40
values education
   in relation to intellectual development, 163
   lesson planning in relation to, 53, 54, 55
   with more able pupils, 174
   teaching approaches for, 63–5

textbooks in relation to, 95
videotapes, 10–11
   with more able pupils, 175
   structured self-observation through, 235
View of the Curriculum, A (HMI), 10
visual aids
   with mixed ability groups, 167
   with more able pupils, 173–5
   see also films; videotapes *etc.*
vocational studies, geography in, 200–2

Waddell Report, 19–20
wall maps, 88, 89
welfare geography, objectives and, 39–40
whiteboards, 87
wordprocessors, resource sheets produced on, 99
worksheets
   in combined studies, 225, 230, 231
   design and production, 96–102
   graded, *169,* 170
   management problems related to, 43
   in mixed ability groups, 53–4, *169,* 170
   for slow learners, 177–8, 179–80
world, knowledge and understanding of, geography's
   contribution to, 10–11
Worldwise Quiz, 12, 143
Writing Across the Curriculum Project, 75
writing skills, 75–7

youth hostel: in map reading lesson, 28–9